CONTEMPORARY MATHEMATICS

451

Frames and Operator Theory in Analysis and Signal Processing

AMS-SIAM Special Session
January 12–15, 2006
San Antonio, Texas

David R. Larson
Peter Massopust
Zuhair Nashed
Minh Chuong Nguyen
Manos Papadakis
Ahmed Zayed
Editors

American Mathematical Society
Providence, Rhode Island

2000 *Mathematics Subject Classification.* Primary 11S80, 20F55, 35S99, 41A30, 42C15, 42C40, 47N40, 47S10, 94A12.

Library of Congress Cataloging-in-Publication Data

Frames and operator theory in analysis and signal processing : AMS Special Session, January 12–15, 2006, San Antonio, Texas / David R. Larson...[et al.], editors.
    p. cm. — (Contemporary mathematics, ISSN 0271-4132 ; v. 451)
    Includes bibliographical references.
    ISBN 978-0-8218-4144-0 (alk. paper)
    1. Operator theory—Congresses. 2. Signal processing—Congresses. I. Larson, David R., 1942–

QA329.F73   2008
515′.724—dc22

2007060586

100546405I

Frames and Operator Theory
in Analysis and Signal Processing

Contents

Preface

The collection of papers in this volume is based on presentations given at the AMS Special Session entitled *Frames and Operator Theory in Analysis and Signal Processing*, which took place at the Joint Mathematics Meetings in San Antonio, Texas, on January 12 – 15, 2006.

One of the goals of this special session was to integrate both the industrial and theoretical aspects of frames and operator theory, in particular their applications to image and signal processing. In addition, the special session contained a strong component of works related to sampling and numerical solutions of partial differential equations. The connections with operator theory (the pure and academic aspects of the special session) reflect a broader group than represented in the previous special sessions dealing with the afore-mentioned topics. Signal and image processing use common mathematical techniques; chief among them are functional analysis, inverse problems, applied harmonic analysis, including wavelets, and transform techniques.

During recent years, the field of frames has undergone a tremendous advancement. Most of the work in these fields has focused on the design and construction of more versatile frames and, more recently, frames tailored towards specific applications, e.g. finite dimensional uniform frames for cellular communication. In addition, frames are now becoming a new hot topic in mathematical research, but surprisingly enough it appears that this recently rediscovered topic is in the heart of many engineering applications, e.g. matching pursuits and greedy algorithms for image and signal processing.

Topics that the special session covered include:

- Application of several branches of analysis (e.g., PDEs, Fourier, wavelet, and harmonic analysis, transform techniques, data representations) to industrial and engineering problems, in particular image and signal processing.
- Theoretical and applied aspects of frames and wavelets.
- Pure aspects of operator theory emphasizing the connections to applied mathematics, frames, and signal processing.

The above-mentioned themes and their scope of applicability are reflected in the papers in this Conference Proceedings volume. A number of interesting and innovative presentations regarding frames and their interactions with operator theory are exhibited and several of these techniques have already been designed for a specific application while others seem to have a potential for future success in applications. For these reasons, we believe that putting together a series of semi-expository and original research papers with the flavor inferred from the above topics will produce a unique volume in the operator and frames literature. We also need to mention that the interest in frames from both the mathematical and the engineering community has recently been rekindled. This renewed interest is mostly due to the foreseen immense potential for applications in the imaging sciences and the burgeoning area of mobile communication. Despite these developments, the current literature on frames and their relation to operators is rather limited and the monographs that are available contain the general theory or parts of it with only minor references to current research trends. A major weakness, in our opinion, of these publications is the lack of information about the applications of frame theory and their connections to operators. The current volume is the first of its kind to simultaneously address theoretical foundations of frame design and operator theory and elucidate their mutual applications.

We believe that this Conference Proceedings volume will be equally attractive to pure mathematicians, working on the foundations of frame and operator theory and their interconnections, to applied mathematicians, who are investigating applications, and to physicists and engineers employing these designs. It thus may be attractive to a wide target group of researchers and may serve as a catalyzer for cross-fertilization of several important areas of mathematics and the applied sciences.

David R. Larson, Peter Massopust, Zuhair Nashed,

Ming Chuong Nguyen, Manos Papadakis, and Ahmed Zayed

September 28, 2007

Speakers and Titles of Talks presented in the AMS - SIAM Special Session on Frames and Operator Theory in Analysis and Signal Processing, in the Joint Mathematics Meetings, San Antonio, Texas, January 12-15, 2006

Radu V. Balan
Siemens Corporate Research
"A Noncommutative Wiener Lemma and a Faithful Tracial State on Banach Algebras of Time-Frequency Shift Operators"

John J. Benedetto
University of Maryland
"Waveform Design and a General Form of Matched Filtering"

Peter G. Casazza
University of Missouri
"The Kadison-Singer Problem in Mathematics and Engineering: Part II"

Ingrid Daubechies
Princeton University
"Two Applications of Frames for the Analysis of Astrophysical Data"

Maarten V. De Hoop
Purdue University
"Curvelets and the One-Way Acoustic Wave Equation"

P.P.B. Eggermont
University of Delaware
"Tikhonov Regularization of Ill-posed Operator Equations with Weakly Bounded Noise"

Adel Faridani
Oregon State University
"Construction of Sampling Theorems for Unions of Shifted Lattices"

Deguang Han
University of Central Florida
"Lattice Tilings, Operator Algebras and Gabor Frames"

Palle Jorgenson
University of Iowa
"Use of Geometry and Operator Algebra Theory in the Computation of Wavelet Coefficients"

Victor Kaftal
University of Cincinnati
"Operator-valued Frames over Hilbert C*-modules"

Costas Karanikas
Aristotle University of Thessaloniki, Greece
"Haar-type Orthonormal Systems, Data Presentation as Riesz Products and a Prediction Method on Symbolic Sequences"

Donald J. Kouri
University of Houston
"New Approaches to the Acoustic Wave Equation"

Mila Nikolova
CMLA UMR 8536–ENS de Cachan, France
"Data Fidelity Terms for Image and Signal Restoration"

Gestur Olafsson
Louisiana State University
"Wavelets, Multiresolution Analysis
and Finite Reflexion Groups"

Judith A. Packer
University of Colorado at Boulder
"The Use of Filters and Direct Limits in
the Construction of Fractal Wavelets"

Vern I. Paulsen
University of Houston
"Frame Paths and Sigma-Delta
Quantization"

Luigi Rodino
Universitá di Torino, Italy
"Pseudo-Differential Operators,
Localization Operators and
Time-Frequency Analysis"

M. Beth Ruskai
Tufts University
"The Relation between Frames and
POVM's in Quantum Information
Theory"

Otmar Scherzer
Universität Innsbruck, Austria
"Regularization Functionals Involving
Discontinuous Operators"

Reinhold Schneider
Christina-Albrechts-Universität Kiel,
Germany
"Multiscale Approximation in
Electronic Structure Calculation"

Xiaopeng Shen
Ohio University
"Prolate Spheroidal Wavelets in a
Periodic Setting"

Hrvoje Šikić
University of Zagreb, Croatia
"The Structure of the Set of Parseval
Frame Wavelets"

Darrin Speegle
St. Louis University
"The Feichtinger Conjecture for Frames
of Translates"

Qiyu Sun
University of Central Florida
"Oversampling a Tight Affine Frame"

Gilbert Walter
University of Wisconsin-Milwaukee
"Multidimensional Prolate Spheroidal
Wavelets"

Yang Wang
Georgia Institute of Technology
"On PCM and Sigma-Delta
Quantization Errors"

Eric Weber
Iowa State University
"The Kadison-Singer Problem in
Mathematics and Engineering: Part II"
"Orthogonal Wavelet Frames"

Contemporary Mathematics
Volume **451**, 2008

Haar-type Orthonormal Systems, data presentation as Riesz Products and a recognition on symbolic sequences

Nikolaos D. Atreas and C. Karanikas

ABSTRACT. For any $p = 2, 3, \ldots$, we provide a sequence $H(n)$ of Haar orthonormal matrices of order $p^n \times p^n$, $n = 1, 2, \ldots$, such that $H(n+1)$ is generated by p-adic dilation and splicing (translations) of block sub-matrices of $H(n)$. Moreover, for any data $\{t_k, k = 1, \ldots, p^n\}$, we get an algorithm to find the coefficients $\{x_j, j = 1, \ldots, p^n\}$ of the Riesz-type product $t_k = \prod_{j=1}^{p^n} \left(1 + x_j h_{j,k}\right)$, where $h_{j,k}$ is the (j,k) entry of the matrix $H(n)$, provided that $\langle t, h_j \rangle \neq 0$ for any j. This work can be extended for data of any length, providing also new tools for fractal sets and singular measure theory. In fact, it enables us to detect the grammar of Cantor-type languages, or the underlying ergodic processes of a class of Cantor type sets. More generally, this approach can be used to approximate in the weak* topology any continuous measure by Riesz-type products, based on unbalanced Haar wavelets.

1. Introduction

The purpose of this work is to introduce new discrete invertible transforms, effective for analyzing symbolic sequences, mimicking the basic biological functionality. Since biological data as genes and proteins can be considered as symbolic sequences in an alphabet of four letters and twenty letters respectively, our attempt is concentrated on mimicking the following biological functionalities/characteristics:

(1) **Replication in multiple copies**. Note that genes written in an alphabet of four letters, A, C, G and T, have repeated copies on certain parts, e.g. the unit 5' CA 3' can be repeated such as 5' CACACACACA 3'. Clearly, the mathematical term of replication is **Dilation**.

(2) **Splicing**, means the union of certain parts of a symbolic sequence to produce a new one (mRNA is spliced from the replica parts of DNA, called exons).

(3) **Local information is important on biological functionality as is the antigen processing**, because the protein's functionality is decided by their peptides (small pieces of the symbolic sequence). Therefore, our

1991 *Mathematics Subject Classification.* Primary 42C10, 65T99; Secondary 65F30, 37N25.
Key words and phrases. Discrete transforms, Haar system, Riesz Products, matrix operators, singular measures, Cantor-type languages.
The first author was supported in part by G.S.R.T. program Pythagoras II.

transforms must have the ability to code local information, cutting data in pieces (as for example does the Haar wavelet transform).

(4) **Usual biological data have an underlying hidden Markov process**, therefore our transforms must have the ability to decode hidden ergodic structure and to deal with Cantor-type sets or languages (see below).

In our attempt to introduce new classes of linear or non-linear discrete transforms inspired by the previous biological meanings, we shall use: **Sparse** matrices (matrices having a small number of non zero elements) with underlying multiresolution structure, suitable to encode local information and to decode hidden ergodic structure. These sparse matrices will be iteratively generated by dilation/replication and union/splicing of block sub-matrices.

First we introduce our matrix operations:

DEFINITION 1.1. Let $p = 2, 3, \ldots$ and let $M_{n,m}$ be the space of all matrices of order $n \times m$, we define the dilation operation $D_p : M_{n,m} \to M_{n,pm}$, such that:

$$D_p(A) = \left\{ A_{i, \lceil \frac{j}{p} \rceil}, \quad i = 1, \ldots, n, \quad j = 1, \ldots, pm \right\},$$

where $\lceil x \rceil$ is the ceiling of x.

EXAMPLE 1.2.

$$D_2\left(\begin{pmatrix} 1\ 2 \\ 3\ 4 \end{pmatrix}\right) = \begin{pmatrix} 1\ 1\ 2\ 2 \\ 3\ 3\ 4\ 4 \end{pmatrix}, \quad D_3\left(\begin{pmatrix} 1\ 2 \\ 3\ 4 \end{pmatrix}\right) = \begin{pmatrix} 1\ 1\ 1\ 2\ 2\ 2 \\ 3\ 3\ 3\ 4\ 4\ 4 \end{pmatrix}.$$

DEFINITION 1.3. Let $T_p : M_{k,l} \to M_{pk,pl}$, be the following translation operator:

$$T_p(A) = \left\{ \begin{cases} A_{\lceil \frac{i}{p} \rceil, Mod(j-1,l)+1}, & whenever\ Mod(i-1,p)+1 = \lceil \frac{j}{l} \rceil \\ 0, & otherwise \end{cases}, i = 1, \ldots, pk,\ j = 1, \ldots, pl \right\}.$$

EXAMPLE 1.4.

$$T_2\left(\begin{pmatrix} b_{11} & b_{12} & b_{13} \\ b_{21} & b_{22} & b_{23} \end{pmatrix}\right) = \begin{pmatrix} b_{11} & b_{12} & b_{13} & 0 & 0 & 0 \\ 0 & 0 & 0 & b_{11} & b_{12} & b_{13} \\ b_{21} & b_{22} & b_{23} & 0 & 0 & 0 \\ 0 & 0 & 0 & b_{21} & b_{22} & b_{23} \end{pmatrix},$$

$$T_3\left(\begin{pmatrix} b_{11}\ b_{12} \\ b_{21}\ b_{22} \end{pmatrix}\right) = \begin{pmatrix} b_{11} & b_{12} & 0 & 0 & 0 & 0 \\ 0 & 0 & b_{11} & b_{12} & 0 & 0 \\ 0 & 0 & 0 & 0 & b_{11} & b_{12} \\ b_{21} & b_{22} & 0 & 0 & 0 & 0 \\ 0 & 0 & b_{21} & b_{22} & 0 & 0 \\ 0 & 0 & 0 & 0 & b_{21} & b_{22} \end{pmatrix}.$$

DEFINITION 1.5. Let $S(.,.) : M_{n,m} \times M_{k,m} \to M_{n+k,m}$ be the following splicing matrix operation:

$$S(A, N) = \left\{ \begin{cases} A_{i,j}, & whenever\ i = 1, \ldots n, \quad j = 1, \ldots m \\ N_{i-n,j}, & whenever\ i = n+1, \ldots, n+k, \quad j = 1, \ldots m \end{cases} \right\}.$$

EXAMPLE 1.6.

$$S\left(\begin{pmatrix} 1\ 1\ 1\ 0 \\ 0\ 0\ 0\ 2 \end{pmatrix},\begin{pmatrix} 1\ 0\ 0\ 0 \\ 0\ 2\ 0\ 0 \\ 0\ 0\ 3\ 0 \end{pmatrix}\right)=\begin{pmatrix} 1\ 1\ 1\ 0 \\ 0\ 0\ 0\ 2 \\ 1\ 0\ 0\ 0 \\ 0\ 2\ 0\ 0 \\ 0\ 0\ 3\ 0 \end{pmatrix}.$$

It is not difficult to check the following:

Observation 1 see [**AK**] For any $p,q=2,3,\ldots$ the operators D_p,T_p satisfy the following properties:

(i): $D_pD_q=D_{pq}$, $D_pT_q=T_qD_p$.
(ii): We suppose that the rows of a matrix $A\in M_{n,m}$ form an orthonormal set, then both operators $\frac{1}{\sqrt{p}}D_p$ and T_p preserve orthonormality.

The first idea of this work derived from the observation that the Gram Schmidt orthonormalization process of the following sparse matrices (see [**AKP1**] and [**AP**]):

$$U(1)=\begin{pmatrix} 1\ 1 \\ 1\ 0 \end{pmatrix},U(2)=\begin{pmatrix} 1\ 1\ 1\ 1 \\ 1\ 1\ 0\ 0 \\ 1\ 0\ 0\ 0 \\ 0\ 0\ 1\ 0 \end{pmatrix},U(3)=\begin{pmatrix} 1\ 1\ 1\ 1\ 1\ 1\ 1\ 1 \\ 1\ 1\ 1\ 1\ 0\ 0\ 0\ 0 \\ 1\ 1\ 0\ 0\ 0\ 0\ 0\ 0 \\ 0\ 0\ 0\ 0\ 1\ 1\ 0\ 0 \\ 1\ 0\ 0\ 0\ 0\ 0\ 0\ 0 \\ 0\ 0\ 1\ 0\ 0\ 0\ 0\ 0 \\ 0\ 0\ 0\ 0\ 1\ 0\ 0\ 0 \\ 0\ 0\ 0\ 0\ 0\ 0\ 1\ 0 \end{pmatrix},\ldots,$$

derives the Haar matrices:

$$H(1)=\begin{pmatrix} \frac{1}{\sqrt{2}}\ \ \frac{1}{\sqrt{2}} \\ \frac{1}{\sqrt{2}}\ -\frac{1}{\sqrt{2}} \end{pmatrix},H(2)=\begin{pmatrix} \frac{1}{2}\ \ \frac{1}{2}\ \ \frac{1}{2}\ \ \frac{1}{2} \\ \frac{1}{2}\ \ \frac{1}{2}\ -\frac{1}{2}\ -\frac{1}{2} \\ \frac{1}{\sqrt{2}}\ -\frac{1}{\sqrt{2}}\ 0\ \ 0 \\ 0\ \ 0\ \ \frac{1}{\sqrt{2}}\ -\frac{1}{\sqrt{2}} \end{pmatrix},$$

$$H(3)=\begin{pmatrix} \frac{1}{2\sqrt{2}}\ \frac{1}{2\sqrt{2}}\ \frac{1}{2\sqrt{2}}\ \frac{1}{2\sqrt{2}}\ \frac{1}{2\sqrt{2}}\ \frac{1}{2\sqrt{2}}\ \frac{1}{2\sqrt{2}}\ \frac{1}{2\sqrt{2}} \\ \frac{1}{2\sqrt{2}}\ \frac{1}{2\sqrt{2}}\ \frac{1}{2\sqrt{2}}\ \frac{1}{2\sqrt{2}}\ -\frac{1}{2\sqrt{2}}\ -\frac{1}{2\sqrt{2}}\ -\frac{1}{2\sqrt{2}}\ -\frac{1}{2\sqrt{2}} \\ \frac{1}{2}\ \ \frac{1}{2}\ -\frac{1}{2}\ -\frac{1}{2}\ 0\ 0\ 0\ 0 \\ 0\ 0\ 0\ 0\ \frac{1}{2}\ \ \frac{1}{2}\ -\frac{1}{2}\ -\frac{1}{2} \\ \frac{1}{\sqrt{2}}\ -\frac{1}{\sqrt{2}}\ 0\ 0\ 0\ 0\ 0\ 0 \\ 0\ 0\ \frac{1}{\sqrt{2}}\ -\frac{1}{\sqrt{2}}\ 0\ 0\ 0\ 0 \\ 0\ 0\ 0\ 0\ \frac{1}{\sqrt{2}}\ -\frac{1}{\sqrt{2}}\ 0\ 0 \\ 0\ 0\ 0\ 0\ 0\ 0\ \frac{1}{\sqrt{2}}\ -\frac{1}{\sqrt{2}} \end{pmatrix},\ldots.$$

Notice that $U(n+1)=S\left(D_2(U(n)),I(2^{n+1},odd)\right)$, $n=2,3,\ldots$, where $I(2^{n+1},odd)$ is obtained from the odd rows of the identity matrix $I(2^{n+1})$. Since sparse matrices $U(n)$ are created by a multiscale construction, we dealt with the problem: can we use a similar construction to create the Haar matrices $H(n)$? In section 2, definition 2.3, we build Haar-type matrices $H(m)$ of order $m\times m$, $m=2,3,\ldots$, using an iteration in scales, determined by the prime integer factorization of m. These matrices endow a generalization of the usual Haar matrices, since the rows h_n, $n=1,\ldots,m$ of $H(m)$ provide unbalanced Haar wavelets, as introduced in [**BP**], [**GS**] and [**S**]. Moreover, we get a multiresolution analysis and a Haar transform:

$$\{t_n:n=1,\ldots,m\}\longleftrightarrow\{\langle t,h_n\rangle,n=1,\ldots,m\}.$$

The second idea of this work arises from the Tree Transform introduced in [**K**] and [**KP**] and the Riesz Products expressed by non-trigonometrical systems as the Rademacher or the Walsh system (see [**B**], [**BK1**], [**BKM**] and [**BK2**]). Tree

Transform provides a representation of any data t of length p^N, ($p = 2, 3, \ldots$ and $N = 1, 2, \ldots$)

$$t = \{t_n, n = 1, \ldots, p^N\} \leftrightarrow \{a_{n,k}, n = 1, \ldots, N, k = 1, \ldots, p^n\}$$

and a presentation of t as a product:

$$t_k = \prod_{n=1}^{N} a_{n, \left\lceil \frac{k}{p^{N-n}} \right\rceil} \left(\sum_{j=1}^{p^N} t_j \right),$$

where $a_{n,k}$, called walks of t, encode the local variability of t. Notice that the walks of t can be stored in the nodes of a tree in which each node has p branches. This transform was shown to be adequate for edge detection of biospots (see [**AKT**]) and for examining hidden Markov processes (see [**AKP2**]). Furthermore, Rademacher Riesz products, as in [**B**], [**BK1**], [**BKM**] and [**BK2**], have an underlying Markovian structure and have been useful to examine certain properties of singular measures as Hausdorff dimension.

In section 3 we deal with multiscale Riesz-type products (see [**BBK**]), based on Haar matrices $H(m)$. We prove that to any step function $f(x)$ on $[0, 1)$ satisfying:

(i): $f(x) = t_n$, $x \in \left[\frac{n-1}{m}, \frac{n}{m} \right)$, $n = 1, \ldots, m$ and
(ii): $\langle f, h_k \rangle \neq 0$ for any $k = 1, \ldots, m$, where $h_k(x) = h_{k,n}$, $x \in \left[\frac{n-1}{m}, \frac{n}{m} \right)$, $n = 1, \ldots, m$ and $h_{k,n}$ is the (k,n) entry of the matrix $H(m)$, corresponds a unique sequence of numbers $\{a_k : k = 1, \ldots, m\}$:

$$f(x) = \prod_{k=1}^{m} \left(1 + a_k h_k(x) \right).$$

In other words, we prove that any data $\{t_n : n = 1, \ldots, m\}$ satisfying $\langle t, h_k \rangle \neq 0$ for any $k = 1, \ldots, m$ can be expressed as a product:

$$t_n = \prod_{k=1}^{m} \left(1 + a_k h_{k,n} \right),$$

called Haar Riesz product associated with the coefficients $\{a_n\}$. As a byproduct of this result, one can express singular measures as weak* limits of Haar Riesz products. In fact, we may present the Cantor measure as the weak* limit of Haar Riesz products.

Cantor set (Cantor language) may be considered as the set of all symbolic sequences of infinite length (respectively of certain length) with entries 0 or 2 in the alphabet of three letters $\{0, 1, 2\}$. A similar definition can be given for Cantor-type sets and languages in an alphabet of more than three letters. There is a great variety of open problems to detect the grammar of a given language (see [**LP**]), especially in the field of bioinformatics for analyzing biological data (see [**KR**]).

In section 4 we make an initial attempt in this direction proving in Theorem 4.2 that the Haar coefficients of the indicator sequence (see definition 4.1) of a Cantor language detect the grammar of this language.

2. Unbalanced Haar orthonormal system

The usual Haar system is created by dyadic dilation and translation. In this section we get an unbalanced Haar orthonormal system by using the dilation, translation and splicing operators defined above.

DEFINITION 2.1. Let $p > 1$ be a prime number, we define the following matrix $\Psi^p = (\psi_{ij}^p)$ of order $(p-1) \times p$:

$$
\psi_{ij}^p = \begin{cases}
\dfrac{1}{\sqrt{p-i}}\dfrac{1}{\sqrt{p-i+1}}, & \text{whenever } 1 \le j \le p-i \\
-\dfrac{\sqrt{p-i}}{\sqrt{p-i+1}}, & \text{whenever } j = p-i+1,\ i = 1,\ldots,p-1\ ,j=1,\ldots,p \\
0, & \text{whenever } p-i < j \le p
\end{cases}
$$

EXAMPLE 2.2.

$$
\Psi^2 = \begin{pmatrix} \frac{1}{\sqrt{2}} & -\frac{1}{\sqrt{2}} \end{pmatrix},\ \Psi^3 = \begin{pmatrix} \frac{1}{\sqrt{6}} & \frac{1}{\sqrt{6}} & -\sqrt{\frac{2}{3}} \\ \frac{1}{\sqrt{2}} & -\frac{1}{\sqrt{2}} & 0 \end{pmatrix},
$$

$$
\Psi^5 = \begin{pmatrix}
\frac{1}{2\sqrt{5}} & \frac{1}{2\sqrt{5}} & \frac{1}{2\sqrt{5}} & \frac{1}{2\sqrt{5}} & \frac{1}{2\sqrt{5}} \\
\frac{1}{2\sqrt{3}} & \frac{1}{2\sqrt{3}} & \frac{1}{2\sqrt{3}} & -\frac{\sqrt{3}}{2} & 0 \\
\frac{1}{\sqrt{6}} & \frac{1}{\sqrt{6}} & -\sqrt{\frac{2}{3}} & 0 & 0 \\
\frac{1}{\sqrt{2}} & -\frac{1}{\sqrt{2}} & 0 & 0 & 0
\end{pmatrix}.
$$

Observation 2 (see [**AK**]) The matrix Ψ^p satisfies the following properties:

(i): $\sum_{j=1}^{p} \psi_{ij}^p = 0,\ i = 1,\ldots,p-1$.

(ii): $\psi_i^p \psi_j^p = \psi_{i,1}^p \psi_j^p$, whenever $i < j,\ i,j = 1,\ldots,p-1$.

(iii): The matrix $S\left(\frac{1}{\sqrt{p}}(1,\ldots,1)_{1\times p}, \Psi^p\right)$ is orthonormal.

DEFINITION 2.3. Let $m = p_1 \cdots p_N$ be the prime integer factorization of m, where $p_1 \ge \ldots \ge p_N$, we define a sequence of block matrices $H^m(n),\ n = 1,\ldots,N$ of order $(\prod_{i=1}^{n} p_i) \times (\prod_{i=1}^{n} p_i)$:

$$
H^m(n) = \begin{cases}
S\left(\frac{1}{\sqrt{p_1}}(1,\ldots,1)_{1\times p_1}, \Psi^{p_1}\right), & n = 1 \\
S\left(\frac{1}{\sqrt{p_n}}D_{p_n}\left(H^m(n-1)\right), T_{p_1\cdots p_{n-1}}\left(\Psi^{p_n}\right)\right), & n = 2,\ldots,N
\end{cases}.
$$

For the case $n = N$ we shall write $H(m)$ instead of writing $H^m(n)$.

EXAMPLE 2.4. Let $m = 6$, then $p_1 = 3$, $p_2 = 2$. We have:

$$
H^6(1) = \begin{pmatrix}
\frac{1}{\sqrt{3}} & \frac{1}{\sqrt{3}} & \frac{1}{\sqrt{3}} \\
\frac{1}{\sqrt{6}} & \frac{1}{\sqrt{6}} & -\sqrt{\frac{2}{3}} \\
\frac{1}{\sqrt{2}} & -\frac{1}{\sqrt{2}} & 0
\end{pmatrix},
$$

$$
H^6(2) = \begin{pmatrix}
\frac{1}{\sqrt{6}} & \frac{1}{\sqrt{6}} & \frac{1}{\sqrt{6}} & \frac{1}{\sqrt{6}} & \frac{1}{\sqrt{6}} & \frac{1}{\sqrt{6}} \\
\frac{1}{2\sqrt{3}} & \frac{1}{2\sqrt{3}} & \frac{1}{2\sqrt{3}} & \frac{1}{2\sqrt{3}} & -\frac{1}{\sqrt{3}} & -\frac{1}{\sqrt{3}} \\
\frac{1}{2} & \frac{1}{2} & -\frac{1}{2} & -\frac{1}{2} & 0 & 0 \\
\frac{1}{\sqrt{2}} & -\frac{1}{\sqrt{2}} & 0 & 0 & 0 & 0 \\
0 & 0 & \frac{1}{\sqrt{2}} & -\frac{1}{\sqrt{2}} & 0 & 0 \\
0 & 0 & 0 & 0 & \frac{1}{\sqrt{2}} & -\frac{1}{\sqrt{2}}
\end{pmatrix}.
$$

THEOREM 2.5. *The matrices $H^m(n),\ n = 1,\ldots,N$ are orthonormal.*

PROOF. The proof is obtained by induction. Observation 2(iii) implies that $H^m(1)$ is orthonormal. Let $H^m(n-1)$ be orthonormal. The set of rows of both matrices $\frac{1}{\sqrt{p_n}}D_{p_n}(H^m(n-1))$, $T_{p_1\ldots p_{n-1}}(\Psi^{p_n})$ is orthonormal (see Observation 1(ii) and the inductive hypothesis), so we have to prove:

$$\frac{1}{\sqrt{p_n}}D_{p_n}(H^m(n-1))\left(T_{p_1\ldots p_{n-1}}(\Psi^{p_n})\right)^* = \mathbf{O},$$

where $\mathbf{O}$ is the zero matrix of order $(p_1\ldots p_{n-1})\times(p_n-1)(p_1\ldots p_{n-1})$ (the symbolism A^* indicates the transpose of the matrix A). Observation 1(i) implies that: $\frac{1}{\sqrt{p_n}}D_{p_n}(H^m(n-1)) = S(A_0,A_1,\ldots,A_{n-2})$, where $A_0 = \frac{1}{\sqrt{p_n\ldots p_2}}D_{p_n\ldots p_2}(H^m(1))$ and $A_k = \frac{1}{\sqrt{p_n\ldots p_{k+2}}}T_{p_1\ldots p_k}\left(D_{p_n\ldots p_{k+2}}(\Psi^{p_{k+1}})\right)$, $k=1,\ldots,n-2$, so it suffices to prove: $A_k\left(T_{p_1\ldots p_{n-1}}(\Psi^{p_n})\right)^* = \mathbf{O}$, $k=0,\ldots,n-2$, where $\mathbf{O}$ is the zero matrix of order $(p_{k+1}-1)(p_1\ldots p_k)\times(p_n-1)(p_1\ldots p_{n-1})$. The key idea for the last equality is the fact that all rows of Ψ^{p_n} have zero mean. For more details see [**AK**]. $\square$

Observation 3 The multiresolution structure arised from $H(m)$, for $m = p^N$.

Let V_m be the space of all real-valued sequences of length m and let h_i be the i-row of the Haar matrix $H(m)$, then any element $t \in V_m$ can be written as:

$$t(n) = \sum_{i=1}^{m}\langle t,h_i\rangle h_{i,n}.$$

Let $j=0,\ldots,N-1$, $k=1,\ldots,p-1$, we define $W_{j,k} = span\{h_{kp^j+s} : s=1,\ldots,p^j\}$. If V_0 is the space of constant sequences, we have the decomposition:

$$V_m = V_0 \oplus_{j=0}^{N-1}\oplus_{k=1}^{p-1}W_{j,k}.$$

EXAMPLE 2.6. Let $m = 3^3$, then $V_m = V_0 + W_{0,1} + W_{0,2} + W_{1,1} + W_{1,2} + W_{2,1} + W_{2,2}$, where

$$V_0 = span\{h_1\}, W_{0,1} = span\{h_2\}, W_{0,2} = span\{h_3\},$$

$$W_{1,1} = span\{h_4,h_5,h_6\}, W_{1,2} = span\{h_7,h_8,h_9\},$$

$$W_{2,1} = span\{h_{10},\ldots,h_{18}\}, W_{2,2} = span\{h_{19},\ldots,h_{27}\}.$$

3. Haar Riesz factorization of data and weak* approximation of singular measures

In the remaining of the text, h_j are rows of the matrix $H(m)$. We start with the following:

EXAMPLE 3.1. Let $H^6(2)$ be the matrix defined in Example 5, we suppose that $t = \{t_1,t_2,\ldots,t_6\} = \prod_{k=1}^{6}(1+a_kh_k)$ and we examine the inner products: $\left\langle t,h_j\right\rangle = \langle\prod_{k=1}^{6}(1+a_kh_k),h_j\rangle$, $j=1,\ldots,6$. We observe that

$$\prod_{k=1}^{6}(1+a_kh_k) = 1 + \sum_{i=1}^{6}a_ih_i + \sum_{i_1=1}^{5}\sum_{i_2=i_1+1}^{6}a_{i_1}a_{i_2}h_{i_1}h_{i_2} + \ldots + (a_1\ldots a_6)h_1\ldots h_6.$$

The fact that $H(m)$ is orthonormal and the equality above imply that:

$$\langle t, h_j \rangle = \left\langle \prod_{k=1}^{j-1} (1 + a_k h_k), h_j \right\rangle a_j, \ j = 1, \ldots, 6.$$

If we denote by $supp(h_j) = \{k = 1, ..., m : h_{j,k} \neq 0\}$, then observation 2(ii) states that if $supp(h_j) \bigcap supp(h_n) \neq \emptyset, j > n$, then for all $k \in supp(h_j)$, $h_{n,k} = h_{n,j_0}$ where h_{j,j_0} is the first non-zero element of the row h_j. Therefore:

$$\langle t, h_j \rangle = \begin{cases} \sqrt{6} + a_1, & j = 1 \\ \prod_{k=1}^{j-1} (1 + a_k h_{k,n_0}) a_j, & j = 2, \ldots, 6 \end{cases},$$

or

$$\frac{t_1 + t_2 + t_3 + t_4 + t_5 + t_6}{\sqrt{6}} = 1 + \frac{a_1}{\sqrt{6}}$$

$$\frac{t_1 + t_2 + t_3 + t_4}{2\sqrt{3}} - \frac{t_5 + t_6}{\sqrt{3}} = \left(1 + \frac{a_1}{\sqrt{6}}\right) a_2$$

$$\frac{t_1 + t_2}{2} - \frac{t_3 + t_4}{2} = \left(1 + \frac{a_1}{\sqrt{6}}\right)\left(1 + \frac{a_2}{2\sqrt{3}}\right) a_3$$

$$\frac{t_1}{\sqrt{2}} - \frac{t_2}{\sqrt{2}} = \left(1 + \frac{a_1}{\sqrt{6}}\right)\left(1 + \frac{a_2}{2\sqrt{3}}\right)\left(1 + \frac{a_3}{2}\right) a_4.$$

$$\frac{t_3}{\sqrt{2}} - \frac{t_4}{\sqrt{2}} = \left(1 + \frac{a_1}{\sqrt{6}}\right)\left(1 + \frac{a_2}{2\sqrt{3}}\right)\left(1 - \frac{a_3}{2}\right) a_5$$

$$\frac{t_5}{\sqrt{6}} - \frac{t_6}{\sqrt{2}} = \left(1 + \frac{a_1}{\sqrt{6}}\right)\left(1 - \frac{a_2}{\sqrt{3}}\right) a_6$$

The system of equations above with unknown $a_n, n = 1, \ldots, 6$ has a unique solution provided that all inner products are non zero. In fact we have:

THEOREM 3.2. *Let* $t = \{t_1, \ldots, t_m\}$ *be a sequence of real numbers such that:*

$$\langle t, h_i \rangle \neq 0, \ i = 1, \ldots, m,$$

then there is a unique sequence of coefficients $\{a_k : k = 1, \ldots, m\}$ *such that:*

$$t_n = \prod_{k=1}^{m} (1 + a_k h_{k,n}).$$

The coefficients $\{a_n : n = 1, \ldots, m\}$ *satisfy:*

$$a_n = \begin{cases} \langle t, h_1 \rangle - \sqrt{m} & n = 1 \\ \frac{\langle t, h_n \rangle}{\prod_{k=1}^{n-1} (1 + a_k h_{k,n_0})}, & n = 2, \ldots, m \end{cases},$$

where h_{n,n_0} *is the first non-zero element of the row* h_n.

PROOF. Based on ideas of previous example and Theorem 1 (more details in [**AK**]). □

Observation 4 The assumption that all inner products must be non zero, can be relaxed as follows: Given t as above and $\varepsilon > 0$, there exists a Haar Riesz Product such that $|t_n - \prod_{k=1}^{m} (1 + a_k h_{k,n})| < \varepsilon$. In fact, replace t with a data t' such that $|t - t'| < \varepsilon$ and whose all inner products are non-zero.

LEMMA 3.3. *Any continuous positive measure μ on $[0,1]$ can be approximated in the weak* topology by a sequence of Haar Riesz products $\{\mu_m, m = 2, 3, \ldots\}$:*

$$d\mu_m(x) = \prod_{k=1}^{m} (1 + a_k h_k(x)) \, dx,$$

where $h_k(x) = h_{k,n}$, $x \in \left[\frac{n-1}{m}, \frac{n}{m}\right)$, $k, n = 1, \ldots, m$, $h_{k,n}$ is the (k,n) entry of the matrix $H(m)$ and a_k are the corresponding coefficients.

PROOF. Apply Theorem 2 and Observation 4 on data $t = \{t_k, k = 1, \ldots, m\}$, where $t_m = \left\{\int_{k/m}^{(k+1)/m} d\mu, k = 1, \ldots, m\right\}$. □

4. Haar coefficients of Cantor-type Languages

Let $p = 3, 5, \ldots$ be a prime number. A Cantor-type language of length N in an alphabet $A = \{a_0, a_1, \ldots, a_{p-1}\}$ of p letters, is the set of all words $\{\varepsilon_1 \varepsilon_2 \ldots \varepsilon_N : \varepsilon_i \in A' \subset A, i = 1, \ldots, N\}$, where A' is a proper subset of A satisfying:

(i): A' contains of at least two letters and
(ii): $a_0 \in A'$, $a_1 \notin A'$.

The corresponding Cantor set on $[0,1)$ is the set $\left\{x = \sum_{n=1}^{N} \varepsilon_n p^{-n} : \varepsilon_n \in B\right\}$, where

$$B = \{i \in \{0, \ldots, p-1\} : a_i \in A'\}.$$

DEFINITION 4.1. We shall call the sequence $t = \{t_1, t_2, \ldots, t_{p^N}\}$:

$$t_n = \begin{cases} 1, & \text{whenever } n = 1 + \sum_{i=1}^{N} \varepsilon_i p^{N-i}, \varepsilon_i \in B \\ 0, & \text{otherwise} \end{cases},$$

indicator sequence of the Cantor-type language.

THEOREM 4.2. *Let t be the indicator sequence of a Cantor language.*

(a): *If $Q_j = \{p^j + 1 \le k \le p^{j+1} : \langle t, h_k \rangle = 0\}$, $j = 1, \ldots, N-1$, then*

$$Q_j = \{s + rp^j : s \in S_j, r = 0, \ldots, (p-2)\},$$

where $S_j = \left\{p^j + 1 \le k \le 2p^j : Mod\left(\left\lceil \frac{k}{p^i} \right\rceil - 1, p\right)\right\} \notin B, i = 0, \ldots, j-1$.
(b):

$$t_n = \sum_{i=0}^{N-1} \sum_{\substack{m = p^i + 1 \\ m \notin Q_i}}^{p^i + 1} \frac{\sqrt{p^i}}{c^i} \langle t, h_{\lceil m/p^i \rceil} \rangle h_{n,m},$$

where c is the cardinality of the set B.

PROOF. The key ideas for the proof are the following:

(i): any non-zero Haar coefficient of the indicator sequence can be written as

$$\langle t, h_{k+1} \rangle = \frac{c^j}{\sqrt{p^j}} \langle t, h_{kp^j+s} \rangle, j = 1, \ldots, N-1, s = 1, \ldots, p^j,$$

where $k = 1, \ldots, p-1$ and c is the cardinality of the set B,
(ii): $\langle t, h_{p^j+s} \rangle = 0 \iff t_n = 0$ for all n satisfying: $n = (s-1)p^{N-j} + 1, \ldots, sp^{N-j}$, where $j = 1, \ldots, N-1$, $s = 1, \ldots, p^j$ and

(iii): $\langle t, h_{p^j+s} \rangle = 0 \Longleftrightarrow \langle t, h_{kp^j+s} \rangle = 0$ for any $k = 1, \ldots, p-1$.

The complete proof is presented in [**AK**]. $\qquad\qquad\qquad\qquad\qquad$ $\square$

References

[AK] Atreas N. D., Karanikas C., *Multiscale Haar unitary matrices with the corresponding Riesz Products and a characterization of Cantor - type languages*, Fourier Anal. Appl., **13**, 2, (2007), 197-210.

[AKP1] Atreas N. D., Karanikas C. and Polychronidou P., *A class of Sparse Unimodular matrices generating Multiresolution and Sampling Analysis for data of any length*, submitted.

[AKP2] Atreas N., Karanikas C., Polychronidou P., *Signal Processing by an Immune Type Tree Transform*, Lecture Notes in Computer Science, **2787**, ICARIS 2003, J.Timmis et al. (Eds), Springer Verlag, Berlin Heildelberg, pp. 111-119.

[AKT] Atreas N., Karanikas C., Tarakanov A., *Signal Analysis on Strings for Immune-type pattern recognition*, Comparative and Functional Genomics, **5**, 1, (2004), 69-74.

[AP] Atreas N. D., Polychronidou P., *A class of sparse invertible matrices and their use for non-linear prediction of nearly periodic time series with fixed period*, submitted.

[BBK] Benedetto J. J., Bernstein E., and Konstantinidis I., *Multiscale Riesz Products and their support properties* , Acta Applicandae Mathematicae, **88**, (2005), 201-227.

[BP] Benedetto J. J. and Pfander G. E., *Periodic wavelet transforms and periodicity detection*, SIAM J. Appl. Math., **62**, (2002), 1329-1368.

[B] Bisbas A., *Singular measures with absolutely continuous convolution squares on locally compact groups*, Proc. Amer. Math. Soc., **127**, no 10, (1999), 2865-2869.

[BK1] Bisbas A., Karanikas C., *Dimension and entropy of a non-ergodic Markovian process and its relation to Rademacher Riesz Products*, Monatsh. Math. **118**, (1994), 21-32.

[BKM] Bisbas A., Karanikas C., Moran W., *Tameness for the distribution of sums of Markov random variables*, Math. Proc. Cambridge Philos. Soc., **121**, no 1, (1997), 115-127.

[BK2] Bisbas A. and Karanikas C., *On the Hausdorff Dimension of Rademacher Riesz Products*, Monatsh. Math., **110**, 15-21, (1990).

[GS] Girardi M., Sweldens W., *A new class of unbalanced Haar wavelets that form an unconditional basis for L_p on general measure spaces*, Fourier Anal. Appl., **3**, 4, (1997), 457-474.

[K] Karanikas C., *The Hausdorff dimension of very weak self-similar fractals described by the Haar wavelet system*, Chaos Solitons Fractals, **11**, No 1-3, (2000), 275-280.

[KP] Karanikas C. and Proios G., *A discrete transform based on the Tree structure of data for pattern recognition of immune type*, Chaos Solitons Fractals **17**, No 2-3, (2002), 195-201.

[KR] Krane D. E., Raymer M. L., *Fundamental Concepts of Bioinformatics*, Pearson Education Inc., San Francinsco, USA, (2003).

[LP] Lewis H. R., and Papadimitriou C. H., *Elements of the Theory of Computation*, Prentice-Hall, Englewood Cliffs, New Jersey, USA, (1981).

[S] Sweldens W., *The lifting scheme: A construction of second generation wavelets*, SIAM J. Appl. Math., **29**, No 2, (1998), 511-546.

TECHNOLOGICAL INSTITUTION OF WEST MACEDONIA, DEPARTMENT OF GENERAL SCIENCES, 501-00 KOILA KOZANIS, KOZANI, GREECE

E-mail address: `natreas@csd.auth.gr`

DEPARTMENT OF INFORMATICS, ARISTOTLE UNIVERSITY OF THESSALONIKI, 54-124, THESSALONIKI, GREECE

E-mail address: `karanika@csd.auth.gr`

Contemporary Mathematics
Volume **451**, 2008

Classes of Finite Equal Norm Parseval Frames

Peter G. Casazza and Nicole Leonhard

ABSTRACT. Finite equal norm Parseval frames are a fundamental tool in applications of Hilbert space frame theory. We will derive classes of finite equal norm Parseval frames for use in applications as well as reviewing the status of the currently known classes.

1. Introduction

Frames for Hilbert spaces were introduced by Duffin and Schaeffer [**20**] in 1952 to study some deep problems in nonharmonic Fourier series. Duffin and Schaeffer abstracted the fundamental notion of Gabor [**22**] for signal processing. These ideas did not generate much interest outside of nonharmonic Fourier series and signal processing until the landmark paper of Daubechies, Grossmannn, and Meyer [**19**] in 1986. After this ground breaking work the theory of frames began to be widely studied.

Frames are redundant sets of vectors in a Hilbert space, which yield one natural representation of each vector in the space, but may have infinitely many different representations for any given vector [**14**]. It is this redundancy that makes frames useful in applications. Frames have traditionally been used in signal processing because of their resilience to additive noise [**18**], resilience to quantization [**24**], as well as their numerical stability of reconstruction [**18**], and their greater freedom to capture important signal characteristics [**14**]. Today, frames play an important role in many applications in mathematics, science, and engineering. Some of these applications include time-frequency analysis [**25**], internet coding [**23**], speech and music processing [**39**], communication [**35**], medicine [**36**], quantum computing [**30**], and many other areas.

Applications generally use equal norm Parseval frames (see Section 2 for the definitions) because of the rapid reconstruction of vectors in addition to giving somewhat equal weight to each vector in the space. Unfortunately, there are a small number of examples of these frames and, far worse, there is no place to go to find the known classes of equal norm Parseval frames. In this paper we will derive several new classes of equal norm Parseval frames as well as reviewing the current known classes.

1991 *Mathematics Subject Classification*. Primary: 42C15, 46C99.
The first author was supported by NSF DMS 0405376.

2. An introduction to Parseval frames

We will denote by ℓ_2^N an N-dimensional real or complex Hilbert space while $\mathbb{R}^N$ (respectively, $\mathbb{C}^N$) will denote an N-dimensional real (respectively, complex) Hilbert space. If the result holds also for infinite dimensional Hilbert spaces, we will denote this $\ell_2(I)$ where I may be a finite or infinite index set.

A family of vectors $\{f_i\}_{i \in I}$ in a Hilbert space $\mathbb{H}$ is a **Riesz basic sequence** if there are constants $A, B > 0$ so that for all scalars $\{a_i\}_{i \in I}$ we have:

$$(1) \qquad A \sum_{i \in I} |a_i|^2 \leq \| \sum_{i \in I} a_i f_i \|^2 \leq B \sum_{i \in I} |a_i|^2.$$

We call A, B the **lower and upper Riesz basis bounds** for $\{f_i\}_{i \in I}$. If the Riesz basic sequence $\{f_i\}_{i \in I}$ spans $\mathbb{H}$ we call it a **Riesz basis** for $\mathbb{H}$. So $\{f_i\}_{i \in I}$ is a Riesz basis for $\mathbb{H}$ means there is an orthonormal basis $\{e_i\}_{i \in I}$ so that the operator $T(e_i) = f_i$ is invertible. In particular, each Riesz basis is **bounded**. That is, $0 < \inf_{i \in I} \|f_i\| \leq \sup_{i \in I} \|f_i\| < \infty$.

Hilbert space frames were introduced by Duffin and Schaeffer [20] to address some very deep problems in nonharmonic Fourier series (see [40]). A family $\{f_i\}_{i \in I}$ of elements of a (finite or infinite dimensional) Hilbert space $\mathbb{H}$ is called a **frame** for $\mathbb{H}$ if there are constants $0 < A \leq B < \infty$ (called the **lower and upper frame bounds**, respectively) so that for all $f \in \mathbb{H}$

$$(2) \qquad A\|f\|^2 \leq \sum_{i \in I} |\langle f, f_i \rangle|^2 \leq B\|f\|^2.$$

A good introduction to frames and Riesz bases is [14]. If we only have the right hand inequality in Equation 2 we call $\{f_i\}_{i \in I}$ a **Bessel sequence with Bessel bound B**. If $A = B$, we call this an A-**tight frame** and if $A = B = 1$, it is called a **Parseval frame**. If all the frame elements have the same norm, this is an **equal norm** frame and if the frame elements are of unit norm, it is a **unit norm frame**. It is immediate that $\|f_i\|^2 \leq B$. If also $\inf \|f_i\| > 0$, $\{f_i\}_{i \in I}$ is a **bounded frame**. The numbers $\{\langle f, f_i \rangle\}_{i \in I}$ are the **frame coefficients** of the vector $f \in \mathbb{H}$. If $\{f_i\}_{i \in I}$ is a Bessel sequence, the **synthesis operator** for $\{f_i\}_{i \in I}$ is the bounded linear operator $T : \ell_2(I) \to \mathbb{H}$ given by $T(e_i) = f_i$ for all $i \in I$. The **analysis operator** for $\{f_i\}_{i \in I}$ is T^* and satisfies: $T^*(f) = \sum_{i \in I} \langle f, f_i \rangle e_i$. In particular,

$$(3) \qquad \|T^* f\|^2 = \sum_{i \in I} |\langle f, f_i \rangle|^2, \text{ for all } f \in \mathbb{H},$$

and hence the smallest Bessel bound for $\{f_i\}_{i \in I}$ equals $\|T^*\|^2$. Comparing this to Equation 2 we have:

THEOREM 2.1. *Let $\mathbb{H}$ be a Hilbert space and $T : \ell_2(I) \to \mathbb{H}$, $Te_i = f_i$ be a bounded linear operator. The following are equivalent:*

(1) *$\{f_i\}_{i \in I}$ is a frame for $\mathbb{H}$.*

(2) *The operator T is bounded, linear, and onto.*

(3) *The operator T^* is an (possibly into) isomorphism.*

Moreover, if $\{f_i\}_{i \in I}$ is a Riesz basis, then the Riesz basis bounds are A, B, the frame bounds for $\{f_i\}_{i \in I}$.

It follows that a Bessel sequence is a Riesz basic sequence if and only if T^* is onto. The **frame operator** for the frame is the positive, self-adjoint invertible operator $S = TT^* : \mathbb{H} \to \mathbb{H}$. That is,

$$(4) \qquad Sf = TT^*f = T\left(\sum_{i \in I} \langle f, f_i \rangle e_i\right) = \sum_{i \in I} \langle f, f_i \rangle Te_i = \sum_{i \in I} \langle f, f_i \rangle f_i.$$

In particular,

$$(5) \qquad \langle Sf, f \rangle = \sum_{i \in I} |\langle f, f_i \rangle|^2.$$

It follows that $\{f_i\}_{i \in I}$ is a frame with frame bounds A, B if and only if $A \cdot I \leq S \leq B \cdot I$. So $\{f_i\}_{i \in I}$ is a Parseval frame if and only if $S = I$. **Reconstruction** of vectors in $\mathbb{H}$ is achieved via the formula:

$$\begin{aligned}
f &= SS^{-1}f = \sum_{i \in I} \langle S^{-1}f, f_i \rangle f_i \\
&= \sum_{i \in I} \langle f, S^{-1}f_i \rangle f_i \\
&= \sum_{i \in I} \langle f, f_i \rangle S^{-1}f_i \\
(6) \qquad &= \sum_{i \in I} \langle f, S^{-1/2}f_i \rangle S^{-1/2}f_i.
\end{aligned}$$

It follows that $\{S^{-1/2}f_i\}_{i \in I}$ is a Parseval frame *equivalent* to $\{f_i\}_{i \in I}$. Two sequences $\{f_i\}_{i \in I}$ and $\{g_i\}_{i \in I}$ in a Hilbert space are *equivalent* if there is an invertible operator T between their spans with $Tf_i = g_i$ for all $i \in I$.

REMARK 2.2. *Any finite set of vectors $\{f_i\}_{i=1}^M$ in a Hilbert space $\mathbb{H}$ has a frame operator $Sf = \sum_{i=1}^M \langle f, f_i \rangle f_i$ associated with it. S is a positive and self-adjoint operator but is not invertible unless $\{f_i\}_{i=1}^M$ spans $\mathbb{H}$.*

PROPOSITION 2.3. *Let $\{f_i\}_{i=1}^M$ be a frame for ℓ_2^N. If $\{g_j\}_{j=1}^N$ is an orthonormal basis of ℓ_2^N consisting of eigenvectors for the frame operator S with respective eigenvalues $\{\lambda_j\}_{j=1}^N$, then for every $1 \leq j \leq N$, $\sum_{i=1}^M |\langle f_i, g_j \rangle|^2 = \lambda_j$. In particular, $\sum_{i=1}^M \|f_i\|^2 = \text{Trace } S$ (= N if $\{f_i\}_{i=1}^M$ is a Parseval frame). Furthermore, if $\{f_i\}_{i \in I}$ is an equal norm Parseval frame for ℓ_2^N then $\|f_i\|^2 = \frac{N}{M}$.*

Another important result is

THEOREM 2.4. *If $\{f_i\}_{i \in I}$ is a frame for $\mathbb{H}$ with frame bounds A, B and P is any orthogonal projection on $\mathbb{H}$, then $\{Pf_i\}_{i \in I}$ is a frame for $P\mathbb{H}$ with frame bounds A, B.*

Proof: For any $f \in P(\mathbb{H})$,

$$\sum_{i \in I} |\langle f, Pf_i \rangle|^2 = \sum_{i \in I} |\langle Pf, f_i \rangle|^2 = \sum_{i \in I} |\langle f, f_i \rangle|^2.$$

$\square$

A fundamental result in frame theory was proved independently by Naimark and Han/Larson [14, 26]. For completeness we include its simple proof.

THEOREM 2.5. *A family $\{f_i\}_{i \in I}$ is a Parseval frame for a Hilbert space $\mathbb{H}$ if and only if there is a containing Hilbert space $\mathbb{H} \subset \ell_2(I)$ with an orthonormal basis $\{e_i\}_{i \in I}$ so that the orthogonal projection P of $\ell_2(I)$ onto $\mathbb{H}$ satisfies $P(e_i) = f_i$ for all $i \in I$.*

Proof. The "only if" part is Theorem 2.4. For the "if" part, if $\{f_i\}_{i \in I}$ is a Parseval frame, then the synthesis operator $T : \ell_2(I) \to \mathbb{H}$ is a partial isometry. So T^* is an isometry and we can associate $\mathbb{H}$ with $T^*\mathbb{H}$. Now, for all $i \in I$ and all $g = T^*f \in T^*(\mathbb{H})$ we have

$$\langle T^*f, Pe_i \rangle = \langle T^*f, e_i \rangle = \langle f, Te_i \rangle = \langle f, f_i \rangle = \langle T^*f, T^*f_i \rangle.$$

It follows that $Pe_i = T^*f_i$ for all $i \in I$. □

Theorem 2.5 helps explain why so few classes of equal norm Parseval frames are known. Namely, to get an equal norm Parseval frame we need to find orthogonal projections which map an orthonormal basis to equal norm vectors. There is very little known about such projections, consequently there lies the challenge. There is a universal method for obtaining Parseval frames given in the next lemma (See [**11**]).

LEMMA 2.6. *There is a unique method for constructing Parseval frames in ℓ_2^N. Let U be an $M \times M$, $M \geq N$, unitary matrix,*

$$U = \begin{bmatrix} u_{11} & . & . & . & u_{1M} \\ . & & & & . \\ . & & & & . \\ . & & & & . \\ u_{M1} & . & . & . & u_{MM} \end{bmatrix}.$$

Define

$$\begin{bmatrix} \varphi_1 \\ . \\ . \\ . \\ \varphi_M \end{bmatrix} = \begin{bmatrix} u_{11} & . & . & . & u_{1N} \\ . & & & & . \\ . & & & & . \\ . & & & & . \\ u_{M1} & . & . & . & u_{MN} \end{bmatrix}, \quad N \leq M.$$

The rows $\{\varphi_i\}_{i=1}^M$ form a Parseval frame for ℓ_2^N.

Another important property of frames comes from [**13**].

THEOREM 2.7. *A family of vectors $\{f_i\}_{i \in I}$ is a frame with frame bounds A and B if and only if forming a matrix C with the $f_i's$ as row vectors, the corresponding column vectors of C form a Riesz basic sequence with Riesz basis bounds A, B.*

There is a classification of the sequence of norms of frame vectors which yield a given frame operator due to Casazza and Leon [**12**]. This result can also be derived from the Schur-Horn Theorem [**2**].

THEOREM 2.8. *Let S be a positive self-adjoint operator on an N-dimensional Hilbert space l_2^N. Let $\lambda_1 \geq \lambda_2 \geq \ldots \geq \lambda_N$ be eigenvalues of S. Fix $M \geq N$ and real numbers $a_1 \geq a_2 \geq \ldots \geq a_M \geq 0$. The following are equivalent:*

(1) There is a frame $\{f_i\}_{i=1}^M$ for l_2^N with frame operator S and $\|f_i\| = a_i$ for all $i = 1, 2, \cdots, M$.

(2) For every $1 \leq k \leq N$ we have

(7)
$$\sum_{j=1}^{k} a_j^2 \leq \sum_{j=1}^{k} \lambda_j,$$

and

(8)
$$\sum_{j=1}^{M} a_j^2 = \sum_{j=1}^{N} \lambda_j.$$

For some reason the following important corollary of Theorem 2.8 has been overlooked until now.

COROLLARY 2.9. *Let S be a positive self-adjoint operator on a N-dimensional Hilbert space l_2^N. For any $M \geq N$ there is an equal norm sequence $\{f_m\}_{m=1}^{M}$ in l_2^N which has S as its frame operator.*

PROOF. Let $\lambda_1 \geq \lambda_2 \geq \ldots \lambda_N \geq 0$ be the eigenvalues of S. Let

(9)
$$a^2 = \frac{1}{M} \sum_{i=1}^{N} \lambda_i.$$

Now we check the conditions of Theorem 2.8 to see that there is a sequence $\{f_i\}_{i=1}^{M}$ in l_2^N with $\|f_i\| = a$ for all $i = 1, 2, \ldots, M$ and the frame operator of $\{f_i\}_{i=1}^{M}$ is precisely S. We are letting $a_1 = a_2 = \ldots a_M = a$. For the second equality in Theorem 2.8, by Equation 9 we have

(10)
$$\sum_{i=1}^{M} \|f_i\|^2 = \sum_{i=1}^{M} a_i^2 = Ma^2 = \sum_{j=1}^{N} \lambda_j.$$

For the first inequality in Theorem 2.8, we note that by Equation 9 we have that

$$a_1^2 = a^2 = \frac{1}{M} \sum_{j=1}^{N} \lambda_j \leq \frac{1}{N} \sum_{j=1}^{N} \lambda_j \leq \lambda_1.$$

So our inequality holds for $i = 1$. Suppose there is an $1 < i \leq N$ for which this inequality fails and i is the first time this fails. So,

$$\sum_{m=1}^{i-1} a_m^2 = (i-1)a^2 \leq \sum_{j=1}^{i-1} \lambda_j,$$

while

$$\sum_{m=1}^{i} a_m^2 = ia^2 > \sum_{j=1}^{i} \lambda_j.$$

It follows that

$$a_i^2 = a^2 > \lambda_i \geq \lambda_{i+1} \geq \lambda_N.$$

Hence,

$$
\begin{aligned}
Ma^2 = \sum_{i=1}^{M} a_i^2 &\geq \sum_{m=1}^{i} a_m^2 + \sum_{m=i+1}^{N} a_m^2 \\
&> \sum_{j=1}^{i} \lambda_j + \sum_{m=i+1}^{N} a_m^2 \\
&\geq \sum_{j=1}^{i} \lambda_j + \sum_{j=i+1}^{N} \lambda_j \\
&= \sum_{j=1}^{N} \lambda_j.
\end{aligned}
$$

But this contradicts Equation 9. $\qquad\qquad\qquad\qquad\qquad\qquad\qquad\square$

We give two more important consequences of Theorem 2.8.

COROLLARY 2.10. *For every $m \geq n$ there is an equal norm Parseval frame for ℓ_2^n containing exactly m-elements.*

COROLLARY 2.11. *Given an N-dimensional Hilbert space l_2^N and a sequence of positive numbers $\{a_i\}_{i=1}^{M}$ with $a_1 \geq a_2 \geq \ldots \geq a_M$, there exists a tight frame $\{f_i\}_{i=1}^{M}$ for l_2^N with $\|f_i\| = a_i$ for all $i = 1, 2, \ldots, M$ if and only if*

$$
a_1^2 \leq \frac{1}{N} \sum_{i=1}^{M} a_i^2.
$$

3. Constructing Tight Frames from sets of vectors

This section will address four methods used to construct tight frames from finite sets of vectors. We also discuss each method's advantages and disadvantages.

Method I: Let $\{f_i\}_{i=1}^{M}$ be a set of norm one vectors in ℓ_2^N. For every $j = 1, \ldots, M$ let $\{f_{ij}\}_{i=1}^{N}$ be an orthonormal basis for ℓ_2^N with $f_{1j} = f_j$. The family $\{f_{ij}\}_{i \in \{1,\ldots,N\},\, j \in \{1,\ldots,M\}}$ is an A-tight frame with tight frame bound A=M.

Note: Method I shows that every finite set of vectors is part of a tight frame for a Hilbert space. But, this technique has the disadvantage that the tight frame bound is exceptionally large, i.e $A = M$.

Method II: Let $\{f_i\}_{i=1}^{M}$ be a set of vectors in ℓ_2^N not all of which are zero. We can add N-1 vectors $\{h_j\}_{j=2}^{N}$ to the family so that $\{f_i\}_{i=1}^{M} \cup \{h_j\}_{j=2}^{N}$ is a tight frame.

Proof of Method II: Let $\{g_j\}_{j=1}^{N}$ be an eigenbasis for the frame operator of $\{f_i\}_{i=1}^{M}$ with respective eigenvalues $\{\lambda_j\}_{j=1}^{N}$, some of which may be zero but one of which must be non-zero. Without loss of generality assume $\lambda_1 \geq \lambda_2 \geq \ldots \geq \lambda_N \geq 0$. For $2 \leq j \leq N$, let $h_j = \sqrt{\lambda_1 - \lambda_j}\, g_j$. If S_1 is the frame operator for $\{f_i\}_{i=1}^{M} \cup \{h_j\}_{j=2}^{N}$ then for all $f \in \ell_2^N$

$$
\begin{aligned}
S_1 f &= \sum_{i=1}^{M} \langle f, f_i \rangle f_i + \sum_{j=2}^{N} \langle f, h_j \rangle h_j \\
&= \sum_{j=1}^{N} \lambda_j \langle f, g_j \rangle g_j + \sum_{j=2}^{N} \sqrt{\lambda_1 - \lambda_j} \langle f, g_j \rangle \sqrt{\lambda_1 - \lambda_j} g_j \\
&= \sum_{j=1}^{N} \lambda_j \langle f, g_j \rangle g_j + \sum_{j=2}^{N} (\lambda_1 - \lambda_j) \langle f, g_j \rangle g_j \\
&= \lambda_1 \langle f, g_1 \rangle g_1 + \sum_{j=2}^{N} \lambda_1 \langle f, g_j \rangle g_j \\
&= \lambda_1 \sum_{j=1}^{N} \langle f, g_j \rangle g_j \\
&= \lambda_1 f
\end{aligned}
$$

Therefore $\{f_i\}_{i=1}^{M} \cup \{h_j\}_{j=2}^{N}$ is a λ_1-tight frame.

Remark: The advantage to this method is that the upper frame bound is the same as the upper frame bound of the original set of vectors and we have to add very few vectors to make the frame tight. However, if the original set consists of equal norm vectors, this method does not ensure that the new frame will be an equal norm frame. In general, even if $\{f_i\}_{i=1}^{M}$ is an equal norm frame for ℓ_2^N we can't make $\{f_i\}_{i=1}^{M} \cup \{h_j\}_{j=2}^{N}$ tight be adding N-1 vectors of the same norm as f_i.

The next method for producing equal norm tight frames comes from [**3**].

Method III: Let $\{f_i\}_{i=1}^{M}$ be a unit norm Bessel sequence in l_2^N with Bessel bound B. There is a unit norm family $\{g_j\}_{j=1}^{K}$ so that $\{f_i\}_{i=1}^{M} \cup \{g_j\}_{j=1}^{K}$ is a unit norm tight frame with tight frame bound $\lambda \leq B + 2$.

Proof of Method III: This theorem and proof are done in the finite dimensional case. The infinite dimensional case also holds by a similar argument using the results of [**10**]. Let $\{f_i\}_{i=1}^{M}$ be a unit norm Bessel sequence with Bessel bound B in l_2^N. Let S be the frame operator for this family and let $\lambda_1 \geq \lambda_2 \geq ..... \geq \lambda_N \geq 0$ be the eigenvalues with respective eigenvectors $\{e_i\}_{i=1}^{N}$ for S. So B=λ_1. If we consider $N(\lambda_1 + 1 + \epsilon) - M$, we see that this equals $N\lambda_1 + N - M$ if $\epsilon = 0$ and it equals $N\lambda_1 + 2N - M$ If $\epsilon = 1$. In particular, there is an $0 \leq \epsilon \leq 1$ so that $N(\lambda_1 + 1 + \epsilon) - M = K \geq N$ where $K \in \mathbb{N}$. Now let S_0 be the positive self-adjoint operator on $\mathbb{H}_N$ given by

$$
(1) \qquad S_0 \left(\sum_{j=1}^{N} c_j e_j \right) = \sum_{j=1}^{N} [(\lambda_1 + 1 + \epsilon) - \lambda_j] c_j e_j.
$$

So S_0 is a positive self-adjoint operator on $\mathbb{H}_N$ with eigenvectors $\{e_j\}_{j=1}^{N}$ having respective eigenvalues $\{\lambda_1 + 1 + \epsilon - \lambda_j\}_{j=1}^{N}$ (which are now in increasing order).

Since each of these eigenvalues is greater than 1, letting $a_i = 1$ for $i = 1, 2, ..., K$ we immediately have the first inequality given in (2) of Theorem 2.8. Also,

$$\sum_{j=1}^{N} [(\lambda_1 + 1 + \epsilon) - \lambda_j] = N(\lambda_1 + 1 + \epsilon) - \sum_{j=1}^{N} \lambda_j = N(\lambda_1 + 1 + \epsilon) - M = K.$$

The last equality above follows from the fact that

$$\sum_{i=1}^{M} \|f_i\|^2 = M = \sum_{j=1}^{N} \lambda_j.$$

Applying Theorerm 2.8, there is a family of unit norm vectors $\{g_j\}_{j=1}^{K}$ in l_2^N having S_0 for its frame operator. If follows that $\{f_i\}_{i=1}^{M} \cup \{g_j\}_{j=1}^{N}$ is a unit norm frame for l_2^N having frame operator $S + S_0$. But $S + S_0$ has eigenvetors $\{e_j\}_{j=1}^{N}$ with respective eigenvalues

$$[(\lambda_1 + 1 + \epsilon) - \lambda_j] + \lambda_j = \lambda + 1 + \epsilon =: \lambda$$

So our unit norm frame is tight with tight frame bound $\lambda \leq \lambda_1 + 2$. $\quad\square$

For our next method, we first need to recall a standard result.

PROPOSITION 3.1. *If S is a positive, self-adjoint bounded operator on $\ell_2(I)$, then $\{S^{1/2}e_i\}_{i \in I}$ is a sequence of vectors with frame operator S for any orthonormal basis $\{e_i\}_{i \in I}$. In particular, if S is also invertible, then we conclude that there is a Riesz basis for $\mathbb{H}$ having frame operator S.*

PROOF. For any $f \in \mathbb{H}$ we have:

$$\sum_{i \in I} \langle f, S^{1/2}e_i \rangle S^{1/2}e_i = S^{1/2} \left(\sum_{i \in I} \langle S^{1/2}f, e_i \rangle e_i \right)$$
$$= S^{1/2} \left(S^{1/2}f \right) = Sf.$$

$\quad\square$

Method IV: Given a Bessel sequence of vectors $\{f_i\}_{i=1}^{M}$ in l_2^N with Bessel bound B and frame operator S, let $\{g_j\}_{j \in J}$ be a family of vectors which has $BI - S$ as its frame operator. Then $\{f_i\}_{i=1}^{M} \cup \{g_j\}_{j \in J}$ is a frame for l_2^N with frame operator $S + (BI - S) = BI$. i.e. This is a tight frame.

Massey and Ruiz [33] generalized Theorem 2.8 to the case where we want to add vectors of prescribed norms to a given family of vectors and end up with a tight frame. There are also variations of this result in [33] including the infinite dimensional case.

THEOREM 3.2 (Massey and Ruiz). *Given vectors $\{f_i\}_{i \in I}$ in $\mathbb{H}_N$ with frame operator S having trace α and eigenvalues $\{\lambda_j\}_{j=1}^{N}$ and a non-increasing sequence $\{a_i\}_{i=1}^{M}$ of positive real numbers, there is a sequence of vectors $\{g_i\}_{i=1}^{M}$ in $\mathbb{H}_N$ with $\|g_i\|^2 = a_i$ and $\{f_i\}_{i \in I} \cup \{g_i\}_{i=1}^{M}$ is a tight frame if and only if*

$$\frac{1}{N} \left(\sum_{i=1}^{M} a_i + \alpha \right) \geq \lambda_1,$$

and

$$\frac{1}{N}\left(\sum_{i=1}^{M} a_i + \alpha\right) \geq \frac{1}{k}\sum(a_i + \lambda_{n-i+1}), \quad 1 \leq k \leq min\{N, M\}.$$

4. Harmonic Frames

In this section we will define three types of harmonic frames and show each type is an equal norm Parseval frame. For results on harmonic frames we refer the reader to [**11, 23, 37, 41**].

4.1. Real Harmonic Frames.

THEOREM 4.1. *The family* $\{\varphi_i\}_{i=0}^{M-1}$ *is an orthonormal basis for* $\mathbb{R}^M$ *where for* $M = 2k+1$

(1)
$$\begin{bmatrix} \varphi_0 \\ \varphi_1 \\ \varphi_2 \\ \vdots \\ \varphi_{M-3} \\ \varphi_{M-2} \\ \varphi_{M-1} \end{bmatrix} = \sqrt{\frac{2}{M}}\begin{bmatrix} \frac{1}{\sqrt{2}} & \frac{1}{\sqrt{2}} & \frac{1}{\sqrt{2}} & \cdots & \frac{1}{\sqrt{2}} \\ 1 & \cos 2\pi\frac{1}{M} & \cos 2\pi\frac{2}{M} & \cdots & \cos 2\pi\frac{(M-1)}{M} \\ 0 & \sin 2\pi\frac{1}{M} & \sin 2\pi\frac{2}{M} & \cdots & \sin 2\pi\frac{M-1}{M} \\ \vdots & \vdots & \vdots & & \vdots \\ 1 & \cos 2\pi\frac{k}{M} & \cos 2\pi\frac{2k}{M} & \cdots & \cos 2\pi\frac{(k(M-1))}{M} \\ 0 & \sin 2\pi\frac{k}{M} & \sin 2\pi\frac{2k}{M} & \cdots & \sin 2\pi\frac{(k(M-1))}{M} \end{bmatrix}$$

and for M=2k

(2)
$$\begin{bmatrix} \varphi_0 \\ \varphi_1 \\ \varphi_2 \\ \vdots \\ \varphi_{M-3} \\ \varphi_{M-2} \\ \varphi_{M-1} \end{bmatrix} = \sqrt{\frac{2}{M}}\begin{bmatrix} \frac{1}{\sqrt{2}} & \frac{1}{\sqrt{2}} & \frac{1}{\sqrt{2}} & \cdots & \frac{1}{\sqrt{2}} \\ 1 & \cos 2\pi\frac{1}{M} & \cos 2\pi\frac{2}{M} & \cdots & \cos 2\pi\frac{(M-1)}{M} \\ 0 & \sin 2\pi\frac{1}{M} & \sin 2\pi\frac{2}{M} & \cdots & \sin 2\pi\frac{M-1}{M} \\ \vdots & \vdots & \vdots & & \vdots \\ 1 & \cos 2\pi\frac{k-1}{M} & \cos 2\pi\frac{2(k-1)}{M} & \cdots & \cos 2\pi\frac{(k-1)(M-1)}{M} \\ 0 & \sin 2\pi\frac{(k-1)}{M} & \sin 2\pi\frac{2(k-1)}{M} & \cdots & \sin 2\pi\frac{((k-1)(M-1))}{M} \\ \frac{1}{\sqrt{2}} & -\frac{1}{\sqrt{2}} & \frac{1}{\sqrt{2}} & \cdots & -\frac{1}{\sqrt{2}} \end{bmatrix}.$$

PROOF. Let $\{\varphi_j\}_{j=0}^{M-1}$ be defined above. First, we want to show that each φ_j has norm one.
For this we just need to check the terms in (1) above. For j=1, $\|\varphi_j\|^2 = \frac{2}{M} \times \frac{M}{2} = 1$.
If $1 \leq 2q - 1 \leq M - 1$,

$$\begin{aligned}
\|\varphi_{2q-1}\|^2 &= \frac{2}{M} \sum_{j=0}^{M-1} \cos^2 2\pi \frac{qj}{M} \\
&= \frac{2}{M} \frac{1}{2} \sum_{j=0}^{M-1} \left(1 + \cos 2\pi 2 \frac{qj}{M} \right) \\
&= \frac{2}{M} \left(\frac{M}{2} + \frac{1}{2} \sum_{j=0}^{M-1} \cos 2\pi 2 \frac{qj}{M} \right) \\
&= \frac{2}{M} \left(\frac{M}{2} + \frac{1}{2} Re \sum_{j=0}^{M-1} \left(\omega^{2q} \right) \right), \ \omega = e^{\frac{2\pi i}{M}} \\
&= \frac{2}{M} \left(\frac{M}{2} + \frac{1}{2} Re \left(\frac{1 - \left(\omega^{2q} \right)^j}{1 - \omega^{2q}} \right) \right) \\
&= 1.
\end{aligned}$$

Similarly, if $0 < 2q \le M - 1$ is even

$$\begin{aligned}
\|\varphi_{2q}\|^2 &= \frac{2}{M} \sum_{j=0}^{M-1} \sin^2 2\pi \frac{qj}{M} \\
&= \frac{2}{M} \left(\frac{M}{2} - \frac{1}{2} \sum_{j=0}^{M-1} \cos 2\pi 2 \frac{qj}{M} \right) \\
&= 1.
\end{aligned}$$

It remains to show that the φ_j's are orthogonal. Again, it suffices to check (1). Let $\omega = e^{\frac{2\pi i}{M}}$. For $0 < 2q$ even

$$\begin{aligned}
\langle \varphi_0, \varphi_{2q-1} \rangle &= \frac{2}{M\sqrt{2}} \sum_{j=0}^{M-1} \cos 2\pi \frac{qj}{M} \\
&= \frac{2}{M\sqrt{2}} Re \sum_{j=0}^{M-1} \left(\omega^q \right)^j \\
&= 0.
\end{aligned}$$

For $2q - 1$ odd

$$
\begin{aligned}
\langle \varphi_0, \varphi_{2q-1} \rangle &= \frac{2}{M\sqrt{2}} \sum_{j=0}^{M-1} \sin 2\pi \frac{qj}{M} \\
&= \frac{2}{M\sqrt{2}} Im \sum_{j=0}^{M-1} (\omega^q)^j \\
&= 0.
\end{aligned}
$$

Finally,

$$
\begin{aligned}
\langle \varphi_{2q}, \varphi_{2\ell-1} \rangle &= \frac{2}{M} \sum_{j=1}^{M-1} \cos \frac{2\pi j (q - \ell)}{M} \sin \frac{2\pi j (q - \ell)}{M} \\
&= \frac{1}{M} \sum_{j=1}^{M-1} \sin \frac{2\pi j 2 (q - \ell)}{M} \\
&= \frac{1}{M} Im \sum_{j=1}^{M-1} \left(\omega^{2(q-\ell)} \right)^j \\
&= \frac{1}{M} Im \left[\sum_{j=1}^{M-1} \left(\omega^{2(q-\ell)} \right)^j - 1 \right] \\
&= \frac{1}{M} Im (0 - 1) \\
&= 0. \quad \square
\end{aligned}
$$

$\square$

By Lemma 2.6 if we take any N-columns, $N < M$, from the matrices given in Theorem 4.1, the corresponding row vectors form a Parseval frame for ℓ_2^N called a **(real) harmonic frame**. Similarly, we could take any N-columns from the transpose of these matrices, then the corresponding row vectors form a Parseval frame for ℓ_2^N.

4.2. Complex Harmonic Frames. In this subsection we look at the complex versions of the harmonic frames.

THEOREM 4.2. *The family* $\{\varphi_i\}_{i=0}^{M-1}$ *in* $\mathbb{C}^M$ *is an orthonormal basis for* $\mathbb{C}^M$ *where for* $\omega = e^{\frac{2\pi i}{M}}$

$$
\begin{bmatrix} \varphi_0 \\ \varphi_1 \\ \varphi_2 \\ \vdots \\ \varphi_{M-2} \\ \varphi_{M-1} \end{bmatrix} = \sqrt{\frac{1}{M}} \begin{bmatrix} 1 & 1 & 1 & \cdots & 1 \\ 1 & \omega^1 & \omega^2 & \cdots & \omega^{M-1} \\ 1 & \omega^2 & \omega^4 & \cdots & \omega^{2(M-1)} \\ \vdots & \vdots & \vdots & & \vdots \\ 1 & \omega^{(M-2)} & \omega^{2(M-2)} & \cdots & \omega^{(M-2)(M-1)} \\ 1 & \omega^{(M-1)} & \omega^{2(M-1)} & \cdots & \omega^{(M-1)(M-1)} \end{bmatrix}.
$$

PROOF. Let $\{\varphi_j\}_{j=0}^{M-1}$ and ω be defined as above. It is obvious that each φ_j has norm one. Now we want to show that the rows are orthogonal. For $\ell \neq j$ we have

$$
\begin{aligned}
\langle \varphi_k, \varphi_\ell \rangle &= \frac{1}{M} \sum_{k=0}^{M-1} \omega^{jk} \overline{\omega^{\ell k}} \\
&= \frac{1}{M} \sum_{k=0}^{M-1} \omega^{jk} \omega^{-\ell k} \\
&= \frac{1}{M} \sum_{k=0}^{M-1} \omega^{k(j-\ell)} \\
&= \frac{1}{M} \left(\frac{1 - \left(\omega^{j-\ell} \right)^M}{1 - \omega^{j-\ell}} \right) \\
&= \frac{1}{M} \left(\frac{1 - \left(\omega^M \right)^{j-\ell}}{1 - \omega^{j-\ell}} \right) \\
&= \frac{1}{M} \left(\frac{1 - 1^{j-\ell}}{1 - \omega^{j-\ell}} \right) = 0 \quad \square
\end{aligned}
$$

$\square$

4.3. General Harmonic Frames.

The material in this section is due to Casazza and Kovačević and comes from [**11**].

DEFINITION 4.3. *Fix $M \geq N$, $|c| = 1$, and $\{b_i\}_{i=1}^N$ with $|b_i| = \frac{1}{\sqrt{M}}$. Let $\{c_i\}_{i=1}^N$ be distinct M^{th} roots of c, and for $0 \leq k \leq M-1$ let $\varphi_k = \left(c_1^k b_1, c_2^k b_2, \cdots, c_N^k b_N \right)$. Then $\{\varphi_i\}_{i=0}^{M-1}$ is a* general harmonic frame *for $\mathbb{C}^N$.*

PROPOSITION 4.4. *Every general harmonic frame for $\mathbb{C}^N$ is unitarily equivalent to a frame of the form $\left\{ c^k \psi_k \right\}_{k=0}^{M-1}$, where $|c| = 1$ and $\{\psi_k\}_{k=0}^{M-1}$ is a harmonic frame.*

PROPOSITION 4.5. *Let $\{\psi_k\}_{k=0}^{M-1}$ be a harmonic frame and let $|c| = 1$. Then $\left\{ c^k \varphi_k \right\}_{k=0}^{M-1}$ is equivalent to $\{\psi_k\}_{k=0}^{M-1}$ if and only if c is an Mth root of unity and there is a permutation σ of $\{1, 2, \cdots, N\}$ so that $\varphi_{kj} = \psi_{k\sigma(j)}$, for all $0 \leq k \leq M-1$ and all $1 \leq j \leq N$. A general harmonic frame is equivalent to a harmonic tight frame if and only if it equals a harmonic tight frame.*

PROPOSITION 4.6. *The family $\left\{ c^k \psi_k \right\}_{k=0}^{M-1}$ is a general harmonic frame for $\mathbb{H}_N$ if and only if there is a vector $\varphi_0 \in \mathbb{C}^N$ with $\|\varphi_0\|^2 = \frac{N}{M}$, an orthonormal basis $\{e_i\}_{i=1}^N$ for $\mathbb{C}^N$ and a unitary operator U on $\mathbb{H}_N$ with $U e_i = c_i e_i$, with $\{c_i\}_{i=1}^N$ distinct M^{th} roots of some $|c| = 1$ so that $\varphi_k = U^k \varphi_0$, for all $0 \leq k \leq M-1$.*

THEOREM 4.7. *Let U be a unitary operator on $\mathbb{C}^N$, $\varphi_0 \in \mathbb{C}^N$ and assume $\left\{ U^k \varphi_0 \right\}_{k=0}^{M-1}$ is a equal norm Parseval frame for $\mathbb{C}^N$. Then $U^M = cI$ for some $|c| = 1$ and $\left\{ U^k \varphi_0 \right\}_{k=0}^{M-1}$ is a general harmonic frame. That is, the general harmonic frames are the only equal norm Parseval frames generated by a group of unitary operators with a single generator.*

A listing of all low dimensional harmonic frames and their properties can be found in [**37**].

4.4. Maximal Robustness to Erasures. The results in this section are due to Kovačević and Puschel and can be found in [**34**].

DEFINITION 4.8. *Let* $\{f_i\}_{i=1}^M$ *be a frame for* ℓ_2^N. *Let* F *be the matrix having the vectors* f_i *as its rows. We call* F **maximally robust to erasures** *if every* $n \times n$ *submatrix of* F *is invertible. If* U *is an* $m \times m$ *matrix and* F *is constructed by keeping all columns with indices in the set* $I \subset \{0, 1, \ldots, m\}$, *then we write*

$$F = U[I].$$

We will work with the **discrete Fourier transform** defined by

$$DFT_m = \frac{1}{\sqrt{m}} \left[\omega_m^{k\ell}\right]_{0 \le k, \ell \le m}, \quad \omega_m = e^{2\pi i/m}.$$

We have from [**34**]

THEOREM 4.9. *For* $n \le m$,

$$F = DFT_m[0, 1, \ldots, n],$$

is an equal norm Parseval frame which is maximally robust to erasures.

The above frames are complex. There are also real versions of these frames given in [**34**]. Define for odd $n = 2k + 1$,

(3)
$$U_n = \begin{bmatrix} 1 & & \\ & I_k & -iJ_k \\ & J_k & iI_k \end{bmatrix}$$

where J_k is I_k with the columns in reversed order.

THEOREM 4.10. *Let* $0 \le k < \frac{m-1}{2}$ *and* $n = 2k + 1$. *Then*

$$\begin{aligned} F &= DFT_m[0, 1, \ldots, k, m-k, m-k+1, \ldots, m-1]U_n \\ &= \left[[cos\,\frac{2j\ell\pi}{n}]_{0 \le j < m, 0 \le \ell \le k}[-sin\,\frac{2j\ell\pi}{n}]_{0 \le j < m, 1 \le \ell \le k}\right]. \end{aligned}$$

is an equal norm (real) tight frame with maximal robustness to erasures.

For even n, we first define

$$\widetilde{DFT_m} = \left[\omega_m^{(k+\frac{1}{2})\ell}\right]_{0 \le k, \ell \le m},$$

and

(4)
$$V_n = \begin{bmatrix} I_k & -iJ_k \\ J_k & iI_k \end{bmatrix}$$

THEOREM 4.11. *Let* $1 \le k \le \frac{m}{2}$ *and let* $n = 2k$. *Then*

$$\begin{aligned} F &= \widetilde{DFT_m}[0, 1, \ldots, k-1, m-k, m-k+1, \ldots m-1]\,V_n \\ &= \left[[cos\,\frac{2(j+\frac{1}{2})\ell\pi}{n}]_{0 \le j < m, o \le \ell < k}[-sin\,\frac{2(j+\frac{1}{2})\ell\pi}{n}]_{0 \le j < m, 1 \le \ell \le k}\right]. \end{aligned}$$

is a equal norm (real) tight frame which is maximally robust to erasures.

5. Structured Parseval Frames

5.1. Using Existing Tight Frames to Construct New Tight Frames.
Fix $N, K, M \in \mathbb{N}$ with $K \leq N$. Let
$$\mathcal{K} = \{A \subset \{1, 2, \ldots, N\} : |A| = K\}.$$
For every $A \in \mathcal{K}$, let $\{f_i^A\}_{i=1}^M$ be a unit norm tight frame for $\mathbb{H}_K$ (so the tight frame bound is M/K). For every $A \in \mathcal{K}$ with $A = \{i_1 < i_2 < \cdots i_K\}$ define $T_A : \mathbb{H}_N \to \mathbb{H}_K$ by $T_A(f)(j) = f(i_j)$.

PROPOSITION 5.1. *The family*
$$\{T_A^* f_i^A\}_{i=1, A \in \mathcal{K}}^M$$
is a unit norm tight frame for $\mathbb{H}_N$ with tight frame bound
$$\frac{M}{N} \binom{N}{K}.$$

PROOF. Each T_A^* is an isometric embedding. Moreover, if $f \in \mathbb{H}_N$ we have
$$
\begin{aligned}
\sum_{A \in \mathcal{K}} \sum_{i=1}^M |\langle f, T_A^* f_i^A \rangle|^2 &= \sum_{A \in \mathcal{K}} \sum_{i=1}^M |\langle T_A f, f_i^A \rangle|^2 \\
&= \sum_{A \in \mathcal{K}} \frac{M}{K} \|T_A f\|^2 \\
&= \frac{M}{K} \sum_{A \in \mathcal{K}} \sum_{i \in A} |F(i)|^2 \\
&= \frac{M}{K} \sum_{i=1}^N \sum_{A \in \mathcal{K}, \ i \in A} |f(i)|^2 \\
&= \frac{M}{K} \sum_{i=1}^N \binom{N}{K} |f(i)|^2 \\
&= \frac{M}{K} \binom{N}{K} \sum_{i=1}^N |f(i)|^2 \\
&= \frac{M}{N} \binom{N}{K} \|f\|^2. \quad \square
\end{aligned}
$$

$\square$

5.2. M-Circle and M-Semicircle Frames.
The results in this section are from Bodmann and Paulsen [**5**]. Bodmann and Paulsen introduced the notion of frame paths to construct M-circle and M-semicircle frames and an alternate frame path definition of real harmonic frames.

DEFINITION 5.2. *A continuous map $f{:}[a, b] \to \mathbb{R}^M$ (respectively, $\mathbb{C}^M$) is called a* **uniform frame path** *iff $\|f(t)\| = 1$ for all t and there are infinitely many choices of N such that $F_N = \{f\left(a + \frac{b-a}{N}\right), f\left(a + \frac{2(b-a)}{N}\right), \cdots, f(b)\}$ is an equal norm $\frac{N}{M}$-tight frame for $\mathbb{R}^M$ (respectively, $\mathbb{C}^M$). We call any such F_N a* **frame obtained by regular sampling of f.**

Examples of such frames would include real and complex harmonic frames [5].

In this section we discuss M-circle frame paths in $\mathbb{R}^M$, $M > 2$, where the image of the frame path is the union of M circles. Let $\{e_i\}_{i=1}^M$ be the canonical orthonormal basis for $\mathbb{R}^M$. The image of $\{e_i\}_{i=1}^M$ will be the union of unit circles in the $e_1 - e_2$-plane, $e_2 - e_3$-plane,$\cdots$,$e_{M-1} - e_M$ plane. For the cases we will consider let $N = 4M$. To define the continuous path one needs only to see that it is possible to traverse this union of M circles in a continuous manner, passing through each quarter circle exactly once. Since the intersection of these circles occur at the 2M points, $\pm e_1, \cdots , \pm e_M$, to define the path it is enough to make clear the order in which one passes through the above points.

PROPOSITION 5.3. *When $M > 2$ is even, the following path traverses each of the quarter circles exactly once,*

$$
\begin{array}{ccccccccc}
+e_1 & \rightarrow & +e_2 & \rightarrow & +e_3 & \rightarrow & \cdots & +e_M & \rightarrow \\
-e_1 & \rightarrow & -e_2 & \rightarrow & -e_2 & \rightarrow & \cdots & -e_M & \rightarrow \\
+e_1 & \rightarrow & -e_2 & \rightarrow & +e_3 & \rightarrow & \cdots & -e_M & \rightarrow \\
-e_1 & \rightarrow & +e_2 & \rightarrow & -e_3 & \rightarrow & \cdots & +e_M & \rightarrow \\
+e_1 &
\end{array}
$$

where the sign in the third and fourth rows alternates.

PROPOSITION 5.4. *When $M > 1$ is odd, the following path traverses each of the quarter circles exactly once,*

$$
\begin{array}{ccccccccc}
+e_1 & \rightarrow & +e_2 & \rightarrow & +e_3 & \rightarrow & \cdots & +e_M & \rightarrow \\
-e_1 & \rightarrow & -e_2 & \rightarrow & -e_2 & \rightarrow & \cdots & -e_M & \rightarrow \\
-e_1 & \rightarrow & +e_2 & \rightarrow & -e_3 & \rightarrow & \cdots & -e_M & \rightarrow \\
+e_1 & \rightarrow & -e_2 & \rightarrow & +e_3 & \rightarrow & \cdots & +e_M & \rightarrow \\
+e_1 &
\end{array}
$$

where the sign in the third and fourth rows alternates.

DEFINITION 5.5. *Now, using the above ordering we will define a piecewise smooth map $f : [0, 4M] \rightarrow \mathbb{R}^M$ so that on ith interval $[i - 1, i]$ the image of f traces the ith quarter circle given in the above ordering. So in the case that $M \geq 4$ is even this is accomplished be setting*

$$
(1) \quad f(t) = \begin{cases}
\left(\cos \frac{\pi t}{2}, \sin \frac{\pi t}{2}, 0, \cdots, 0\right) & 0 \leq t \leq 1 \\
\left(0, \cos \frac{\pi(t-1)}{2}, \sin \frac{\pi(t-1)}{2}, 0, \cdots, 0\right) & 1 \leq t \leq 2 \\
\left(0, \cdots, 0, \cos \frac{\pi(t-M+1)}{2}, \sin \frac{\pi(t-M+1)}{2}\right) & M - 1 \leq t \leq M \\
\left(-\sin \frac{\pi(t-M+1)}{2}, 0, \cdots, \cos \frac{\pi(t-M+1)}{2}\right) & M \leq t \leq M + 1 \\
\left(\sin \frac{\pi(t-4M+1)}{2}, 0, \cdots, \cos \frac{\pi(t-4M+1)}{2}\right) & 4M - 1 \leq t \leq 4M
\end{cases}
$$

[5].

THEOREM 5.6. *(The Circle Frames). If $n \in \mathbb{N}$, $N = 4nM$ with $M \geq 3$ and let $f : [0, 4M] \rightarrow \mathbb{R}^M$ denote the M-circle path defined above, then $\{f\left(\frac{4Mj}{N}\right) : j = 1, \cdots, N\}$ is an equal norm $\frac{N}{M}$-tight frame for $\mathbb{R}^M$.*

THEOREM 5.7. *(The Semicircle Frames).* *If* $f : [0,1] \to \mathbb{R}^2$ *be defined by* $f(t) = (\cos(\pi t), \sin(\pi t))$, *then for any* $N > 2$, *the set* $F_N = \{f\left(\frac{j}{N}\right) : 1 \le j \le N\}$ *is an equal norm,* $\frac{N}{2}$*-tight frame for* $\mathbb{R}^2$.

The construction of the **M-semicircles path** will be similar to the construction of the M-circles path in that the construction of the map $f : [0, 2M] \to \mathbb{R}^M$ is identical. The construction of these paths differs because to construct the M-semicircles path we need only choose a path that exhausts a connected semicircle on each of the M-circles.

PROPOSITION 5.8. *When* $M > 2$ *is even, the following path traverses a connected semicircle on each of the* M *circles exactly once,*

$$
\begin{array}{ccccccccc}
+e_1 & \to & +e_2 & \to & +e_3 & \to & \cdots & +e_M & \to \\
-e_1 & \to & +e_2 & \to & -e_3 & \to & \cdots & +e_M & \to \\
+e_1 & & & & & & & &
\end{array}
$$

When $M > 1$ *is odd, the following path traverses a connected semicircle on each of the* M *circles exactly once,*

$$
\begin{array}{ccccccccc}
+e_1 & \to & +e_2 & \to & +e_3 & \to & \cdots & +e_M & \to \\
-e_1 & \to & +e_2 & \to & -e_3 & \to & \cdots & -e_M & \to \\
+e_1 & & & & & & & &
\end{array} .
$$

[**5**]

THEOREM 5.9. *If* $n \in \mathbb{N}$, $N = 2nM$ *with* $M \ge 3$ *and* $f : [0, 2M] \to \mathbb{R}^M$ *denote the M-semicircle path defined above, then* F_N *is an equal norm* $\frac{N}{M}$*-tight frame obtained by regular sampling.*

6. Frames of Translates

Translates of a single function play a fundamental role in frame theory, time-frequency analysis, sampling theory and more [**1, 17, 18**].

There is a simple classification of which functions give (tight) frames of translates. For this we need a definition. For $x \in \mathbb{R}$ we define *translation by x* by

$$
\tau_x : L^2(\mathbb{R}) \to L^2(\mathbb{R}), (\tau_x f)(y) = f(y - x), \ y \in \mathbb{R}.
$$

We first introduce some notation. For a function $\phi \in L^1(\mathbb{R})$ we denote by $\hat{\phi}$ the *Fourier transform of* ϕ

$$
\hat{\phi}(\xi) = \int \phi(x) e^{-2\pi i \xi x}(x) dx.
$$

As usual the definition of the Fourier transform extends to an isometry $\phi \to \hat{\phi}$ on $L^2(\mathbb{R})$.

Now suppose $\phi \in L^2(\mathbb{R})$ and that $b > 0$. Let us identify the circle $\mathbb{T}$ with the interval $[0, 1)$ via the standard map $\xi \to e^{2\pi i \xi}$. We define the function $\Phi_b : \mathbb{T} \to \mathbb{R}$ by

$$
\Phi_b(\xi) = \sum_{n \in \mathbb{Z}} |\hat{\phi}(\frac{\xi + n}{b})|^2.
$$

Note that $\Phi_b \in L^1(\mathbb{T})$.

For any $n \in \mathbb{Z}$ we note that

$$\langle \tau_{nb}\phi, \phi \rangle = \langle e^{-2\pi i n \xi b} \hat{\phi}, \hat{\phi} \rangle = \frac{1}{b} \int_0^1 \Phi_b(\xi) e^{-2\pi i n \xi} d\xi = \frac{1}{b} \hat{\Phi}_b(n).$$

We now have a classification of (tight) frames of translates originally due to Benedetto and Li [4]. Casazza, Christensen and Kalton [9] and Kim and Lim [29] removed an unnecessary assumption from the results in [4] as well as generalizing these results.

THEOREM 6.1. *If* $\phi \in L^2(\mathbb{R})$, *and* $b > 0$ *then:*
(1) $(\tau_{nb}\phi)_{n\in\mathbb{Z}}$ *is an orthonormal sequence if and only if*

$$\Phi_b(\gamma) = b \quad a.e.$$

(2) $(\tau_{nb}\phi)_{n\in\mathbb{Z}}$ *is a Riesz basic sequence with frame bounds* A, B *if and only if*

$$bA \leq \Phi_b(\gamma) \leq bB \quad a.e.$$

(3) $(\tau_{nb}\phi)_{n\in\mathbb{Z}}$ *is a frame sequence with frame bounds* A, B *if and only if*

$$bA \leq \Phi_b(\gamma) \leq bB \quad a.e.$$

on $\mathbb{T} \setminus N_b$ *where* $N_b = \{\xi \in \mathbb{T} : \Phi_b(\xi) = 0\}$.

We also mention a surprising result from [9].

THEOREM 6.2. *Let* $I \subset \mathbb{Z}$ *be bounded below,* $a > 0$ *and* $g \in L^2(\mathbb{R})$. *Then* $\{T_{na}g\}_{n\in I}$ *is a frame sequence if and only if it is a Riesz basic sequence.*

7. Gabor Frames

An excellent reference for time-frequency analysis is Gröchenig's book [25]. Although we are only dealing with finite tight frames in this note, since we now discuss discrete Gabor frames we briefly mention the infinite dimensional equivalent for comparison.

Given a function $g \in L^2(\mathbb{R}^d)$, for any $f \in L^2(\mathbb{R}^d)$ we define **translation of f by** $x \in \mathbb{R}^d$ and **modulation of f by** $y \in \mathbb{R}^d$ respectively as

$$T_x(f)(t) = f(t - x) \quad \text{and} \quad M_y(f)(t) = e^{2\pi i y \cdot t} f(t).$$

DEFINITION 7.1. *Given a non-zero* **window function** $g \in L^2(\mathbb{R}^d)$ *and lattice parameters* $\alpha, \beta > 0$, *the set of time-frequency shifts*

$$\mathcal{G}(g, \alpha, \beta) = \{T_{\alpha k} M_{\beta n} g\}_{k,n\in\mathbb{Z}^d}$$

is called a **Gabor system**. *If this family forms a frame for* $L^2(\mathbb{R}^d)$, *it is called a* **Gabor frame** *or* **Weyl-Heisenberg frame**.

It is a very deep question when g, α, β yields a Gabor frame [25]. It can be shown [25] that the frame operator for a Gabor frame commutes with translation and modulation which yields:

THEOREM 7.2. *For any Gabor frame* $\mathcal{G}(g, \alpha, \beta)$ *with frame operator* S, *the family* $\mathcal{G}(S^{-1/2}g, \alpha, \beta)$ *is an equal norm Parseval frame for* $L^2(\mathbb{R}^d)$.

The Wexler-Raz biorthogonality relations give an exact calculation for determining when a Gabor frame is Parseval (See [25]).

THEOREM 7.3 (Wexler-Raz). *A Gabor frame $\mathcal{G}(g, \alpha, \beta)$ is a Parseval frame if and only if*

$$(\alpha\beta)^{-1}\langle g, M_{\frac{k}{\alpha}}T_{\frac{n}{\beta}}\rangle = \delta_{k0}\delta n0, \quad \text{for all } k, n \in \mathbb{Z}.$$

Now let us look at finite Gabor systems. Let $\omega = e^{2\pi i/n}$. The **translation operator** T is the unitary operator on $\mathbb{C}^n$ given by

$$Tx = T(x_0, x_1, \ldots, x_{n-1}) = (x_{n-1}, x_0, x_1, \ldots, x_{n-2}),$$

and the **modulation operator** M is the unitary operator defined by

$$Mx = M(x_0, x_1, \ldots, x_{n-1}) = (\omega^0 x_0, \omega^1 x_1, \ldots, \omega^{n-1} x_{n-1}).$$

Given a vector $g \in \mathbb{C}^n$, the **finite Gabor system with window function g** is the family

$$\left\{M^\ell T^k g\right\}_{\ell, k \in \mathbb{Z}_n}.$$

In the discrete case, Gabor systems always form tight frames. Although this has been folklore for quite some time, the first formal proof we have seen is in [**32**].

THEOREM 7.4. *For any $0 \neq g \in \mathbb{C}^n$, the collection $\{M^\ell T^k g\}_{\ell, k \in \mathbb{Z}_n}$ is an equal norm tight frame for $\mathbb{C}^n$ with tight frame bound $n^2\|f\|^2$.*

Lawrence, Pfander and Walnut [**32**] examined the linear independence of discrete Gabor systems.

THEOREM 7.5. *If n is prime, there is a dense open set E of full measure (i.e. The Lebesgue measure of $\mathbb{C}^n \setminus E$ is 0) in $\mathbb{C}^n$ such that for every $f \in E$, every subset of the Gabor system $\{M^\ell T^k g\}_{\ell, k \in \mathbb{Z}_n}$ containing n-elements is linearly independent.*

8. Filter Bank Frames

Important results on filter bank frames can be found in [**16, 31**]. The results here are due to Casazza, Chebira, and Kovačević [**8**].

DEFINITION 8.1. *Given $k, N, M \in \mathbb{N}$, a **Filter bank frame** for $\ell_2(\mathbb{Z})$ is a frame for $\ell_2(\mathbb{Z})$, say $\{\varphi_m\}_{m\in\mathbb{Z}}$, satisfying the following:*
1. For $0 \leq i \leq kN - 1$ and $j \notin \{0, 1, \ldots, kN - 1\}$ we have that $\varphi_i(j) = 0$.
2. For $j = 0, 1, \ldots, M - 1$ and $i \in \mathbb{Z}$, $\varphi_{iM+j} = T_{iN}\varphi_j$, where T_{iN} is translation by iN.

NOTATION 8.2. *Throughout this section we will let $\{e_i\}_{i\in\mathbb{Z}}$ be the natural orthonormal basis for $\ell_2(\mathbb{Z})$.*

PROPOSITION 8.3. *Let $N < L - 1$ be natural numbers and let $\{\varphi_j\}_{j=0}^{M-1}$ be a frame for the span of $\{e_i\}_{i=0}^{L-1}$ with frame bounds A, B. Let $\varphi_{iM+j} = T_{iN}\varphi_j$ for all $0 \leq j \leq M - 1$ and all $i \in \mathbb{Z}$. Then $\{\varphi_i\}_{i\in\mathbb{Z}}$ is a frame for $\ell_2(\mathbb{Z})$ with frame bounds $A\lfloor\frac{L}{N}\rfloor, B\lceil\frac{L}{N}\rceil$.*

PROOF. Let $\varphi \in \ell_2(\mathbb{Z})$ and compute

$$
\begin{aligned}
\sum_{m \in \mathbb{Z}} |\langle \varphi, \varphi_m \rangle|^2 &= \sum_{i \in \mathbb{Z}} \sum_{j=0}^{M-1} |\langle \varphi, \varphi_{iM+j} \rangle|^2 \\
&= \sum_{i \in \mathbb{Z}} \sum_{j=0}^{M-1} |\langle \varphi, T_{iN}\varphi_j \rangle|^2 \\
&\leq \sum_{i \in \mathbb{Z}} B \sum_{n=1}^{iN+L-1} |\varphi(n)|^2 \\
&\leq B \left\lceil \frac{L}{N} \right\rceil \sum_{n \in \mathbb{Z}} |\varphi(n)|^2 \\
&= B \left\lceil \frac{L}{N} \right\rceil \|\varphi\|^2.
\end{aligned}
$$

The lower frame bound follows similarly. $\square$

Next we see when we can get a tight frame from a filter bank frame.

PROPOSITION 8.4. *Let $0 < L < N$ be natural numbers and Let $\{\varphi_j\}_{j=0}^{M-1}$ be a frame for span $\{e_i\}_{i=0}^{KN+L}$, with frame operator S having eigenvectors $\{e_i\}_{i=0}^{KN+L}$ and respective eigenvalues $\{\lambda_i\}_{i=0}^{KN+L}$. Let $\varphi_{iM+j} = T_{iN}\varphi_j$ for all $0 \leq j \leq M-1$ and all $i \in \mathbb{Z}$. The following are equivalent:*

1. The family $\{\varphi_i\}_{i \in \mathbb{Z}}$ is a λ-tight frame for $\ell_2(\mathbb{Z})$.

2. We have

$$
\lambda = \begin{cases} \sum_{j=0}^{K} \lambda_{jN+m} : & : \quad 0 \leq m \leq L \\ \sum_{j=0}^{K-1} \lambda_{jN+m} : & : \quad L < m \leq N \end{cases}
$$

PROOF. Let $\varphi \in \ell_2(\mathbb{Z})$ and compute

$$
\begin{aligned}
\sum_{m \in \mathbb{Z}} |\langle \varphi, \varphi_m \rangle|^2 &= \sum_{i \in \mathbb{Z}} \sum_{j=0}^{M-1} |\langle \varphi, \varphi_{iM+j} \rangle|^2 \\
&= \sum_{i \in \mathbb{Z}} \sum_{j=0}^{M-1} |\langle \varphi, T_{iN}\varphi_j \rangle|^2 \\
&= \sum_{i \in \mathbb{Z}} \sum_{n=0}^{(i+K)N+L} \lambda_{n-iN} |\varphi(n)|^2 \\
&= \sum_{i \in \mathbb{Z}} \sum_{n=0}^{KN+L} \lambda_n |\varphi(n+iN)|^2 \\
&= \sum_{m=0}^{L} \sum_{i \in \mathbb{Z}} \left(\sum_{j=0}^{K} \lambda_{jN+m} \right) |\varphi(iN+m)|^2 \\
&\quad + \sum_{m=L+1}^{N} \sum_{i \in \mathbb{Z}} \left(\sum_{j=0}^{K-1} \lambda_{jN+m} \right) |\varphi(iN+m)|^2.
\end{aligned}
$$

The result follows immediately from here. $\square$

Since a filter bank frame is an equal norm frame if and only if we start with an equal norm frame before translation, the next corollary also classifies which filter bank frames are equal norm tight frames.

COROLLARY 8.5. *Let $\{\varphi_j\}_{j=0}^{M-1}$ be an A-tight frame for the span of $\{e_i\}_{i=0}^{KN+L}$ for $0 \leq L < N$. Let $\varphi_{iM+j} = T_{iN}\varphi_j$, for all $0 \leq j \leq M-1$ and all $i \in \mathbb{Z}$. Then $\{\varphi_i\}_{i \in \mathbb{Z}}$ is a λ-tight frame for $\ell_2(\mathbb{Z})$ if and only if $L = 0$. Moreover, in this case $\lambda = (K+1)A$.*

PROOF. Since $\{\varphi\}_{i=0}^{M-1}$ is an A-tight frame, every orthonormal basis consists of eigenvectors of the frame operator for this frame. By Proposition 8.4, $\{\varphi_i\}_{i \in \mathbb{Z}}$ is a frame with $\{e_i\}_{i \in \mathbb{Z}}$ eigenvectors for the frame operator having eigenvalues

$$\sum_{j=0}^{K} A = (K+1)A, \quad \text{for } e_{iN+m}, \;\; 0 \leq m \leq L,$$

and

$$\sum_{j=0}^{K-1} A = KA, \quad \text{for } e_{iN+m}, \;\; L+1 \leq m \leq N.$$

It follows that this is a tight frame if and only if $L = 0$. □

THANKS: The referee did a very serious job of greatly improving the presentation in this manuscript.

References

[1] A. Aldroubi, *p frames and shift-invariant subspaces of L^p*, Journal of Fourier Analysis and Applications **7** (2001) 1–21.

[2] J. Antezana, P. Massey, M. Ruiz and D. Stojanoff, *The Schur-Horn Theorem for operators and frames with prescribed norms and frame operator*, Illinois Journal of Mathematics, To appear.

[3] Balan, R., Casazza, P.G., Edidin, D., and Kutyniok, G., *Decomposition of Frames and a New Frame Identity*, Proceedings of SPIE, **5914** (2005), 1-10.

[4] Benedetto, J. and Li, S., *The Theory of Multiresolution Analysis Frames and Applications to Filter Banks*, ACHA, **4**, No. 4 (1998) 389-427.

[5] Bodmann, B.G and Paulsen, V.I, *Frame Paths and Error Bounds for Sigma-Delta Quantization*, Applied and Computational Harmonic Analysis, to appear (2006).

[6] Bodmann, B.G, Paulsen, V.I, and Abdulbaki, Soha A., *Smooth Frame-Path Termination for Higher Order Sigma-Delta Quantizations*, to appear (2006).

[7] M. Bownik and D. Speegle, *The Feichtinger conjecture for wavelet frames, Gabor frames and frames of translates*, Preprint.

[8] Casazza, P.G., Chebira, A. and Kovačević, J., *Filterbank frames*, In progress.

[9] Casazza, P.G., Christensen, O., and Kalton, N.J. *Frames of translates*, Collect. Math. **52** (2001) 35-54.

[10] Casazza, P.G., Fickus, M., Leon, M., and Tremain, J.C., *Constructing infinite tight frames*, Preprint.

[11] Casazza, P.G. and Kovačević, J., *Uniform Tight Frames with Erasures*, Adv. in Comp. Math. **18** No. 2-4 (2003), 387-430.

[12] Casazza, P.G. and Leon, M., *Constructing Frames with a given Frame Operator*, preprint.

[13] Casazza, P.G., Kutyniok, G., and Lammers, M.C., *Duality Principles in Frame Theory*, J. of Fourier Anal. and App. **10** (2004), 383-408.

[14] Christensen, O., *An introduction to frames and Riesz bases*, Birkhauser, Boston, 2003.

[15] Christensen, O., Deng,B and Heil, C., *Density of Gabor frames*, Appl. and Com. Har. Anal. **7** (1999) 292-304.

[16] Cvetković, Z. and Vetterli, M., *Oversampled filter banks*, IEEE Trans. Signal Process. **46** No. 5 (1998)1245–1255.

[17] M. Bownik and D. Speegle, *The Feichtinger conjecture for wavelet frames, Gabor frames and frames of translates*, Preprint.

[18] Daubechies, I., *Ten Lectures on Wavelets*, SAIM, Philadelphia, PA, 1992.

[19] Daubechies, I., Grossmannn, A., and Meyer, Y., *Painless nonorthogonal expansions*, J. Math. Phys. **27** (1986), 1271-1283.

[20] Duffin, R.J and Schaeffer, A.C., *A class of nonharmonic Fourier series*, Trans. Amer. Soc. **72** (1952), 341-366.

[21] Feng, D.J., Wang, L. and Wang, Y., *Generation of finite tight frames by Householder transformations*, Advances in Computational Mathematics **24** (2006) 297-309.

[22] Gabor, D., *Theory of communication* J. Inst. Elec. Engrg. **93**(1946),429-457.

[23] Goyal, V.K., Kovačević, J., and Kelner, J.A., *Quantized frame expansions with erasures*, Appl. Comput. Harmon. Anal. **10** (2001), 203-233.

[24] Goyal, V.K, Vetterli, M., and Thoa, N.T., *Quantized overcomplete expansions in $\mathbb{R}^N$: Analysis, synthesis, and algorithms*, IEEE Trans. Inform. Th. **44(1)** (January 1998), 16-31.

[25] Gröchening, K., *Foundations of Time Frequency Analysis*, Birkhauser, Boston, 2001.

[26] Han, D. and Larson, D.R., *Frames, bases and group representations*, Memoirs AMS **697** (2000).

[27] Hernández, E. and Weiss, G., *A First Course on Wavelets*, CRC, Inc., 1996.

[28] Hernández, E., Labate, D., and Weiss, G., *A Unified Characterization of Reproducing Systems Generated by a Finite Family, II*, 2002.

[29] H.O. Kim and J.K. Lim, *Frame multiresolution analysis*, Comm. Korean Math. Soc. Vol. **15**, No. 2 (2000) 285-308.

[30] Klapperenecker, A. and Rötteler, M., *Mutually unbiased bases, spherical designs and frames* in "Wavelets XI" (San Diego, CA 2005), Proc. SPIE **5914**, M. Papadakis et al., eds., Bellingham, WA (2005) 5914 P1-13.

[31] Kovačević, J., Dragotti, P.L., Goyal, V.K., *Finter bank frame expansions with erasures*, IEEE Trans. Inform. Theory, **48** No. 6 (2002) 1439–1450.

[32] Lawrence, G., Pfander, E. and Walnut, D., *Linear independence of Gabor systems in finite dimensional vector spaces*, preprint.

[33] P. Massey and R. Ruiz, *Tight frame conpletions with prescribed norms*, Preprint.

[34] Puschel, M. and Kovačević, J., *Real, tight frames with maximal robustness to erasures*, Preprint.

[35] Strohmer, T. and Heath, Jr., R.W., *Grassmannian frames with applications to coding and communications*, Appl. Comput. Harmon. Anal., **13** (2003), 257-275.

[36] Unser, M., Aldroubi, A., and Laine, A., eds., em Special Issue on Wavelets in Medical Imaging, IEEE Trans. Medical Imaging, **22** (2003).

[37] Waldron, S. and Hay, N, *On computing all harmonic frames of n vectors in $\mathbb{C}^d$*, Applied and Computational Harmonic Analysis, to appear (2006).

[38] Walnut, D.F., *An introduction to wavelet analysis*, Birkhäuser, Boston (2002).

[39] Wolfe, P.J., Godsill, S.J., and Ng, W.J., *Bayesain variable selection and regularization for time-frequency surface estimation*, J. R. Stat. Soc. Ser. B Stat. Methodol., **66** (2004), 575-589.

[40] Young, R., *An Introduction to Nonharmonic Fourier Series*, Revised First Edition, Academic Press, San Diego, 2001.

[41] Zimmermann, G., *Normalized Tight Frames in Finite Dimensions*, in "Recent Progress in Multivariate Approximation, ISNM 137" (W. Haussmann and K. Jetter Eds.), 249-252, Birkhäuser, Basel, 2001.

DEPARTMENT OF MATHEMATICS, UNIVERSITY OF MISSOURI-COLUMBIA, COLUMBIA, MO 65211
E-mail address: `pete,leonhard@math.missouri.edu`

Contemporary Mathematics
Volume **451**, 2008

P-ADIC PSEUDODIFFERENTIAL OPERATORS AND WAVELETS

NGUYEN MINH CHUONG AND NGUYEN VAN CO

ABSTRACT. Some most recent results on Archimedean and non- Archimedean pseudodifferential operators and wavelets are shortly introduced without proofs and some new ones are presented in detail.

I. Introduction

In [14] it is interesting to see the relations between operator theory, frame theory on a separable Hilbert space, orthonormal wavelet theory and group representations. The three-volume book [25] provides various useful relations of wavelets and singular, pseudodifferential operators, Fourier integral operators in many important functional spaces such as Sobolev, Besov, Hardy, BMO, VMO. Recently, one can discover more and more, exciting, profound connections between such above mentioned operators with p-adic wavelet analysis and p-adic analysis, tree, graph theory which are very needed to almost all branches of sciences, mathematics, theoretical as well as practical, deterministic as well as stochastic, more precisely, say, to engineering, economy, cognitive sciences etc. By means of such discovery , we hope, not only pure mathematics but mathematics for physics, industry, economy etc, may be developed more and more intensively nowadays and even in future (se e.g. [22-24]) .

In this note we will present some most recent results and some new ones, which did not appear elsewhere, on the above mentioned relations.

II. Pseudodifferential operators and wavelets

As in the introduction, the famous book [25] gives us various deep and beautiful aspects of pseudodifferential operators and wavelets. For wavelet approximation mehods for pseudodifferential operators, we refer to [13] and the references therein.

2000 *Mathematics Subjet Classification.* Primary 35S10, 11S80; Secondary 47G30, 58J40

Key words and phrases. Pseudodifferential operators over the p-adic field, p-adic wavelet analysis, p-adic analysis non-Archimedean.

Supported by the National Basic Research Program, Vietnamese Academy of Science and Technology.

Here will be stated some most recent approximation results from [12] for these operators.

1. Let us consider a Cauchy problem for a class of interesting complicated pseudodifferential equations as follows:

$$(2.1) \qquad \frac{\partial u}{\partial t} = -\mu A u(x,t) - \int_0^t a(t-\tau) A u(x,\tau) d\tau + b(x,t),$$

$$(2.2) \qquad u(x,0) = u_0(x),$$

where $x \in \mathcal{J}^n = \mathbb{R}^n/\mathbb{Z}^n, \mu > 0, A = \sigma(D)$ is a psedodifferential operarator, defined by

$$\sigma(D)u(x) = \sum_{\xi \in \mathbb{Z}^n} e^{2\pi i x \xi} \sigma(\xi) \tilde{u}(\xi), u \in C^\infty(\mathbb{Z}^n),$$

where the symbol $\sigma(\xi)$ belongs to $S^{2m}(\mathbb{Z}^n)$, i.e., $\sigma \in C^\infty(\mathbb{Z}^n)$ and satisfies

$$|\Delta^\alpha \sigma(\xi)| \leqslant C_\alpha (1+|\xi|)^{2m-|\alpha|} \quad \text{for all } \xi \in \mathbb{Z}^n \text{ and for all multi - indexes } \alpha,$$

and Δ is difference operator, defined by $\Delta := (\tau_1 - 1, \tau_2 - 1, \cdots, \tau_n - 1)^T$, where $\tau_j f(x) := f(x + e^j)$, $e^j = (\delta_{j,l})_{l=1}^n$ is the jth coordinate vector.

Here it is asuumed also that $\sigma \in C^\infty(\mathbb{R}^n) \setminus \{0\}$ and $\sigma(t\xi) = t^{2m}\sigma(\xi)$, with $t > 0$, $\sigma(0) = 1$.

Moreover
$$(2.3) \qquad \sigma(\xi) \geq c \left(1 + |\xi|^2\right)^m, |\xi| \geq \lambda > 0.$$

The function $a(t), b(x,t)$ are given. Note that when $\mu = 0$, under some assumpotions on $a(t)$ the type of equation (2.1) may be changed.

2. Let us recall here some usual notations, definitions and some basic facts. For a function $u(x) \in L^2(\mathcal{J}^n)$, the discrete Fourier transform is defined by

$$\mathcal{F}(u)(\xi) = \tilde{u}(\xi) = \int_{\mathcal{J}^n} e^{-2\pi i x \xi} u(x) dx = \int_{[0,1]^n} e^{-2\pi i x \xi} u(x) dx, \xi \in \mathbb{Z}^n.$$

The discrete inverse Fourier transform is then $u(x) := \sum_{\xi \in \mathbb{Z}^n} \tilde{u}(\xi) e^{2\pi i x \xi}$.

The Fourier transform of the function $f \in L^2(\mathbb{R}^n)$ is

$$\hat{f}(\xi) = \int_{\mathbb{R}^n} e^{-2\pi i x \xi} f(x) dx, \xi \in \mathbb{R}^n.$$

with the inverse Fourier transform $f(x) = \int_{\mathbb{R}^n} e^{-2\pi i x \xi} \hat{f}(\xi) d\xi, x \in \mathbb{R}^n$. The Laplace transform of a function $u(x)$ increasing less than an exponent is

$$L(u)(s) = \breve{u}(s) := \int_0^\infty e^{-2\pi s t} u(t) dt, s \in \mathbb{C}.$$

We consider now the space $H^s(\mathcal{J}^n), (s \in \mathbb{R})$ of functions $u \in \mathcal{D}'(\mathcal{J}^n)$ (Schwartz space of distributions), such that $\langle D \rangle^s u \in L^2(\mathcal{J}^n)$ and having the finite norm

$$\|u\|_s = \left(\sum_{\xi \in \mathbb{Z}^n} \langle \xi \rangle^{2s} \left| \tilde{u}(\xi) \right|^2 \right)^{\frac{1}{2}},$$

where

$$\langle \xi \rangle = \left\{ \begin{array}{l} 1, \text{ if } \xi = 0, \\ |\xi|, \text{ if } \xi \neq 0. \end{array} \right.$$

Definition 1. Let $L^2(H^{q,s_0}), (q \geq 0, s_0 > 0)$, be the space of all functions $u(x,t), x \in \mathcal{J}^n, t \geq 0$, satifying the following conditions: for each $0 \leqslant \alpha \leqslant q$,
  i) $u(x,t) \in H^\alpha(\mathcal{J}^n)$ uniformly in $t \in [0, +\infty)$,
  ii) The following series converges:

$$\sum_{\xi \in \mathbb{Z}^n} \langle \xi \rangle^{2s} \int_0^\infty e^{-4\pi s_0 t} |\tilde{u}(\xi, t)|^2 \, dt.$$

The norm of the function $u \in L^2(H^{q,s_0})$ is defined by

(2.4)
$$\|u\|_{L^2(H^{q,s_0})} = \left(\sum_{\xi \in \mathbb{Z}^n} \langle \xi \rangle^{2s} \int_0^\infty e^{-4\pi s_0 t} |\tilde{u}(\xi, t)|^2 \, dt \right)^{\frac{1}{2}}.$$

Next we introduce some notation and definitions regarding wavelets.

Definition 2. An multiresolution approximation (M.R.A.) of $L^2(\mathbb{R}^n)$ is, by definition, an sequence $V_j, j \in \mathbb{Z}$, of closed linear subspaces of $L^2(\mathbb{R}^n)$ with the following properties:

(1)
$$\cdots \subset V_{-1} \subset V_0 \subset V_1 \subset \cdots \subset V_n \cdots,$$

(2)
$$\bigcap_{j \in \mathbb{Z}} V_j = \{0\}, \overline{\bigcup_{j \in \mathbb{Z}} V_j} = L^2(\mathbb{R}^n),$$

(3)   for all $f \in L^2(\mathbb{R}^n)$ and all $j \in \mathbb{Z}$
$$f(x) \in V_j \Leftrightarrow f(2x) \in V_{j+1},$$

(4)   for all $f \in L^2(\mathbb{R}^n)$ and $k \in \mathbb{Z}^n$
$$f(x) \in V_0 \Leftrightarrow f(x-k) \in V_0,$$

(5)   there exists a function, which is called scaling function (S.F.) $\phi(x) \in V_0$, such that the sequence
$$\{\phi(x-k), \, k \in \mathbb{Z}^n\},$$
is a Riesz basic of V_0.

An M.R.A. of $L^2(\mathbb{R}^n)$ is said to be r-regular $(r \in \mathbb{N})$ if the function ϕ is r-regular, that is for each $m \in \mathbb{N}$ there exists c_m such that for all multi-index $\alpha, |\alpha| \leqslant r$ the following condition holds

$$|D^\alpha \phi(x)| \leqslant c_m (1 + |x|)^{-m}.$$

Let us denote by

$$\phi_{jk}(x) = 2^{nj/2} \phi(2^j x - k), k \in \mathbb{Z}^n,$$

Obviously
$$V_j = \overline{span}\{\phi_{jk}(x), k \in \mathbb{Z}^n\}, \text{ (the } L^2\text{-closure)}.$$

The notation $[u]$ stands for the periodization operator of a function u, that is
$$[u](x) = \sum_{k \in \mathbb{Z}^n} u(x + k).$$

Denote by
$$\phi_j^k(x) := [\phi_{jk}](x) = 2^{\frac{nj}{2}} \sum_{l \in \mathbb{Z}^n} \phi(2^j(x + l) - k).$$

Similarly let us define a M.R.A. of $L^2(\mathcal{J}^n)$ as follows

$$[V_j] := \overline{span}\{\phi_j^k(x), k \in \mathbb{Z}^{nj}\}, j \geq 0,$$

where $\mathbb{Z}^{nj} = \mathbb{Z}^n/2^j\mathbb{Z}^n$.

It is easy to check
$$[V_0] \subset [V_1] \subset ... \subset [V_n] \subset ..., \overline{\bigcup_{j \geq 0} V_j} = L^2(\mathcal{J}^n).$$

Furthermore if $(\phi_{jk}, \phi_{jl}) = \delta_{kl}, k, l \in \mathbb{Z}^n$, then
$$(\phi_k^j, \phi_l^j) = \delta_{kl}; \; k, l \in \mathbb{Z}^{nj},$$

and dim $[V_j] = 2^{nj}$.

Definition 3. A C^1-mapping in t $\;$ $u : [0, \infty) \to H^m(\mathcal{J}^n)$ satisfying

(2.5) $$\left(\frac{\partial u}{\partial t}, v\right) = -\mu(Au, v) - \int_0^t a(t - \tau)(Au(\tau), v) \, d\tau + (b, v),$$

(2.6) $$(u(x, 0), v) = (u_0, v), \forall v \in L^2(\mathcal{J}^n),$$

is called a weak solution of problem $(2.1) - (2.2)$.

Definition 4. A C^1-mapping in t $u_h : [0, \infty) \to V_h$ satisfying

(2.7) $$\left(\frac{\partial u_h}{\partial t}, v\right) = -\mu(Au_h, v) - \int_0^t a(t - \tau)(Au(\tau), v) \, d\tau + (b, v),$$

(2.8) $$(u_h(x, 0), v) = (u_{0h}, v), \forall v \in V_h,$$

where $u_{0h} := \mathcal{R}u_0$, $\mathcal{R}$ is a linear approximation of u_0 in V_h, is called a Galerkin-wavelet solution of problem $(2.1) - (2.2)$.

The following theorem asserts the stability of the weak solution of problem $(2.1) - (2.2)$.

Theorem 1. *Let* $u(x, t) \in L^2(H^{2m, s_0})$ *be a weak solution of problem* $(2.1) - (2.2)$, *with* $b = 0$. *Assume, morever that there exsits a number* $s_0 > 0$ *such that*

(2.9) $$\mu + \mathrm{Re}\breve{a}(s_0 + i\delta) \geq 0, \; \forall \delta \in \mathbb{R}.$$

Then

$$\|u\|^2_{L^2(H^{0,s_0})} \leqslant \frac{1}{4\pi s_0} \|u_0\|^2.$$

Let us consider the equation

$$Au(x) = f(x),$$

where A is the pseudodifferential operator introduced as above.

Theorem 2. *Let $u(x,t)$ be the solution of $(2.5) - (2.6)$ and $u_h(x,t)$ a solution of $(2.7) - (2.8)$. If (2.9) is satisfied and $\dfrac{\partial^j u}{\partial t^j} \in H^q(\mathcal{J}^n)$ $(j = 0,1, q \geq 2m)$ then*

$$\|u - u_h\|^2_{L^2\left(H^{0,s_0}\right)} \leqslant$$

$$\leqslant C \left\{ \|u_0 - u_{0,h}\|^2 + h^{2(q-2m)} \left[\|u_0\|^2_q + \int_0^\infty e^{-4\pi s_0 t} \left(\|u_0\|^2_q + \left\| \frac{\partial u}{\partial t} \right\|^2_q \right) dt \right] \right\}.$$

III. p-adic Pseudodifferential operators and wavelets

1. Some basic facts (see also [26]).

a) Any p-adic number $x, x \neq 0$ has the canonical form

$$x = p^\gamma \left(x_0 + x_1 p + x_2 p^2 + \cdots \right),$$

where $\gamma = \gamma(x) \in \mathbb{Z}$ is an integer set, $x_j = 0, 1, ..., p-1$, $x_0 \neq 0, (j = 0, 1, \cdots)$. The norm $|x|_p$ on Q_p is defined by $|0|_p = 0$, $|x|_p = p^{-\gamma(x)}$ if $x \neq 0$. The norm is non-Archimedean, i.e.

i) $|x|_p \geq 0, |x|_p = 0 \Leftrightarrow x = 0$,

ii) $|xy|_p = |x|_p |y|_p$,

iii) $|x + y|_p \leqslant \max\left\{ |x|_p, |y|_p \right\}$, if $|x|_p \neq |y|_p$ then $|x + y|_p = \max\left\{ |x|_p, |y|_p \right\}$.

For each $a \in Q_p, \gamma \in \mathbb{Z}$, let

$$B_\gamma(a) = \left\{ x \in Q_p : |x - a|_p \leqslant p^\gamma \right\}, \ B_\gamma(0) = B_\gamma.$$

We let also

$$S_\gamma(a) = \left\{ x \in Q_p : |x - a|_p = p^\gamma \right\}, S_\gamma(0) = S_\gamma.$$

b) The Haar measure on Q_p is normalized by $\int_{B_0} dx = 1$ and $d(ax + b) = |a|_p, a \in Q_p^*$. For the additive normalized character $\chi_p(x)$ on Q_p we have

$$\int_{B_\gamma} \chi_p(\xi x) dx = p^\gamma \Omega\left(p^\gamma |\xi|_p \right),$$

where

$$\Omega(t) = \begin{cases} 1, & \text{if } 0 \leqslant t \leqslant 1 \\ 0, & \text{if } t > 1; t \in \mathbb{R}. \end{cases}$$

c) The Fourier transform of a test function $\varphi \in \mathcal{D}(Q_p)$, $(\mathcal{D}(Q_p)$ is the space of local constant functions from Q_p to $\mathbb{C}$ with compact supports) is defined by

$$\widetilde{\varphi}(\xi) = \int\limits_{Q_p} \varphi(x)\chi_p(\xi x)\, dx.$$

We have $\widetilde{\varphi}(\xi) \in \mathcal{D}(Q_p)$ and $\varphi(x) = \int\limits_{Q_p} \widetilde{\varphi}(\xi)\chi_p(-\xi x)\, d\xi$.

$L^2(Q_p)$ is the set of measurable complex-valued functions f on Q_p such that

$$\|f\|_{L^2(Q_p)} = \left(\int\limits_{Q_p} |f(x)|^2\, dx \right)^{\frac{1}{2}} < \infty.$$

This is a Hilbert space with an inner product defined by

$$\langle f, g \rangle = \int\limits_{Q_p} f(x)\overline{g(x)}dx;\, f, g \in L^2(Q_p),$$

and

$$\|f\|^2_{L^2(Q_p)} = \langle f, f \rangle.$$

If $f \in L^2(Q_p)$, then

$$\widetilde{f}(\xi) = \lim_{\gamma \to +\infty} \int\limits_{B_\gamma} f(x)\chi_p(\xi x)dx \text{ in } L^2(Q_p), \widetilde{f} \in L^2(Q_p).$$

We have

$$\mathcal{D}(Q_p) \subset L^2(Q_p),$$

and the Parseval equality is

$$\langle f, g \rangle = \left\langle \widetilde{f}, \widetilde{g} \right\rangle;\, \|f\|_{L^2(Q_p)} = \left\| \widetilde{f} \right\|_{L^2(Q_p)};\, f, g \in L^2(Q_p).$$

d) For $\alpha > 0$, the operator D^α is defined by V.S. Vladimirov in [26] as

$$(D^\alpha \varphi)(x) = \int\limits_{Q_p} |\xi|_p^\alpha\, \widetilde{\varphi}(\xi)\, \chi_p(-\xi x)\, d\xi, \varphi \in \mathcal{D}(Q_p).$$

For $\varphi \in \mathcal{D}'(Q_p)$, $(\mathcal{D}'(Q_p)$ is the dual space of $\mathcal{D}(Q_p))$, if the convolution $f_{-\alpha} * \varphi$ exists, with $\alpha \neq -1, \alpha$ is a real number, then

$$D^\alpha \varphi = f_{-\alpha} * \varphi \text{ with } f_\alpha(x) = \frac{|x|_p^{\alpha-1}}{\Gamma_p(\alpha)}, \Gamma_p(\alpha) = \frac{1 - p^{\alpha-1}}{1 - p^{-\alpha}}.$$

2. An orthonormal basis in $L^2(Q_p)$

Here and in the sequel as in [23-24] we will present an orthonormal basis in $L^2(Q_p)$ consisting of test functions in Q_p. Such test functions are eigenfunctions of $D^\alpha, \alpha > 0$. However, the basis we construct is different from the one in [23-24-26].

1. Let $\psi \in L^2(Q_p)$. The function ψ is eigenfunction of the operator $D^\alpha, \alpha > 0$ with eigenvalue $\lambda \neq 0$

$$D^\alpha \psi(x) = \lambda \psi(x), \quad \psi \neq 0.$$

By the Fourier transform, we have

$$\left(|\xi|_p^\alpha - \lambda \right) \widetilde{\psi}(\xi) = 0, \quad \lambda = p^{\alpha N}, \quad N \in \mathbb{Z}.$$

Therefore

$$\widetilde{\psi}\left(\xi\right) = \delta\left(\left|\xi\right|_p - p^N\right)\rho\left(\xi\right), \quad \text{(see [26])},$$

where

$$\delta(t) = \begin{cases} 0, & \text{if} \quad t = 0 \\ 1, & \text{if} \quad t \neq 0; t \in \mathbb{R}. \end{cases}$$

a) Let us choose

$$\widetilde{\psi_0}\left(\xi\right) = \delta\left(\left|\xi\right|_p - p^N\right)\delta\left(\xi_0 - 1\right).$$

This function has the following properties.

Property 1.

i) $\widetilde{\psi_0} \in \mathcal{D}\left(Q_p\right),$

ii) $\psi_0(x) = p^{N-1}\chi_p\left(-xp^{-N}\right)\Omega\left(p^{N-1}\left|x\right|_p\right),$

iii) $D^\alpha\psi_0(x) = p^{\alpha N}\psi_0(x).$

Proof. i) It is obvious that $\widetilde{\psi_0} \in \mathcal{D}_N^{N-1}\left(Q_p\right)$, consisting of functions $\varphi \in \mathcal{D}\left(Q_p\right)$, which are local constant functions on B_{N-1} and have supports on B_N, $N \in \mathbb{Z}$.

ii) Since $\widetilde{\psi_0}\left(\xi\right) \in \mathcal{D}\left(Q_p\right)$, by using the Fourier transform we get

$$\psi_0(x) = \int\limits_{Q_p} \widetilde{\psi_0}\left(\xi\right)\chi_p\left(-\xi x\right)d\xi = \int\limits_{S_N} \delta\left(\xi_0 - 1\right)\chi_p\left(-x\xi\right)d\xi.$$

Put $\xi = p^{-N} + \xi'$, with $\left|\xi'\right|_p \leqslant p^{N-1}$. It is easy to see that

$$\psi_0(x) = \chi_p\left(-xp^{-N}\right)\int\limits_{B_{N-1}} \chi_p\left(-x\xi'\right)d\xi' = p^{N-1}\chi_p\left(-xp^{-N}\right)\Omega\left(p^{N-1}\left|x\right|_p\right).$$

iii) $D^\alpha\psi_0(x) = \int\limits_{Q_p} \left|\xi\right|_p^\alpha \widetilde{\psi_0}\left(\xi\right)\chi_p\left(-x\xi\right)d\xi$

$$= \int\limits_{Q_p} \left|\xi\right|_p^\alpha \delta\left(\left|\xi\right|_p - p^N\right)\delta\left(\xi_0 - 1\right)\chi_p\left(-x\xi\right)d\xi = p^{\alpha N}\int\limits_{S_N} \delta\left(\xi_0 - 1\right)\chi_p\left(-x\xi\right)d\xi.$$

Put $\xi = p^{-N} + \xi', \left|\xi'\right|_p \leqslant p^{N-1},$

we get

$$D^\alpha\psi_0(x) = p^{\alpha N}\chi_p\left(-xp^{-N}\right)\int\limits_{B_{N-1}} \chi_p\left(-x\xi'\right)d\xi' =$$

$$= p^{\alpha N}\chi_p\left(-xp^{-N}\right)p^{N-1}\Omega\left(p^{N-1}\left|x\right|_p\right) = p^{\alpha N}\psi_0(x).$$

$\square$

b) Since $\psi_0 \in \mathcal{D}\left(Q_p\right)$ we have $\psi_0 \in L^2\left(Q_p\right)$. After normalizing the functions ψ_0 in $L^2\left(Q_p\right)$, with $N = 0$, we obtain

$$\psi(x) = p^{\frac{-1}{2}}\chi_p\left(-x\right)\Omega\left(p^{-1}\left|x\right|_p\right).$$

Consider now the class of functions $\psi_{Nkr}(x) = p^{\frac{N}{2}}\psi\left(p^{-N}kx - kr\right).$

Obviously

$$\psi_{Nkr}(x) = p^{\frac{N-1}{2}} \chi_p(kr) \chi_p\left(-xp^{-N}k\right) \Omega\left(p^{-1}\left|p^{-N}x - r\right|_p\right),$$

or

$$(3.1) \qquad \psi_{Nkr}(x) = p^{\frac{N-1}{2}} \chi_p(kr) \chi_p\left(-xp^{-N}k\right) \Omega\left(p^{N-1}\left|x - p^N r\right|_p\right),$$

where $x \in Q_p$, $N \in \mathbb{Z}$, $k = 1, 2, \cdots, p-1$ and $r \in \frac{Q_p}{\frac{1}{p}Z_p}$, $Z_p = \left\{x \in Q_p, |x|_p \leqslant 1\right\}$, i.e. $r = 0$ or $r = p^{-m}\left(r_0 + r_1 p + \cdots + r_{m-2}p^{m-2}\right)$, $m \in \mathbb{Z}$, $m \geq 2$, with $r_0 = 1, 2, \cdots, p-1$; $r_j = 0, 1, \cdots, p-1$; $1 \leqslant j \leqslant m-2$.

Property 2. $\psi_{Nkr} \in \mathcal{D}(Q_p)$, *or more precisely*

i) $\psi_{Nk0} \in \mathcal{D}_{1-N}^{-N}(Q_p)$; $(r = 0)$,

ii) $\psi_{Nkr} \in \mathcal{D}_{m-N}^{-N}(Q_p)$; $(r \neq 0)$.

Proof. i) If $r = 0$, then

$$\psi_{Nk0}(x) = p^{\frac{N-1}{2}} \chi_p\left(-xp^{-N}k\right) \Omega\left(p^{N-1}|x|_p\right) \in \mathcal{D}_{1-N}^{-N}(Q_p).$$

Indeed, from $|x|_p > p^{1-N}$, it follows that $\Omega\left(p^{N-1}|x|_p\right) = 0$, hence $\psi_{Nk0} = 0$, consequently $\psi_{Nk0}(x)$ has a compact support. Moreover it is locally constant, because

by $|x'|_p \leqslant p^{-N}$, we obtain

$$\psi_{Nk0}(x + x') = p^{\frac{N-1}{2}} = \psi_{Nk0}(x); \quad \text{with} \quad |x|_p \leqslant p^{-N},$$

$$\psi_{Nk0}(x + x') = 0 = \psi_{Nk0}(x); \quad \text{with} \quad |x|_p \geq p^{2-N},$$

and

$$\psi_{Nk0}(x + x') = p^{\frac{N-1}{2}} \chi_p\left(-(x+x')p^{-N}k\right) \Omega\left(p^{N-1}|x+x'|_p\right)$$
$$= p^{\frac{N-1}{2}} \chi_p\left(-xp^{-N}k\right) \Omega\left(p^{N-1}|x|_p\right) = \psi_{Nk0}(x); \quad \text{with} \quad |x|_p = p^{1-N}.$$

ii) If $r \neq 0$, then

$$r = p^{-m}\left(r_0 + r_1 p + \cdots + r_{m-2}p^{m-2}\right), r_0 \neq 0, |r|_p = p^m \geq p^2.$$

Indeed, similarly to above, from $|x|_p > p^{m-N}$, it follows that $\left|p^N r\right|_p = p^{m-N}$. So $\left|x - p^N r\right|_p = |x|_p > p^{m-N}$. Hence $\left(p^{N-1}\left|x - p^N r\right|_p\right) > p^{m-1} > 1$. So $\psi_{Nkr}(x) = 0$, and by $|x'|_p \leqslant p^{-N}$, we get

$$\psi_{Nkr}(x + x') = 0 = \psi_{Nkr}(x); \quad \text{with} \quad |x|_p < p^{m-N},$$

$$\psi_{Nkr}(x + x') = 0 = \psi_{Nkr}(x); \quad \text{with} \quad |x|_p > p^{m-N},$$

$\psi_{Nkr}(x + x') = 0 = \psi_{Nkr}(x)$; with $|x|_p = p^{m-N}$ and there exists j, $j = 0, 1, \cdots, m-2$, so that $x_j - r_j \neq 0$,

$\psi_{Nkr}(x+x') = p^{\frac{N-1}{2}} \chi_p\left(-xp^{-N}k\right) = \psi_{Nkr}(x)$; with $|x|_p = p^{m-N}$ and $x_j - r_j = 0, \forall j = 0, 1, \cdots, m-2$. $\qquad\square$

Property 3.

$$\widetilde{\psi}_{Nkr}(\xi) = p^{\frac{1-N}{2}} \chi_p\left(p^N r\xi\right) \delta\left(|\xi|_p - p^N\right) \delta(\xi_0 - k).$$

Proof. By property 2, we have $\widetilde{\psi}_{Nkr} \in \mathcal{D}\left(Q_p\right)$, consequently, by (3.1) we obtain

$$\widetilde{\psi}_{Nkr}\left(\xi\right) = \int\limits_{Q_p} p^{\frac{N-1}{2}} \chi_p(kr) \chi_p\left(-xp^{-N}k\right) \Omega\left(p^{N-1}\left|x - p^N r\right|_p\right) \chi_p\left(\xi x\right) dx.$$

Put $x = p^N r + x'$. It is obvious that

$$\widetilde{\psi}_{Nkr}(\xi) = p^{\frac{N-1}{2}} \chi_p\left(p^N r\xi\right) \int\limits_{Q_p} \chi_p\left(-x'p^{-N}k\right) \Omega\left(p^{N-1}\left|x'\right|_p\right) \chi_p\left(\xi x'\right) dx'$$

$$= p^{\frac{N-1}{2}} \chi_p\left(p^N r\xi\right) \int\limits_{B_{1-N}} \chi_p\left(\left(\xi - p^{-N}k\right) x'\right) dx'$$

$$= p^{\frac{1-N}{2}} \chi_p\left(p^N r\xi\right) \Omega\left(p^{1-N}\left|\xi - p^{-N}k\right|_p\right)$$

$$= p^{\frac{1-N}{2}} \chi_p\left(p^N r\xi\right) \delta\left(\left|\xi\right|_p - p^N\right) \delta\left(\xi_0 - k\right).$$

$\square$

2. Theorem 2. *System of the functions $\{\psi_{Nkr}\}$ is an orthonormal basis of $L^2(Q_p)$ consisting of eigenfunctions of the Vladimirov operator $D^\alpha, (\alpha > 0)$ corresponding to the eigenvalues $p^{\alpha N}$*

$$(3.2) \qquad\qquad D^\alpha \psi_{Nkr} = p^{\alpha N} \psi_{Nkr}.$$

Proof. The following steps are needed to prove the Theorem.

a) It will be proved that $\{\psi_{Nkr}\}$ is an orthonormal system .

Let us use the Parseval equality

$$\langle \psi_{Nkr}, \psi_{N'k'r'} \rangle = \left\langle \widetilde{\psi}_{Nkr}\left(\xi\right), \widetilde{\psi}_{N'k'r'}\left(\xi\right) \right\rangle =$$

$$= b \int\limits_{Q_p} \chi_p\left(p^N r\xi\right) \delta\left(\left|\xi\right|_p - p^N\right) \delta\left(\xi_0 - k\right) \chi_p\left(-p^{N'}r'\xi\right) \delta\left(\left|\xi\right|_p - p^{N'}\right) \delta\left(\xi_0 - k'\right) d\xi,$$

where $b = p^{\frac{1-N}{2}} p^{\frac{1-N'}{2}}$.

Obviously

$$\delta\left(\left|\xi\right|_p - p^N\right) \delta\left(\left|\xi\right|_p - p^{N'}\right) \delta\left(\xi_0 - k\right) \delta\left(\xi_0 - k'\right) =$$

$$= \delta_{NN'} \delta_{kk'} \delta\left(\left|\xi\right|_p - p^N\right) \delta\left(\xi_0 - k\right),$$

with δ_{ij} is the Kronecker symbol.

Put $\xi = p^{-N}k + \xi'$, with $\left|\xi'\right|_p \leqslant p^{N-1}$. We have

$$\langle \psi_{Nkr}, \psi_{N'k'r'} \rangle = p^{1-N} \delta_{NN'} \delta_{kk'} \chi_p\left(\left(r - r'\right) k\right) \int\limits_{B_{N-1}} \chi_p\left(p^N\left(r - r'\right)\xi'\right) d\xi'.$$

For $r = r'$, we get $\chi_p\left((r - r')k\right) = 1$, and $\int\limits_{B_{N-1}} \chi_p\left(p^N(r - r')\xi'\right)d\xi' = p^{N-1}$.

For $r \neq r'$, we obtain $|r - r'|_p \geq p^2$ so $\int\limits_{B_{N-1}} \chi_p\left(p^N(r - r')\xi'\right)d\xi' = 0$.

Finally we get

$$\left\langle \widetilde{\psi}_{Nkr}, \widetilde{\psi}_{N'k'r'} \right\rangle = \delta_{NN'}\delta_{kk'}\delta_{rr'}.$$

Thus $\{\psi_{Nkr}\}$ is an orthonormal system.

b) It is sufficient to prove the completeness of $\left\{\widetilde{\psi}_{Nkr}\right\}$ in $L^2(S_N)$ for the character set

$$\left\{\chi_p(\sigma\xi) : |\sigma|_p \geq p^{-N}\right\}.$$

Let us consider

Case 1. For $|\sigma|_p = p^{-N+h}$, with $h = 0$, or $h = 1$. The Fourier coefficient of the expansion of $\chi_p(\sigma\xi)$ in $\left\{\widetilde{\psi}_{Nkr}(\xi)\right\}$ is

$$C_{Nkr} = \int\limits_{Q_p} \chi_p(\sigma\xi)\overline{\widetilde{\psi}_{Nkr}(\xi)}d\xi =$$

$$= p^{\frac{1-N}{2}} \int\limits_{Q_p} \chi_p(\sigma\xi)\chi_p\left(-p^N r\xi\right)\delta\left(|\xi|_p - p^N\right)\delta\left(\xi_0 - k\right)d\xi.$$

Put $\xi = p^{-N}k + \xi'$ with $|\xi'| \leqslant p^{N-1}$. We obtain

$$C_{Nkr} = p^{\frac{1-N}{2}}\chi_p\left(\sigma p^{-N}k\right)\chi_p(-rk)\int\limits_{B_{N-1}}\chi_p\left((\sigma - p^N r)\xi'\right)d\xi'.$$

For $r = 0$, we have $\chi_p(-rk) = 1$ and $\left|\sigma - p^N 0\right|_p = |\sigma|_p = p^{-N+h} \leqslant p^{-N+1}$.

So

$$\int\limits_{B_{N-1}} \chi_p\left((\sigma - p^N 0)\xi'\right)d\xi' = p^{N-1}.$$

Consequently

$$C_{Nk0} = p^{\frac{N-1}{2}}\chi_p\left(\sigma p^{-N}k\right) = p^{\frac{N-1}{2}}\chi_p\left(\frac{\sigma_0 k}{p^h}\right); \text{ with } h = 0, \text{ or } h = 1.$$

For $r \neq 0$, we have $\left|p^N r\right|_p > |\sigma|_p$.

Hence $C_{Nkr} = 0$.

So in case 1, we get

$$C_{Nkr} = p^{\frac{N-1}{2}}\chi_p\left(\frac{\sigma_0 k}{p^h}\right)\delta_{r0}; \text{ with } h = 0, \text{ or } h = 1.$$

Further it is clear that

$$\sum_{k,r} |C_{Nkr}|^2 = \sum_k p^{N-1} = (p-1)p^{N-1} = \int\limits_{S_N} |\chi_p(\sigma\xi)|^2 d\xi = \|\chi_p(\sigma\xi)\|_{L^2(S_N)},$$

that is we obtain the expansion

$$\chi_p\left(\sigma\xi\right)=\sum_{k,r}C_{Nkr}\widetilde{\psi}_{Nkr}\left(\xi\right)\text{ in }L^2\left(S_N\right).$$

Case 2. $|\sigma|_p=p^{-N+h}, h$ is an integer ≥ 2

Similarly to the case 1, we have

$$C_{Nkr}=p^{\frac{1-N}{2}}\chi_p\left(\sigma p^{-N}k\right)\chi_p\left(-rk\right)\int_{B_{N-1}}\chi_p\left(\left(\sigma-p^Nr\right)\xi'\right)d\xi'.$$

For $r=0$, obviously

$$\left|\sigma-p^N0\right|_p=|\sigma|_p=p^{-N+h}\geq p^{-N+2}.$$

So $C_{Nk0}=0.$

For $r\neq 0$, we have

$$r=p^{-m}\left(r_0+r_1p+\cdots+r_{m-2}p^{m-2}\right),r_0\neq 0.$$

Thus

$$p^Nr=p^{N-m}\left(r_0+r_1p+\cdots+r_{m-2}p^{m-2}\right),$$

and

$$\sigma=p^{N-h}\left(\sigma_0+\sigma_1p+\cdots+\sigma_{m-2}p^{m-2}\right)+\sigma'.$$

For $m\neq h$ we have

$$\left|\sigma-p^Nr\right|_p=\max\left(p^{-N+m},p^{-N+h}\right)\geq p^{-N+2},$$

consequently $C_{Nkr}=0.$

For $m=h$, if there exists $j,j=0,1,\cdots,m-2$, so that $r_j\neq\sigma_j$, then

$$\left|\sigma-p^Nr\right|_p\geq p^{-N+2}.$$

So $C_{Nkr}=0.$

If $r_j=\sigma_j,\forall j=0,1,\cdots,m-2$, then

$$\sigma-p^Nr=\sigma',|\sigma'|_p\leqslant p^{-N+1},$$

implying

$$C_{Nkr}=p^{\frac{1-N}{2}}\chi_p\left(\sigma p^{-N}k\right)\chi_p\left(-rk\right)p^{N-1}=p^{\frac{N-1}{2}}\chi_p\left(\frac{k\sigma_{h-1}}{p}\right).$$

Setting

$$\overline{\delta}_{ij}=1-\delta_{ij}=\left\{\begin{array}{l}0,\text{ if }i=j\\1,\text{ if }i\neq j,\end{array}\right.$$

we get for the case 2

$$C_{Nkr}=p^{\frac{N-1}{2}}\chi_p\left(\frac{k\sigma_{h-1}}{p}\right)\overline{\delta_{r0}}\delta_{mh}\delta_{r_j\sigma_j}.$$

Similarly to case 1, we obtain also

$$\sum_{k,r}|C_{Nkr}|^2=\sum_{k=1}^{p-1}p^{N-1}=\left(p-1\right)p^{N-1}=\|\chi_p\left(\sigma\xi\right)\|_{L^2(S_N)}.$$

Thus
$$\chi_p(\sigma\xi) = \sum_{k,r} C_{Nkr} \widetilde{\psi}_{Nkr}(\xi) \text{ in } L^2(S_N), \text{ (see [26])}.$$

c) Let us now prove (3.2). Obviously
$$D^\alpha \psi_{Nkr}(x) = \int_{Q_p} |\xi|_p^\alpha \, \widetilde{\psi}_{Nkr}(\xi) \chi_p(-x\xi) \, d\xi$$

$$= p^{\frac{1-N}{2}} \int_{Q_p} |\xi|_p^\alpha \chi_p(p^N r\xi) \delta\left(|\xi|_p - p^N\right) \delta(\xi_0 - k) \chi_p(-x\xi) \, d\xi.$$

We have to consider the integral only on S_N and $\xi_0 = k$. Put $\xi = p^{-N}k + \xi'$ with $|\xi'|_p \leqslant p^{N-1}$.

It this clear that

$$D^\alpha \psi_{Nkr}(x) = p^{\frac{1-N}{2}} p^{\alpha N} \chi_p(rk) \chi_p\left(-xp^{-N}k\right) \int_{B_{N-1}} \chi_p\left(-\left(x - p^N r\right)\xi'\right) d\xi'$$

$$= p^{\alpha N} p^{\frac{1-N}{2}} p^{N-1} \chi_p(rk) \chi_p\left(-xp^{-N}k\right) \Omega\left(p^{N-1}\left|x - p^N r\right|_p\right) = p^{\alpha N} \psi_{Nkr}(x).$$

$\square$

3. **Further discussions.** After establishing a correspondence between an orthonormal wavelet basis in $L^2(\mathbb{R}_+)$ and a basis of eigenvectors of the Vladimirov pseudodifferential operaror in [22], S.V. Kozyrev possibly together with A.Yu. Khrennikov have discussed such a problem in the so called ultrametric spaces, for much more general p-adic pseudodifferential operators (see [16, 22, 24] and references therein). Note that an ultrametric space is metric space with the metric $||.||$ satisfying the strongly triangle inequality

$$||xz|| \leqslant \max(||xy||, ||yz||), \forall x, y, z,$$

and the simplest example of an ultrametric space is the field Q_p.

Acknowledgment

The authors are grateful to the referee for helpful comments.

References

[1] M.S. Agranovich and M.I.Vishik, *Elliptic problem with parameter and parabolic problem of general form*, Uspehi Math, Nauk, 19 (1964), 2 (17), 53-161.

[2] Nguyen Minh Chuong and Ha Duy Hung, *Boundedness of p-adic maximal operators and p-adic weighted Hardy-Littlewood averages in L^l and BMO*, preprint 03/22, Inst. of Math, Hanoi, 2003 (submitted to Rossia Izv.Akad. Nauk).

[3] Nguyen Minh Chuong, *Parabolic pseudodifferential operators of variable order*, Dokl. Akad. Nauk SSSR 258 (1981), No 6, 1308-1312.

[4] Nguyen Minh Chuong, *Parabolic pseudodifferential operators of variable order in Sobolev spaces with weighted norms*, Dokl. Akad. Nauk SSSR 264 (1982), No 2, 299-302.

[5] Nguyen Minh Chuong, *Degenerate parabolic pseudodifferential operators of variable order*, Dokl. Akad. Nauk SSSR 268 (1983), No 5, 1055-1058.

[6] Nguyen Minh Chuong and Bui Kien Cuong, *Galerkin-wavelet approximation for a class partial integro-differential equations*, Fractional Calculus and Applied Analysis, 2001.

[7] Nguyen Minh Chuong and Ta Ngoc Tri, *The integral wavelet transform in $L_p(\mathbb{R})$*, Fractional Calculus and Applied Analysis, 3 (2000), No 2, 133-140.

[8] Nguyen Minh Chuong and Bui Kien Cuong, *The convergence estimates for Galerkin-wavelet solution for periodic pseudodifferential initial value problems*, Intern. J. of Math. and Math. Sci. Vol. 2003, No 14, March 2003, 857-867.

[9] Nguyen Minh Chuong (Author in Chief), Ha Tien Ngoan, Nguyen Minh Tri and Le Quang Trung, *Partial Differential Equations*, Publishing House "Education", Hanoi 2000.

[10] Nguyen Minh Chuong and Nguyen Van Co, *p -adic multidimensional Green function*, Proc. Americ. Math. Soc. 127 (1999). No 3, 685-694.

[11] Nguyen Minh Chuong and Nguyen Van Co, *Parabolic equation of Vladimirov type over 2 dimensional p-adic space*, preprint 05/25, Ins. of Math, Hanoi, 2005.

[12] Nguyen Minh Chuong and Bui Kien Cuong, *Convergence estimates of Galerkin-wavelet solutions to a Cauchy problem for a class of periodic pseudodifferential equations*, Proc. Americ. Math. Soc. 132 (2004), No 12, 3589-3597.

[13] W.Dahmen, S.Prössdorf, R.Schneider, *Wavelet approximation methods for pseudodiffirential equation I: Stability and convergence*, Math. Zeit. Springer-Verlag, 215 (1994), 583-620.

[14] Dequang Han, David R. Larson, *Frames, Bases, and Group Representations*, Memoirs of the AMS, 147 (2000), No 697.

[15] A. Yu. Khrennikov, *p-adic valued distributions in mathematical physics*, Kluwer Acdemic publishers, Dordrect, Boston, London, 1994.

[16] A. Yu. Khrennikov, and S.V. Kozyrev, *Wavelet on ultrametric space*, Appl/ Comput. Harmon. Anal. 19 (2005), No 1, 61-76.

[17] A.N. Kochubei, *Parabolic equations over the field of p-adic numbers*, Math. USSR. Izv. Vol. 39 (1992), No 3, 1263-1280.

[18] A.N. Kochubei, *Pseudodifferential equations and stochastics over non-archimedean fields*, Marcel Dekker, Inc. New York-Basel, 2001.

[19] A.N. Kochubei. *Stochastic integrals and stochastic differential equations over the field of p-adic numbers*, Potential Anal., 6 (1997), No 2, 105-125.

[20] A.N. Kochubei, *Analysis and Probability over infinite extension of local field*, Potential Anal. 10 (1999), No 3, 305-525.

[21] A.N. Kochubei, *Pseudo-differential equations and Stochastics over Non-Archimedean Field*, Marcel Dekker, 2001.

[22] S.V. Kozyrev, and A.Yu. Khrenikov, *Pseudo-differential operators on ultrametric spaces and ultrametric wavelets*, Izv. Ross. Akad. Nauk. Ser. Math. 69 (2005), No 5, 133-148.

[23] S.V. Kozyrev, *Wavelet theory as p-adic spectral analysis*, Izv. Ross. Akad. Nauk. Ser. Math. 66 (2002), No 2, 149-158.

[24] S.V. Kozyrev, *p-adic pseudodifferential operators and p- adic wavelets*, Tooret. Math. Fiz. 138 (2004), No 3, 322-332.

[25] Y. Meyer, *Ondelettes et Opérators*, Herman, Paris, 1990.

[26] S. V. Vladimirov, L. V. Volovich, and E. L. Zelenov, *p-adic analysis and mathematical physics*, Russian, Nauka, Moscow, 1994.

INSTITUTE OF MATHEMATICS, VIETNAMESE ACADEMY OF SCIENCE AND TECHNOLOGY, HANOI, VIETNAM

E-mail address : nmchuong@math.ac.vn

DEPARTMENT OF MATHEMATICS, HANOI UNIVERSITY OF EDUCATION, VIETNAM

E-mail: nguyenvancodhsp@yahoo.com

Contemporary Mathematics
Volume **451**, 2008

Short-Time Fourier Transform Analysis of Localization Operators

Elena Cordero and Luigi Rodino

ABSTRACT. We perform a detailed study of the boundedness properties for
localization operators. The language and the tools employed are provided
by time-frequency analysis. In particular, the time-frequency representation
named short-time Fourier transform (STFT) is used both to define and to obtain the boundedness properties of localization operators. Our results widen
most part of the ones known in the literature. Moreover, the sufficient conditions are worked out by STFT estimates, rather than the Weyl connections as
done, e.g., in [**4**, **8**]. Besides, the necessary boundedness conditions are referred
to a fixed choice of window functions, so that we can claim the *optimality* of
our results. Finally, we deal with the problem of finding symbol conditions for
Weyl operators that guarantee to rewrite them as localization operators.

CONTENTS

1. Introduction and Definitions

Localization operators belong to the family of pseudodifferential operators. This paper shows that they can be introduced and studied only using tools and techniques from time-frequency analysis. Besides, this approach provides the best boundedness results. The definition of localization operators can be given by means of a time-frequency representation: the short-time Fourier transform (STFT). Indeed, we consider the linear operators of translation and modulation (so-called

2000 *Mathematics Subject Classification.* 47G30,35S05,46E35,47B10.

Key words and phrases. Localization operator, modulation space, Weyl calculus, convolution
relations, Wigner distribution, short-time Fourier transform, Schatten class.

time-frequency shifts) given by

$$(1.1) \qquad T_x f(t) = f(t-x) \quad \text{and} \quad M_\omega f(t) = e^{2\pi i \omega t} f(t) \,.$$

Next, let g be a non-zero window function in the Schwartz class $\mathcal{S}(\mathbb{R}^d)$, then the short-time Fourier transform (STFT) of a signal $f \in L^2(\mathbb{R}^d)$ with respect to the window g is given by

$$(1.2) \qquad V_g f(x,\omega) = \langle f, M_\omega T_x g \rangle = \int_{\mathbb{R}^d} f(t) \, \overline{g(t-x)} \, e^{-2\pi i \omega t} \, dt \,.$$

We have $V_g f \in L^2(\mathbb{R}^{2d})$. This definition can be extended to pairs of dual topological vector spaces, whose duality, denoted by $\langle \cdot, \cdot \rangle$, extends the inner product on $L^2(\mathbb{R}^d)$. For instance, it may be suited to the framework of distributions and ultra-distributions (see Section 2). Just few words to explain the intuitive meaning of the previous "time-frequency" expression. If $f(t)$ represents a signal varying in time, its Fourier transform $\hat{f}(\omega)$ shows the distribution of its frequency ω, without any additional information about "when" these frequencies appear. To overcome this problem, one may choose a non-negative window function g well localized around the origin. Then, the information of the signal f at the instant x can be obtained by shifting the window g till the instant x under consideration, and by computing the Fourier transform of the product $f(x)g(t-x)$, that localizes f around the instant time x.

Once the analysis of the signal f is terminated, we can reconstruct the original signal f by a suitable inversion procedure. Namely, the reproducing formula related to the STFT, for every pairs of windows $\varphi_1, \varphi_2 \in \mathcal{S}(\mathbb{R}^d)$ with $\langle \varphi_1, \varphi_2 \rangle \neq 0$, reads as follows

$$(1.3) \qquad \int_{\mathbb{R}^{2d}} V_{\varphi_1} f(x,\omega) M_\omega T_x \varphi_2 \, dx d\omega = \langle \varphi_2, \varphi_1 \rangle f \,.$$

The function φ_1 is called the *analysis* window, because the STFT $V_{\varphi_1} f$ gives the time-frequency distribution of the signal f, whereas the window φ_2 permits to come back to the original f and, consequently, is called the *synthesis* window. We address to Section 2 below for other properties of the STFT.

The signal analysis often requires to highlight some features of the time-frequency distribution of f. This is achieved by first multiplying the STFT $V_{\varphi_1} f$ by a suitable function $a(x,\omega)$ and secondly by re-constructing $\tilde{f}$ from the product $aV_{\varphi_2} f$. In other words, we recover a filtered version of the original signal f which we denote by $A_a^{\varphi_1, \varphi_2} f$. This intuition motivates the definition of time-frequency localization operators.

DEFINITION 1.1. The localization operator $A_a^{\varphi_1, \varphi_2}$ with symbol $a \in \mathcal{S}(\mathbb{R}^{2d})$ and windows $\varphi_1, \varphi_2 \in \mathcal{S}(\mathbb{R}^d)$ is defined to be

$$(1.4) \qquad A_a^{\varphi_1, \varphi_2} f(t) = \int_{\mathbb{R}^{2d}} a(x,\omega) V_{\varphi_1} f(x,\omega) M_\omega T_x \varphi_2(t) \, dx d\omega \,, \quad f \in L^2(\mathbb{R}^d).$$

The preceding definition makes sense also if we assume $a \in L^\infty(\mathbb{R}^{2d})$, see below. In particular, if $a = \chi_\Omega$ for some compact set $\Omega \subset \mathbb{R}^{2d}$, then $A_a^{\varphi_1, \varphi_2}$ is interpreted as the part of f that "lives on the set Ω" in the time-frequency plane. This is why $A_a^{\varphi_1, \varphi_2}$ is called a *localization* operator.

Often it is more convenient to interpret the definition of $A_a^{\varphi_1,\varphi_2}$ in a weak sense, then (1.4) can be recast as

$$(1.5) \qquad \langle A_a^{\varphi_1,\varphi_2} f, g \rangle = \langle a V_{\varphi_1} f, V_{\varphi_2} g \rangle = \langle a, \overline{V_{\varphi_1} f} V_{\varphi_2} g \rangle, \quad f, g \in \mathcal{S}(\mathbb{R}^d) \,.$$

If we enlarge the class of symbols to the tempered distributions, i.e., we take $a \in \mathcal{S}'(\mathbb{R}^{2d})$ whereas $\varphi_1, \varphi_2 \in \mathcal{S}(\mathbb{R}^d)$, then (1.4) is a well-defined continuous operator from $\mathcal{S}(\mathbb{R}^d)$ to $\mathcal{S}'(\mathbb{R}^d)$. The previous assertion can be proven directly using the weak definition. For every window $\varphi_1 \in \mathcal{S}(\mathbb{R}^d)$ the STFT V_{φ_1} is a continuous mapping from $\mathcal{S}(\mathbb{R}^d)$ into $\mathcal{S}(\mathbb{R}^{2d})$ (see, e.g., [22, Theorem 11.2.5]). Since also $V_{\varphi_2} g \in \mathcal{S}(\mathbb{R}^{2d})$, the brackets $\langle a, \overline{V_{\varphi_1} f} V_{\varphi_2} g \rangle$ are well-defined in the duality between $\mathcal{S}'(\mathbb{R}^{2d})$ and $\mathcal{S}(\mathbb{R}^{2d})$. Consequently, the left-hand side of (1.5) can be interpreted in the duality between $\mathcal{S}'(\mathbb{R}^d)$ and $\mathcal{S}(\mathbb{R}^d)$ and this shows that $A_a^{\varphi_1,\varphi_2}$ is a continuous operator from $\mathcal{S}(\mathbb{R}^d)$ to $\mathcal{S}'(\mathbb{R}^d)$. The continuity of the mapping $A_a^{\varphi_1,\varphi_2}$ is achieved by using the continuity of both the STFT and the brackets $\langle \cdot, \cdot \rangle$. Similar arguments can be applied for tempered ultra-distributions, as we are going to see later on.

If $\varphi_1(t) = \varphi_2(t) = e^{-\pi t^2}$, then $A_a = A_a^{\varphi_1,\varphi_2}$ is the so-called (anti-)Wick operator. Actually, the name "localization operator" first appeared in 1988, when Daubechies [10] used these operators as a mathematical tool to localize a signal on the time-frequency plane. But localization operators with Gaussian windows were already known in physics: they were introduced as a quantization rule by Berezin [1] in 1971 (the Wick operators cited above). Since their first appearance, they have been extensively studied as an important mathematical tool in signal analysis and other applications (see [11, 29, 35] and references therein). Beyond signal analysis and the Wick quantization procedure [1, 30], we recall their employment as approximation of pseudodifferential operators ("wave packets") [9, 20]. Besides, in other branches of mathematics, localization operators are also named Toeplitz operators (see, e.g., [12]) or short-time Fourier transform multipliers [18].

In [4] localization operators are viewed as a multilinear mapping

$$(1.6) \qquad\qquad\qquad (a, \varphi_1, \varphi_2) \mapsto A_a^{\varphi_1,\varphi_2},$$

acting on products of symbol and window spaces. The dependence of the localization operator $A_a^{\varphi_1,\varphi_2}$ on all three parameters has been studied there in different functional frameworks. The start is given by subspaces of the tempered distributions. The basic subspace is $L^2(\mathbb{R}^d)$, but many other Banach and Hilbert spaces, as well as topological vector spaces, are considered. The results in [4] enlarge the ones in the literature, concerning L^p spaces [2, 35], potential and Sobolev spaces [3], modulation spaces [18, 33, 34].

On the footprints of [4], the study of localization operators can be carried to Gelfand-Shilov spaces and spaces of ultra-distributions [8].

Our present setting includes both of the previous studies. Depending on the weights we are dealing with, we shall find ourselves in the field of either distributions or ultra-distributions. Besides, the techniques employed come from the very definition of localization operators and use the STFT, instead of evoking Weyl operators results as in [4, 8].

Let us give the guidelines of the study. The continuity of the mapping in (1.6) can be expressed by an inequality of the form

$$(1.7) \qquad\qquad\qquad \|A_a^{\varphi_1,\varphi_2}\|_{op} \le C \|a\|_{B_1} \|\varphi_1\|_{B_2} \|\varphi_2\|_{B_3} \,,$$

where B_1, B_2, B_3 are suitable spaces of symbols and windows. For example, if $a \in L^\infty(\mathbb{R}^d)$ and $\varphi_1, \varphi_2 \in L^2(\mathbb{R}^d)$, then

$$
\begin{aligned}
\|A_a^{\varphi_1,\varphi_2}\|_{B(L^2)} &= \sup_{\|f\|_{L^2}=1} \sup_{\|g\|_{L^2}=1} |\langle A_a^{\varphi_1,\varphi_2} f, g \rangle| \\
&= \sup_{\|f\|_{L^2}=1} \sup_{\|g\|_{L^2}=1} |\langle a, \overline{V_{\varphi_1} f}\, V_{\varphi_2} g \rangle| \\
&\leq \sup_{\|f\|_{L^2}=1} \sup_{\|g\|_{L^2}=1} \|a\|_{L^\infty} \|\overline{V_{\varphi_1} f}\, V_{\varphi_2} g\|_{L^1} \\
&\leq \sup_{\|f\|_{L^2}=1} \sup_{\|g\|_{L^2}=1} \|a\|_{L^\infty} \|V_{\varphi_1} f\|_{L^2} \|V_{\varphi_2} g\|_{L^2} \\
&= \|a\|_{L^\infty} \|\varphi_1\|_{L^2} \|\varphi_2\|_{L^2},
\end{aligned}
$$

where the last inequality is achieved by using the orthogonality relations for the STFT

$$\|V_\varphi f\|_{L^2(\mathbb{R}^{2d})} = \|\varphi\|_{L^2(\mathbb{R}^d)} \|f\|_{L^2(\mathbb{R}^d)}, \quad \forall \varphi, f \in L^2(\mathbb{R}^d).$$

Thus for this particular choice of symbol classes and window spaces we obtain the L^2 boundedness. The previous easy proof gives just a flavour of the boundedness results for localization operators known in literature, we shall see that the symbol class L^∞ can be enlarged significantly. Even a tempered distribution like δ may give the boundedness of the corresponding localization operator.

Among the many function/(ultra-)distribution spaces employed, modulation spaces reveal to be the *optimal choice* for handling localization operators, see Section 4 below. As special case we mention Feichtinger's algebra $M^1(\mathbb{R}^d)$ defined by the norm

$$\|f\|_{M^1} := \|V_g f\|_{L^1(\mathbb{R}^{2d})}$$

for some (hence all) non-zero $g \in \mathcal{S}(\mathbb{R}^d)$ [**16, 22**]. Its dual space $M^\infty(\mathbb{R}^{2d})$ is a very useful subspace of tempered distributions and possesses the norm

$$\|f\|_{M^\infty} := \sup_{(x,\omega) \in \mathbb{R}^{2d}} |V_g f(x,\omega)|.$$

With these spaces the estimate (1.7) reads as follows:

THEOREM 1.2. *If $a \in M^\infty(\mathbb{R}^{2d})$, and $\varphi_1, \varphi_2 \in M^1(\mathbb{R}^d)$, then $A_a^{\varphi_1,\varphi_2}$ is bounded on $L^2(\mathbb{R}^d)$, with operator norm at most*

$$\|A_a^{\varphi_1,\varphi_2}\|_{B(L^2)} \leq C \|a\|_{M^\infty} \|\varphi_1\|_{M^1} \|\varphi_2\|_{M^1}.$$

The striking fact is the converse of the preceding result.

THEOREM 1.3. *Let $a \in \mathcal{S}'(\mathbb{R}^{2d})$ and $\varphi_1, \varphi_2 \in \mathcal{S}(\mathbb{R}^d)$. If the operators $A_{M_\zeta a}^{\varphi_1,\varphi_2}$ are uniformly bounded on $L^2(\mathbb{R}^d)$, i.e., there exists a constant $C = C(a, \varphi_1, \varphi_2) > 0$ such that,*

$$\|A_{M_\zeta a}^{\varphi_1,\varphi_2}\|_{B(L^2)} \leq C, \qquad \forall \zeta \in \mathbb{R}^{2d},$$

then $a \in M^\infty$.

The observation $a \in M^\infty(\mathbb{R}^{2d})$ then $M_\zeta a \in M^\infty(\mathbb{R}^{2d})$, with $\|M_\zeta a\|_{M^\infty} = \|a\|_{M^\infty}$, for every $\zeta \in \mathbb{R}^{2d}$, shows that condition $\|A_{M_\zeta a}^{\varphi_1,\varphi_2}\|_{B(L^2)} \leq C$ is not a *further, technical assumption* and that the necessary results are optimal, indeed. These topics shall be detailed in Section 4.

In the last section shall work on the Weyl-Wick connection. Whereas a Wick operator A_a can always be written in a Weyl form $L\sigma$ [**20, 30**], with symbol relation

$\sigma = a * 2^d e^{-2\pi(\cdot)^2}$, the converse is not true in general. We find conditions on the Weyl symbol σ guaranteeing the equality $L_\sigma = A_a$.

Notation. To be definite, let us fix some notation we shall use later on (and have already used in this Introduction). We define $t^2 = t \cdot t$, for $t \in \mathbb{R}^d$, and $xy = x \cdot y$ is the scalar product on $\mathbb{R}^d$. For $z = (x, \omega) \in \mathbb{R}^{2d}$, we denote by $\pi(z) = M_\omega T_x$ the time-frequency shift operator.

The Schwartz class is denoted by $\mathcal{S}(\mathbb{R}^d)$, the space of tempered distributions by $\mathcal{S}'(\mathbb{R}^d)$. We use the brackets $\langle f, g \rangle$ to denote the extension to $\mathcal{S}(\mathbb{R}^d) \times \mathcal{S}'(\mathbb{R}^d)$ of the inner product $\langle f, g \rangle = \int f(t) \overline{g(t)} dt$ on $L^2(\mathbb{R}^d)$. The Fourier transform is normalized to be $\hat{f}(\omega) = \mathcal{F}f(\omega) = \int f(t) e^{-2\pi i t \omega} dt$.

We write J for the symplectic matrix

$$J = \begin{bmatrix} 0 & I_d \\ -I_d & 0 \end{bmatrix}.$$

For functions F on the phase space (x, ω), the symplectic operator T_J is defined by

(1.8) $$(T_J F)(x, \omega) = F({}^t J(x, \omega)) = F(-\omega, x), \quad (x, \omega) \in \mathbb{R}^{2d}.$$

Throughout the paper, we shall use the notation $A \lesssim B$ to indicate $A \le cB$ for a suitable constant $c > 0$, whereas $A \asymp B$ if $A \le cB$ and $B \le kA$, for suitable $c, k > 0$.

2. Time-Frequency Methods

First we summarize some concepts and tools of time-frequency analysis, for an extended exposition we refer to the textbooks [**20, 22**].

The time-frequency representations required for localization operators and the Weyl calculus are the *short-time Fourier transform* and the *Wigner distribution*.

The short-time Fourier transform (STFT) is defined in (1.2). The *cross-Wigner distribution* $W(f, g)$ of $f, g \in L^2(\mathbb{R}^d)$ is given by

(2.1) $$W(f, g)(x, \omega) = \int f(x + \frac{t}{2}) \overline{g(x - \frac{t}{2})} e^{-2\pi i \omega t} dt.$$

The quadratic expression $Wf = W(f, f)$ is usually called the Wigner distribution of f.

Both the STFT $V_g f$ and the Wigner distribution $W(f, g)$ are defined for f, g in many possible pairs of Banach spaces or topological vector spaces. For instance, they both map $L^2(\mathbb{R}^d) \times L^2(\mathbb{R}^d)$ into $L^2(\mathbb{R}^{2d})$ and $\mathcal{S}(\mathbb{R}^d) \times \mathcal{S}(\mathbb{R}^d)$ into $\mathcal{S}(\mathbb{R}^{2d})$. Furthermore, they can be extended to a map from $\mathcal{S}'(\mathbb{R}^d) \times \mathcal{S}'(\mathbb{R}^d)$ into $\mathcal{S}'(\mathbb{R}^{2d})$ or, e.g., from $\mathcal{S}_{1/2}^{1/2}(\mathbb{R}^d) \times \mathcal{S}_{1/2}^{1/2}(\mathbb{R}^d)$ into $\mathcal{S}_{1/2}^{1/2}(\mathbb{R}^{2d})$. The crucial properties of the STFT (for proofs, see [**22**, Ch. 3] and [**23**]).

LEMMA 2.1. *Let $f, g, f_j, g_j \in L^2(\mathbb{R}^d), j = 1, 2$, then we have*
(i) (STFT of time-frequency shifts) For $y, \xi \in \mathbb{R}^d$, we have

(2.2) $$V_g(M_\xi T_y f)(x, \omega) = e^{-2\pi i (\omega - \xi) y} (V_g f)(x - y, \omega - \xi)$$

(2.3) $$V_{(M_\xi T_y g)}(M_\xi T_y f)(x, \omega) = e^{2\pi i (\xi x - \omega y)} (V_g f)(x, \omega)$$

(ii) (Fourier transform of a product of STFTs),

(2.4) $$\left(\widehat{V_{g_1} f_1 \overline{V_{g_2} f_2}}\right)(x, \omega) = (V_{f_2} f_1 \overline{V_{g_2} g_1})(-\omega, x) = T_J(V_{f_2} f_1 \overline{V_{g_2} g_1})(x, \omega).$$

(iii) (STFT of the Fourier transforms),

$$(2.5) \qquad (V_g f)(x, \omega) = -2\pi i x \omega V_{\hat{g}} \hat{f}(\omega, -x).$$

Note that (2.2) and (2.3) can be read backwards and yield a formula for the $2d$-dimensional time-frequency shift $M_\zeta T_z(V_g f), z, \zeta \in \mathbb{R}^{2d}$.

To investigate the local properties of the STFT, we will need to compute the STFT of a STFT.

LEMMA 2.2. *Fix a nonzero $\varphi \in \mathcal{S}(\mathbb{R}^d)$ and let $f, g \in \mathcal{S}(\mathbb{R}^d)$.*

Set $\Phi = V_\varphi \varphi \in \mathcal{S}(\mathbb{R}^{2d})$. Then the STFT of $V_g f$ with respect to the window Φ is given by

$$(2.6) \qquad V_\Phi(V_g f)(z, \zeta) = e^{-2\pi i z_2 \zeta_2} \overline{V_\varphi g(-z_1 - \zeta_2, \zeta_1)} \, V_\varphi f(-\zeta_2, z_2 + \zeta_1),$$

where $z = (z_1, z_2) \in \mathbb{R}^{2d}$ and $\zeta = (\zeta_1, \zeta_2) \in \mathbb{R}^{2d}$.

For a non-zero $g \in L^2(\mathbb{R}^d)$, we write V_g^* for the *adjoint* of V_g, given by

$$\langle V_g^* F, f \rangle = \langle F, V_g f \rangle, \quad f \in L^2(\mathbb{R}^d), \ F \in L^2(\mathbb{R}^{2d}).$$

In particular, for $F \in \mathcal{S}(\mathbb{R}^{2d})$, $g \in \mathcal{S}(\mathbb{R}^d)$, we have

$$(2.7) \qquad V_g^* F(t) = \int_{\mathbb{R}^{2d}} F(x, \omega) M_\omega T_x g(t) \, dx \, d\omega \in \mathcal{S}(\mathbb{R}^d).$$

Take $f \in \mathcal{S}(\mathbb{R}^d)$ and set $F = V_g f$, then

$$(2.8) \quad f(t) = \frac{1}{\|g\|_{L^2}^2} \int_{\mathbb{R}^{2d}} V_g f(x, \omega) M_\omega T_x g(t) \, dx \, d\omega \in \mathcal{S}(\mathbb{R}^d) = \frac{1}{\|g\|_{L^2}^2} V_g^* V_g f(t).$$

We refer to [**22**, Proposition 11.3.2] for a detailed treatment of the adjoint operator.

Localization operators as Weyl operators. Let $W(g, f)$ be the cross-Wigner distribution as defined in (2.1). Then the Weyl operator L_σ of symbol $\sigma \in \mathcal{S}'(\mathbb{R}^{2d})$ is defined by

$$(2.9) \qquad \langle L_\sigma f, g \rangle = \langle \sigma, W(g, f) \rangle, \qquad f, g \in \mathcal{S}(\mathbb{R}^d).$$

Every linear continuous operator from $\mathcal{S}(\mathbb{R}^d)$ to $\mathcal{S}'(\mathbb{R}^d)$ can be represented as a Weyl operator, and a calculation in [**3, 20, 30**] reveals that

$$(2.10) \qquad A_a^{\varphi_1, \varphi_2} = L_{a * W(\varphi_2, \varphi_1)},$$

so the (Weyl) symbol of $A_a^{\varphi_1, \varphi_2}$ is given by

$$(2.11) \qquad \sigma = a * W(\varphi_2, \varphi_1).$$

Then same arguments might be adapted for the Gelfand-Shilov spaces and related ultra-distribution spaces [**8**].

3. Function Spaces

Gelfand-Shilov spaces. The Gelfand-Shilov spaces were introduced by Gelfand and Shilov in [**21**]. They have been applied by many authors in different contexts, see, e.g. [**24, 26, 32**]. For a comprehensive treatment of Gelfand-Shilov spaces we refer to [**21**]. We limit ourselves to highlight those features our study requires. For $\alpha \in \mathbb{R}_+^d$, the Gelfand-Shilov spaces $\mathcal{S}_\alpha^\alpha$ and Σ_α^α are introduced via the following characterization:

PROPOSITION 3.1. The next statements are equivalent [8, 21, 23]:
(i) $f \in \mathcal{S}_\alpha^\alpha(\mathbb{R}^d)$ (resp. $f \in \Sigma_\alpha^\alpha(\mathbb{R}^d)$).
(ii) $f \in \mathcal{C}^\infty(\mathbb{R}^d)$ and there exist (resp. for every) constants $h > 0, k > 0$,

$$(3.1) \qquad \|f e^{h|x|^{1/\alpha}}\|_{L^\infty} < \infty \quad \text{and} \quad \|\mathcal{F}f e^{k|\omega|^{1/\alpha}}\|_{L^\infty} < \infty,$$

where $|x|^{1/\alpha} = |x_1|^{1/\alpha_1} + \cdots + |x_d|^{1/\alpha_d}$, $|\omega|^{1/\alpha} = |\omega_1|^{1/\alpha_1} + \cdots + |\omega_d|^{1/\alpha_d}$.

(iii) $f \in \mathcal{C}^\infty(\mathbb{R}^d)$ and there exists (resp. for every) $C > 0, h > 0$,

$$(3.2) \qquad \|(\partial^q f) e^{h|x|^{1/\alpha}}\|_{L^\infty} \le C^{|q|+1}(q!)^\alpha, \quad \forall q \in \mathbb{N}_0^d.$$

Gelfand-Shilov spaces enjoy the following embeddings:
(i) For $\alpha > 0$ [21],

$$(3.3) \qquad \Sigma_\alpha^\alpha \hookrightarrow \mathcal{S}_\alpha^\alpha \hookrightarrow \mathcal{S}.$$

(ii) For every $0 < \alpha_1 < \alpha_2$ [8],

$$(3.4) \qquad \mathcal{S}_{\alpha_1}^{\alpha_1} \hookrightarrow \Sigma_{\alpha_2}^{\alpha_2}.$$

Furthermore, $\mathcal{S}_\alpha^\alpha$ is not trivial if and only if $\alpha \ge 1/2$ whereas $\Sigma_\alpha^\alpha \ne \emptyset$ if and only if $\alpha > 1/2$ (see Pilipović [26]).

The Gelfand-Shilov spaces $\mathcal{S}_\alpha^\alpha$ are invariant under Fourier transform $\mathcal{F}$ and time-frequency shifts:

$$(3.5) \qquad \mathcal{F}(\mathcal{S}_\alpha^\alpha) = \mathcal{S}_\alpha^\alpha, \quad T_x(\mathcal{S}_\alpha^\alpha) = \mathcal{S}_\alpha^\alpha \quad \text{and} \quad M_\omega(\mathcal{S}_\alpha^\alpha) = \mathcal{S}_\alpha^\alpha.$$

The same holds true for the spaces Σ_α^α. Therefore, the spaces $\mathcal{S}_\alpha^\alpha$ and Σ_α^α are a family of *Fourier transform and time-frequency shift invariant spaces* contained in the Schwartz class $\mathcal{S}$; the *smallest*, non-trivial one being $\mathcal{S}_{1/2}^{1/2}$. Functions in $\mathcal{S}_{1/2}^{1/2}$ are, e.g., the Gaussian $f(x) = e^{-\pi x^2}$ or the Hermite h_α detailed in Section 5.

Another useful characterization of the space $\mathcal{S}_\alpha^\alpha$ and Σ_α^α involves the STFT and is proven in [24, Proposition 4.3]: $f \in \mathcal{S}_\alpha^\alpha(\mathbb{R}^d)$ if and only if $V_g f \in \mathcal{S}_\alpha^\alpha(\mathbb{R}^{2d})$, and the same for Σ_α^α. We shall heavily use the case $\alpha = 1$: for a non-zero window $g \in \Sigma_1^1$ we have

$$(3.6) \qquad V_g f \in \Sigma_1^1(\mathbb{R}^{2d}) \Leftrightarrow f \in \Sigma_1^1(\mathbb{R}^d).$$

The strong duals of Gelfand-Shilov classes $\mathcal{S}_\alpha^\alpha$ and Σ_α^α are spaces of tempered ultra-distributions of Roumieu and Beurling type and will be denoted by $(\mathcal{S}_\alpha^\alpha)'$ and $(\Sigma_\alpha^\alpha)'$, respectively.

Modulation Spaces. We do not use the traditional definition of these spaces as subspaces of tempered distributions and related to polynomial weights, for which we refer to [13], [22, Ch. 11-13] and the original literature quoted there. Ultra-distributions as symbols for localization operators require a more general definition, including weights of exponential growth. Thus, we first introduce the class of weights we shall work with.

Weight Functions. In this paper v is always a continuous, positive, even, sub-multiplicative function (submultiplicative weight), i.e., $v(0) = 1$, $v(z) = v(-z)$, and $v(z_1 + z_2) \le v(z_1)v(z_2)$, for all $z, z_1, z_2 \in \mathbb{R}^{2d}$. Moreover, v is *invariant* under symplectic transformations: $(T_J v)(x, \omega) = v(x, \omega)$, $(x, \omega) \in \mathbb{R}^{2d}$ and the symplectic operator T_j is defined in (1.8).

Continuity and submultiplicativity imply that $v(z)$ is *dominated* by an exponential function, i.e.

$$(3.7) \qquad \exists\, C, k > 0 \quad \text{such that} \quad v(z) \le C e^{k|z|}, \quad z \in \mathbb{R}^{2d}.$$

For example, every weight of the form $v(z) = e^{a|z|^b}(1 + |z|)^s \log^r(e + |z|)$ for parameters $a, r, s \ge 0$, $0 \le b \le 1$ satisfies the above conditions.

Associated to every submultiplicative weight we consider the class of so-called *v-moderate* weights $\mathcal{M}_v(\mathbb{R}^{2d})$. A positive, even weight function m on $\mathbb{R}^{2d}$ belongs to $\mathcal{M}_v(\mathbb{R}^{2d})$ if it satisfies the condition

$$m(z_1 + z_2) \le C v(z_1) m(z_2) \quad \forall z_1, z_2 \in \mathbb{R}^{2d}.$$

We note that this definition implies that $\frac{1}{v} \lesssim m \lesssim v$, $m \ne 0$ everywhere, and that $1/m \in \mathcal{M}_v(\mathbb{R}^{2d})$.

For the investigation of localization operators the weights mostly used are defined by

$$(3.8) \quad v_s(z) \;=\; v_s(x, \omega) = \langle z \rangle^s = (1 + x^2 + \omega^2)^{s/2}, \quad z = (x, \omega) \in \mathbb{R}^{2d}$$

$$(3.9) \quad w_s(z) \;=\; w_s(x, \omega) = e^{s|(x,\omega)|}, \quad z = (x, \omega) \in \mathbb{R}^{2d},$$

$$(3.10) \quad \tau_s(z) \;=\; \tau_s(x, \omega) = \langle \omega \rangle^s$$

$$(3.11) \quad \mu_s(z) \;=\; \mu_s(x, \omega) = e^{s|\omega|}.$$

DEFINITION 3.2. Let m be a weight in $\mathcal{M}_v(\mathbb{R}^{2d})$, and g a non-zero *window* function in $\Sigma_1^1(\mathbb{R}^d)$. For $1 \le p, q \le \infty$ and $f \in \Sigma_1^1(\mathbb{R}^d)$ we define the modulation space norm (on $\Sigma_1^1(\mathbb{R}^d)$) by

$$\|f\|_{M_m^{p,q}} = \|V_g f\|_{L_m^{p,q}} = \left(\int_{\mathbb{R}^d} \left(\int_{\mathbb{R}^d} |V_g f(x, \omega)|^p m(x, \omega)^p \, dx \right)^{q/p} d\omega \right)^{1/q},$$

(with obvious changes if either $p = \infty$ or $q = \infty$). If $p, q < \infty$, the modulation space $M_m^{p,q}$ is the norm completion of Σ_1^1 in the $M_m^{p,q}$-norm. If $p = \infty$ or $q = \infty$, then $M_m^{p,q}$ is the completion of Σ_1^1 in the weak* topology. If $p = q$, $M_m^p := M_m^{p,p}$, and, if $m \equiv 1$, then $M^{p,q}$ and M^p stand for $M_m^{p,q}$ and $M_m^{p,p}$, respectively.

Notice that:
(i) If $f, g \in \Sigma_1^1(\mathbb{R}^d)$, the above integral is convergent thanks to (3.1) and (3.6). Namely, for a given $m \in \mathcal{M}_v$ there exist $l > 0$ such that $m(x, \omega) \le C e^{l|(x,\omega)|}$ and therefore

$$\left| \int_{\mathbb{R}^d} \left(\int_{\mathbb{R}^d} |V_g f(x, \omega)|^p m(x, \omega)^p \, dx \right)^{q/p} d\omega \right|$$

$$\le C \left| \int_{\mathbb{R}^d} \left(\int_{\mathbb{R}^d} |V_g f(x, \omega)|^p e^{lp|(x,\omega)|} \, dx \right)^{q/p} d\omega \right| < \infty$$

in view of (3.6). This implies $\Sigma_1^1 \subset M_m^{p,q}$.
(ii) By definition, $M_m^{p,q}$ is a Banach space. Besides, it is proven for the subexponential case in [13] and for the exponential one in [8] that their definition does not depend on the choice of the window g, that can be enlarged to the modulation algebra M_v^1.

(iii) For $m \in \mathcal{M}_v$ of at most polynomial growth, $M_m^{p,q} \subset \mathcal{S}'$ and the definition 3.2 reads as [4, 22]:
$$M_m^{p,q}(\mathbb{R}^d) = \{f \in \mathcal{S}'(\mathbb{R}^d) : V_g f \in L_m^{p,q}(\mathbb{R}^{2d})\}.$$
(iv) For every weight $m \in \mathcal{M}_v$, $M_m^{p,q}$ is the subspace of ultra-distribution $(\Sigma_1^1)'$ defined in [8, Definition 2.1].

(iv) If m belongs to $\mathcal{M}_v$ and fulfills the GRS-condition $\lim_{n \to \infty} v(nz)^{1/n} = 1$, for all $z \in \mathbb{R}^{2d}$, the definition of modulation spaces is the same as in [6] (because the "space of special windows" $\mathcal{S}_\mathcal{C}$ is a subset of Σ_1^1).

(v) For related constructions of modulation spaces, involving the theory of coorbit spaces, we refer to [15, 17].

(vi) A function $f \in M^{p,q}$ behaves like $f \in L^p$ and $\hat{f} \in L^q$; so the weight's x-variable is related to the decay of f at infinity, whereas the ω-variable gives information on the smoothness of f. In the sequel, it will be useful to consider weights $m(x,\omega) = (\nu \otimes \mu)(x,\omega) = \nu(x) \otimes \mu(\omega)$, where the two different types of information can be handled separately.

(vii) *Duality of modulation spaces* [22, Thm. 11.3.6.]: If $1 \le p, q < \infty$, then $(M_m^{p,q})^* = M_{1/m}^{p',q'}$ under the duality

(3.12) $$\langle f, h \rangle = \int_{\mathbb{R}^{2d}} V_g f(z) \overline{V_g h(z)}\, dz, \quad g \in \Sigma_1^1.$$

The class of modulation spaces contains the following well-known function spaces:

Weighted L^2-spaces: $M_{\langle x \rangle^s}^2(\mathbb{R}^d) = L_s^2(\mathbb{R}^d) = \{f : f(x)\langle x \rangle^s \in L^2(\mathbb{R}^d)\}, s \in \mathbb{R}$.

Sobolev spaces: $M_{\langle \omega \rangle^s}^2(\mathbb{R}^d) = H^s(\mathbb{R}^d) = \{f : \hat{f}(\omega)\langle \omega \rangle^s \in L^2(\mathbb{R}^d)\}, s \in \mathbb{R}$.

Shubin-Sobolev spaces [30, 3]: $M_{\langle (x,\omega) \rangle^s}^2(\mathbb{R}^d) = L_s^2(\mathbb{R}^d) \cap H^s(\mathbb{R}^d) = Q_s(\mathbb{R}^d)$.

Feichtinger's algebra: $M^1(\mathbb{R}^d) = S_0(\mathbb{R}^d)$.

For comparison, we list the following embeddings between potential and modulation spaces [4].

LEMMA 3.3. *We have*
(i) *If* $p_1 \le p_2$ *and* $q_1 \le q_2$, *then* $M_m^{p_1,q_1} \hookrightarrow M_m^{p_2,q_2}$.
(ii) *For* $1 \le p \le \infty$ *and* $s \in \mathbb{R}$
$$W_s^p(\mathbb{R}^d) \hookrightarrow M_{\tau_s}^{p,\infty}(\mathbb{R}^d).$$

Consequently, $L^p \subseteq M^{p,\infty}$, and in particular, $L^\infty \subseteq M^\infty$. But M^∞ contains all bounded measures on $\mathbb{R}^d$ and other tempered distributions. For instance, the point measure δ belongs to M^∞, because for $g \in \mathcal{S}$ we have
$$|V_g \delta(x,\omega)| = |\langle \delta, M_\omega T_x g \rangle| = |\bar{g}(-x)| \le \|g\|_{L^\infty}, \quad \forall (x,\omega) \in \mathbb{R}^{2d}.$$

But one might go far beyond the tempered distributions and still find elements in suitable weighted modulation spaces. An interesting example is provided by ultra-distributions with compact support, denoted by $\mathcal{E}'_t$, $t > 1$. Recall the embeddings:
$$\mathcal{E}'_t \subset (\mathcal{S}_t^t)' \subset (\Sigma_t^t)', \quad t > 1.$$
We skip the precise definition of $\mathcal{E}'_t$, which can be found in many places, see e.g. [28, Definition 1.5.5] and subsequent anisotropic generalization. The following structure theorem, obtained by a slight generalization of [28, Theorem 1.5.6] to the anisotropic case, will be sufficient for our purposes.

THEOREM 3.4. *Let $t \in \mathbb{R}^d$, $t > 1$, i.e. $t = (t_1, \ldots, t_d)$, with $t_1 > 1, \ldots, t_d > 1$. Every $u \in \mathcal{E}'_t$ can be represented as*

$$(3.13) \qquad u = \sum_{\alpha \in \mathbb{N}_0^d} \partial^\alpha \mu_\alpha,$$

where μ_α is a measure satisfying

$$(3.14) \qquad \int_K |d\mu_\alpha| \leq C_\epsilon \epsilon^{|\alpha|} (\alpha!)^{-t},$$

for every $\epsilon > 0$ and a suitable compact set $K \subset \mathbb{R}^d$, independent of α.

Using the preceding characterization, the STFT of an ultra-distribution with compact support is estimated as follows. [**8**, Proposition 4.2].

PROPOSITION 3.5. *Let $t \in \mathbb{R}^d$, $t > 1$, and $a \in \mathcal{E}'_t(\mathbb{R}^d)$. Then its STFT with respect to any window $g \in \Sigma_1^1$ satisfies the estimate*

$$|V_g a(x, \omega)| \lesssim e^{-h|x|} e^{t|2\pi\omega|^{1/t}},$$

for every $h > 0$,.

The STFT estimate given in Proposition 3.5 shows that ultra-distributions with compact support are a subset of fitting modulation spaces. Namely, for all $\epsilon > 0$, there exists $C_\epsilon > 0$ such that,

$$t \cdot |2\pi\omega|^{1/t} = \sum_{i=1}^d t_i |2\pi\omega_i|^{1/t_i} \leq \epsilon|\omega|, \quad \text{for } |\omega| > C_\epsilon,$$

then the estimate of Proposition 3.5 gives $a \in M_{1/\mu_\epsilon}^{1,\infty}(\mathbb{R}^{2d})$.

The characterization of the Schwartz class and the tempered distributions is given in [**23**]: we have $\mathcal{S}(\mathbb{R}^d) = \bigcap_{s \geq 0} M_{\langle \cdot \rangle^s}^1(\mathbb{R}^d)$ and $\mathcal{S}'(\mathbb{R}^d) = \bigcup_{s \geq 0} M_{1/\langle \cdot \rangle^s}^\infty(\mathbb{R}^d)$. A similar characterization for Gelfand-Shilov spaces and tempered ultra-distributions is in [**8**, Proposition 2.3]: Let $1 \leq p, q \leq \infty$, and let w_s be given by (3.9), then,

$$(3.15) \qquad \Sigma_1^1 = \bigcap_{s \geq 0} M_{w_s}^{p,q}, \qquad (\Sigma_1^1)' = \bigcup_{s \geq 0} M_{1/w_s}^{p,q}.$$

$$(3.16) \qquad \mathcal{S}_1^1 = \bigcup_{s > 0} M_{w_s}^{p,q}, \qquad (\mathcal{S}_1^1)' = \bigcap_{s > 0} M_{1/w_s}^{p,q}.$$

Nevertheless, in order to obtain the sharpest results for boundedness of localization operators, we shall present another characterization of the preceding spaces, using weights with time and frequency variables decoupled. For the space Σ_1^1 and its dual we have:

$$(3.17) \qquad \Sigma_1^1 = \bigcap_{r,s \geq 0} M_{w_r \otimes w_s}^{p,q}, \quad (\Sigma_1^1)' = \bigcup_{r,s \geq 0} M_{(1/w_r) \otimes (1/w_s)}^{p,q}.$$

and, analogously, for the other spaces. This is proven using the following simple embeddings: for every $r, s > 0$,

$$M_{w_l}^{p,q} \hookrightarrow M_{w_r \otimes w_s}^{p,q} \hookrightarrow M_{w_t}^{p,q}, \quad l \geq 2^{1/2} \max\{r, s\}, \quad 0 < t \leq \min\{r, s\},$$

where $w_l(x, \omega) = e^{l|(x,\omega)|}$, $(w_r \otimes w_s)(x, \omega) = e^{r|x|} e^{s|x|}$, $w_t(x, \omega) = e^{t|(x,\omega)|}$.

More generally, using the arguments above, one can recover Σ_1^1 as the intersection of all the $M_{v \otimes w}^{p,q}$, with v, w submultiplicative weights on $\mathbb{R}^d$ and, by duality, the space $(\Sigma_1^1)'$.

Wiener Amalgam Spaces. They were introduced by Feichtinger to overcome the inability of the L^p norm to distinguish between local and global properties of functions. We refer to [**14, 25**] for a complete treatment of the subject. Here we present a definition similar to what done above for modulation spaces.

Let $\mathcal{F}L^p$ denote the image of L^p under the Fourier transform, i.e.,

$$\mathcal{F}L^p(\mathbb{R}^d) = \{f \in \mathcal{S}'(\mathbb{R}^d), \text{ such that } \exists\, h \in L^p \text{ with } \hat{h} = f\}$$

equipped with the norm $\|f\|_{\mathcal{F}L^p} = \|h\|_p$, with $\hat{h} = f$. Here we present their definition for a general weight $m \in \mathcal{M}_v$ and using the STFT. Namely, for $1 \le p, q \le \infty$, the Wiener amalgam space $W(\mathcal{F}L^p, L_m^q)$ is the space of all functions/tempered (ultra-)distributions for which the norm

$$(3.18) \qquad \|f\|_{W(\mathcal{F}L^p, L_m^q)} = \left(\int_{\mathbb{R}^d} \left(\int_{\mathbb{R}^d} |V_g f(z, \zeta)|^p \, d\zeta \right)^{q/p} m(z)^q dz \right)^{1/q}$$

is finite. As for modulation spaces [**8, 32**], it is straightforward to check that different choices of $g \in \Sigma_1^1$ generate the same space and yield equivalent norms. Modulation spaces and are closely related: for $p = q$, we have

$$(3.19) \qquad \|f\|_{W(\mathcal{F}L^p, L_m^p)} = \left(\int_{\mathbb{R}^d} \int_{\mathbb{R}^d} |V_g f(z, \zeta)|^p m(z)^p dz \, d\zeta \right)^{1/p} \asymp \|f\|_{M_{m \otimes 1}^p},$$

that is, $W(\mathcal{F}L^p, L_m^p) = M_{m \otimes 1}^p$. Among the properties of Wiener amalgam spaces, we shall recall their *pointwise multiplication*, formulated for the spaces object of our study:

$$(3.20) \qquad \|f \cdot g\|_{W(\mathcal{F}L^1, L_m^1)} \lesssim \|f\|_{W(\mathcal{F}L^1, L_m^1)} \|g\|_{W(\mathcal{F}L^1, L^\infty)}.$$

3.1. Estimates for the STFT. In [**4, 8**] the investigation of optimal sufficient boundedness results for localization operators was led exploiting boundedness results for Weyl operators. Here we shall prove them by the very definition of localization operators. The following local regularity properties of the STFT play the central role for our results. They are obtained by using two different convolution relations in the last step of the proof of [**4**, Lemma 4.1]. For sake of clarity, we report the whole proof.

LEMMA 3.6. *Let* $1 \le p, q \le \infty$. *Then,*
(i) If $f \in M_m^{p,q}(\mathbb{R}^d)$ *and* $g \in M_v^1(\mathbb{R}^d)$, *then* $V_g f \in W(\mathcal{F}L^1, L_m^{p,q})(\mathbb{R}^{2d})$ *with norm estimate*

$$(3.21) \qquad \|V_g f\|_{W(\mathcal{F}L^1, L_m^{p,q})} \lesssim \|f\|_{M_m^{p,q}} \|g\|_{M_v^1}.$$

(ii) If $f \in M^{p,q}(\mathbb{R}^d)$ *and* $g \in M^{p',q'}(\mathbb{R}^d)$, *then* $V_g f \in W(\mathcal{F}L^1, L^\infty)(\mathbb{R}^{2d})$ *with norm estimate*

$$(3.22) \qquad \|V_g f\|_{W(\mathcal{F}L^1, L^\infty)} \lesssim \|f\|_{M^{p,q}} \|g\|_{M^{p',q'}}.$$

PROOF. We shall use the amalgam norm (3.18), Lemma 2.2 and Young's Inequality to to reach the desired results. Let $\varphi \in \Sigma_1^1(\mathbb{R}^d) \setminus \{0\}$ and set $\Phi = V_\varphi \varphi \in \Sigma_1^1(\mathbb{R}^{2d})$. We only consider $f, g \in \Sigma_1^1(\mathbb{R}^d)$ and extend to arbitrary f, g by approximation argument.

(i) In the follwing computations we first perform the change of variables: $w_1 = -\zeta_2$, $w_2 = z_2 + \zeta_1$, that let us express the Wiener amalgam norm as a convolution product of two STFTs.

$$\|V_g f\|_{W(\mathcal{F}L^1, L_m^{p,q})}$$

$$\asymp \left(\int_{\mathbb{R}^d} \left(\int_{\mathbb{R}^d} \left(\int_{\mathbb{R}^{2d}} |\mathcal{V}_{V_\varphi \varphi}(V_g f)(z_1, z_2, \zeta_1, \zeta_2)| d\zeta_1 d\zeta_2 \right)^p m(z_1, z_2)^p \, dz_1 \right)^{q/p} dz_2 \right)^{1/q}$$

$$= \left(\int_{\mathbb{R}^d} \left(\int_{\mathbb{R}^d} \left(\int_{\mathbb{R}^{2d}} |V_\varphi g(-z_1 - \zeta_2, \zeta_1)| \, |V_\varphi f(-\zeta_2, z_2 + \zeta_1)| d\zeta_1 d\zeta_2 \right)^p \times \right. \right.$$
$$\left. \left. \times m(z_1, z_2)^p \, dz_1 \right)^{q/p} dz_2 \right)^{1/q}$$

$$= \left(\int_{\mathbb{R}^d} \left(\int_{\mathbb{R}^d} \left(\int_{\mathbb{R}^{2d}} |V_\varphi f(w_1, w_2)| |V_\varphi g(w_1 - z_1, w_2 - z_2)| \, dw_1 dw_2 \right)^p \times \right. \right.$$
$$\left. \left. \times m(z_1, z_2)^p \, dz_1 \right)^{q/p} dz_2 \right)^{1/q}$$

$$= \left(\int_{\mathbb{R}^d} \left(\int_{\mathbb{R}^d} (|V_\varphi f| * |V_\varphi g|^*)(z_1, z_2))^p m(z_1, z_1)^p \, dz_1 \right)^{q/p} dz_2 \right)^{1/q}$$

$$= \||V_\varphi f| * |V_\varphi g|^*\|_{L_m^{p,q}}.$$

Now Young's theorem for mixed-norm spaces (see, e.g., [**22**, Proposition 11.1.3]) yields the desired estimate:

$$\|V_g f\|_{W(\mathcal{F}L^1, L_m^{p,q})} \lesssim \|V_\varphi f\|_{L_m^{p,q}} \|V_\varphi g\|_{L_v^1} \asymp \|f\|_{M_m^{p,q}} \|g\|_{M_v^1}.$$

(ii) The convolution relation $L^{p,q} * L^{p',q'} \hookrightarrow L^\infty$ and calculations similar to step (i) yield

$$\|V_g f\|_{W(\mathcal{F}L^1, L^\infty)} = \||V_\varphi f| * |V_\varphi g|^*\|_{L^\infty} \lesssim \|V_\varphi f\|_{L^{p,q}} \|V_\varphi g\|_{L^{p',q'}} \asymp \|f\|_{M^{p,q}} \|g\|_{M^{p',q'}},$$

that is the estimate (3.22). ∎

4. Boundedness Results

In this section, we first give general sufficient conditions for boundedness of localization operators. Then we show these are the best conditions we can get at, providing negative results when enlarging the class of symbols. At last, we shall state a compactness result for localization operators due to [**19**].

4.1. Sufficient Conditions. We use a different approach from the standard way of study the boundedness of localization operators [**4, 8**]. We shall not come back to the Weyl form of a localization operator, but use the weak definition (1.5), together with STFT properties and Wiener amalgam estimates.

To start with, we provide a necessary boundedness condition to be satisfied by the Weyl symbol of a localization operator with Gaussian windows (Wick operator).

PROPOSITION 4.1. Let a be a positive measure on $\mathbb{R}^{2d}$. Then, if $\varphi(t) = 2^{d/4} e^{-\pi t^2}$ and $\sigma(z) = 2^d (a * e^{-2\pi z^2})$ is the Weyl symbol of the Wick operator $A_a^{\varphi,\varphi}$, we have

(4.1) $$\sigma(z) \le 2^d \|A_a^{\varphi,\varphi}\|_{B(L^2)}, \qquad \forall z \in \mathbb{R}^{2d}.$$

PROOF. For $\varphi(t) = 2^{d/4}e^{-\pi t^2}$, we have $V_\varphi\varphi(x,\omega) = e^{\pi i x\omega}e^{-pi(x^2+\omega^2)/2}$. For $z = (x,\omega)$ and using the time-frequency shift operator $\pi(z) = M_\omega T_x$, the positivity of a let us write the following lower bound

$$
\begin{aligned}
\langle A_a^{\varphi,\varphi}\pi(z)\varphi, \pi(z)\varphi\rangle &= \langle a, \overline{V_\varphi(\pi(z)\varphi)}V_\varphi(\pi(z)\varphi)\rangle \\
&= \langle a, T_z(\overline{V_\varphi\varphi}V_\varphi\varphi)\rangle \\
&= (a * e^{-\pi(\cdot)^2})(z) \\
&\geq \int_{\mathbb{R}^{2d}} a(z-y)e^{-2\pi y^2}\,dy = 2^{-d}\sigma(z).
\end{aligned}
$$

Hence,

$$
\sigma(z) \leq 2^d\langle A_a^{\varphi,\varphi}\pi(z)\varphi, \pi(z)\varphi\rangle \leq 2^d\|A_a^{\varphi,\varphi}\|_{B(L^2)}\|\pi(z)\varphi\|_2^2 = 2^d\|A_a^{\varphi,\varphi}\|_{B(L^2)},
$$

that is, the estimate expected.   ∎

The straightforward consequence of (4.1) is

COROLLARY 4.2. A necessary condition for the Wick operator $A_a^{\varphi,\varphi}$ being bounded on $L^2(\mathbb{R}^d)$ is that the corresponding Weyl symbol σ lives in $L^\infty(\mathbb{R}^{2d})$.

REMARK 4.3. The preceding necessary condition extends the classical necessary result for smooth symbols in the Shubin classes [**30**, Chp. IV].

PROPOSITION 4.4. Every symbol $a(x,\omega) = v(x)$, $x,\omega \in \mathbb{R}^d$, where v is a submultiplicative weight on $\mathbb{R}^d$ with $v \notin L^\infty(\mathbb{R}^d)$, gives rise to a Wick operator $A_a^{\varphi,\varphi}$ that is unbounded on $L^2(\mathbb{R}^d)$.

PROOF. The proof uses the Wick-Weyl connection, the submultiplicativity of v, and Corollary 4.2. In details,

$$
\begin{aligned}
\sigma(z) &= 2^d(a * e^{-2\pi(\cdot)^2})(z) = 2^d\int_{\mathbb{R}^{2d}} v(z_1 - y_1)e^{-2\pi y_1^2}e^{-2\pi y_2^2}\,dy_1 dy_2 \\
&\geq 2^d\int_{\mathbb{R}^{2d}} \frac{v(z_1)}{v(y_1)}e^{-2\pi y_1^2}e^{-2\pi y_2^2}\,dy_1 dy_2 = Cv(z_1),
\end{aligned}
$$

with $C = \int_{\mathbb{R}^{2d}}(1/v(y_1))e^{-2\pi y_1^2}e^{-2\pi y_2^2}\,dy_1 dy_2$. Hence the Weyl symbol σ is unbounded on $\mathbb{R}^{2d}$ and, consequently, $A_a^{\varphi,\varphi}$ is unbounded on L^2.   ∎

COROLLARY 4.5. Every modulation space $M_{(1/v)\otimes 1}^\infty(\mathbb{R}^{2d})$ with v submultiplicative and $v \notin L^\infty(\mathbb{R}^{2d})$ contains symbols a for unbounded localization operators $A_a^{\varphi_1,\varphi_2}$ on $L^2(\mathbb{R}^d)$.

PROOF. Take $a(x,\omega) = v(x)$, then we shall verify $a \in M_{(1/v)\otimes 1}^\infty$. For every not null $g \in \Sigma_1^1(\mathbb{R}^{2d})$, we have

$$
\begin{aligned}
\|V_g a\|_{L_{(1/v)\otimes 1}^\infty} &= \sup_{(z,\zeta)\in\mathbb{R}^{4d}} \frac{1}{v(z)}\left|\int_{\mathbb{R}^{2d}} v(t)\overline{g(t-z)}e^{-2\pi i t\zeta}\,dt\right| \\
&\leq \sup_{(z,\zeta)\in\mathbb{R}^{4d}} \frac{1}{v(z)}\int_{\mathbb{R}^{2d}} v(t-z)v(z)|g(t-z)|\,dt \\
&= \sup_{(z,\zeta)\in\mathbb{R}^{4d}} \int_{\mathbb{R}^{2d}} v(t-z)|g(t-z)|\,dt < \infty.
\end{aligned}
$$

Next, pick up the Wick operator $A_a^{\varphi,\varphi}$, that is unbounded in virtue of Proposition 4.4.   ∎

On the other hand, we shall presently see that symbols a in the modulation spaces $M_{1\otimes(1/v)}^\infty$ give L^2-bounded localization operators. Therefore, weights that have time-frequency variables decoupled appear to be the right setting for the study of boundedness property of localization operators. We start with analysing the behavior of modulation spaces under the action of the Fourier transform $\mathcal{F}$ symplectic operator T_J defined in (1.8). It is well-known that, for even, symplectic invariant weights m, the spaces M_m^p are invariant under the Fourier transform [**22**, Thm. 11.3.5]. For a tensor product of even weights we have:

LEMMA 4.6. *Let ν, μ be even weights on $\mathbb{R}^d$. Then, the Fourier transform $\mathcal{F}$ is an isometry from $M_{\nu\otimes\mu}^p(\mathbb{R}^d)$ onto $M_{\mu\otimes\nu}^p(\mathbb{R}^d)$, where $1 \leq p \leq \infty$.*

PROOF. For $f \in M_{\nu\otimes\mu}^p(\mathbb{R}^d)$, $g \in \Sigma_1^1(\mathbb{R}^d) \setminus \{0\}$, the indipendence of the $M_{\mu\otimes\nu}^p$-norm of the choice of $g \in \Sigma_1^1(\mathbb{R}^d) \setminus \{0\}$ and formula (2.5) help us to achieve our purpouse:

$$
\begin{aligned}
\|\mathcal{F}f\|_{M_{\mu\otimes\nu}^p(\mathbb{R}^d)} &\asymp \|V_{\hat{g}}\hat{f}\|_{L_{\mu\otimes\nu}^p} \\
&= \left(\int_{\mathbb{R}^{2d}} |V_{\hat{g}}\hat{f}(x,\omega)|^p (\mu\otimes\nu)^p(x,\omega)\,dxd\omega\right)^{1/p} \\
&= \left(\int_{\mathbb{R}^{2d}} |V_g f(-\omega,x)|^p \mu^p(x)\nu^p(\omega)\,dxd\omega\right)^{1/p} \\
&= \left(\int_{\mathbb{R}^{2d}} |V_g f(x,\omega)|^p \mu^p(\omega)\nu^p(-x)\,dxd\omega\right)^{1/p} \\
&= \left(\int_{\mathbb{R}^{2d}} |V_g f(x,\omega)|^p \nu^p(x)\mu^p(\omega)\,dxd\omega\right)^{1/p} \\
&\asymp \|f\|_{M_{\nu\otimes\mu}^p}.
\end{aligned}
$$

∎

LEMMA 4.7. *Let ν, μ be even weights on $\mathbb{R}^{2d}$ such that $T_J\mu = \mu$ and $T_J\nu = \nu$. Then, the symplectic operator T_J is an isometry on $M_{\nu\otimes\mu}^{p,q}(\mathbb{R}^{2d})$.*

PROOF. We simply use the symplectic invariance and the even property of ν and μ. Namely, take the window $g = \varphi\otimes\varphi$, with $\varphi(t) = e^{-\pi t^2}$. Starting with a function $F \in \Sigma_1^1(\mathbb{R}^{2d})$,

$$\|T_J F\|_{M_{\mu\otimes\nu}^{p,q}} \asymp \left(\int_{\mathbb{R}^{2d}} \left(\int_{\mathbb{R}^{2d}} \left| \int_{\mathbb{R}^{2d}} T_J F(z,\zeta)\overline{g(z-\eta_1,\zeta-\eta_2)} e^{-2\pi(z,\zeta)(y_1,y_2)} dzd\zeta \right|^p \right. \right.$$
$$\left. \left. \cdot\, \mu(y_1,y_2)^p dy_1 dy_2 \right)^{q/p} \nu(\eta_1,\eta_2)^q d\eta_1 d\eta_2 \right)^{1/q}$$

$$= \left(\int_{\mathbb{R}^{2d}} \left(\int_{\mathbb{R}^{2d}} \left| \int_{\mathbb{R}^{2d}} F(-\zeta,z)\varphi(z-\eta_1)\varphi(\zeta-\eta_2) e^{-2\pi(z,\zeta)(y_1,y_2)} dzd\zeta \right|^p \right. \right.$$
$$\left. \left. \cdot\, \mu(y_1,y_2)^p dy_1 dy_2 \right)^{q/p} \nu(\eta_1,\eta_2)^q d\eta_1 d\eta_2 \right)^{1/q}$$

$$= \left(\int_{\mathbb{R}^{2d}} \left(\int_{\mathbb{R}^{2d}} \left| \int_{\mathbb{R}^{2d}} F(z,\zeta)\varphi(-(z+\eta_2))\varphi(\zeta-\eta_1) e^{-2\pi(z,\zeta)(-y_2,y_1)} dzd\zeta \right|^p \right. \right.$$
$$\left. \left. \cdot\, \mu(y_1,y_2)^p dy_1 dy_2 \right)^{q/p} \nu(\eta_1,\eta_2)^q d\eta_1 d\eta_2 \right)^{1/q}$$

$$= \left(\int_{\mathbb{R}^{2d}} \left(\int_{\mathbb{R}^{2d}} \left| \int_{\mathbb{R}^{2d}} F(z,\zeta)\varphi(z-\eta_1)\varphi(\zeta-\eta_2) e^{-2\pi(z,\zeta)(y_1,y_2)} dzd\zeta \right|^p \right. \right.$$
$$\left. \left. \cdot\, \mu(y_2,-y_1)^p dy_1 dy_2 \right)^{q/p} \nu(\eta_2,-\eta_1)^q d\eta_1 d\eta_2 \right)^{1/q}$$

$$= \left(\int_{\mathbb{R}^{2d}} \left(\int_{\mathbb{R}^{2d}} \left| \int_{\mathbb{R}^{2d}} F(z,\zeta)\varphi(z-\eta_1)\varphi(\zeta-\eta_2) e^{-2\pi(z,\zeta)(y_1,y_2)} dzd\zeta \right|^p \right. \right.$$
$$\left. \left. \cdot\, \mu(y_1,y_2)^p dy_1 dy_2 \right)^{q/p} \nu(\eta_1,\eta_2)^q d\eta_1 d\eta_2 \right)^{1/q}$$

$$\asymp \|F\|_{M_{\mu\otimes\nu}^{p,q}},$$

then, by density argument, we get the isometry for every $F \in M_{\mu\otimes\nu}^{p,q}$. ∎

THEOREM 4.8. *If $a \in M_{1\otimes(1/v)}^\infty(\mathbb{R}^{2d})$, $\varphi_1,\varphi_2 \in M_v^1(\mathbb{R}^d)$, the operator $A_a^{\varphi_1,\varphi_2}$ is bounded on $M^{p,q}(\mathbb{R}^d)$, for all $1 \le p,q \le \infty$, and the operator norm satisfies the uniform estimate*

$$\|A_a^{\varphi_1,\varphi_2}\|_{B(M^{p,q})} \lesssim \|a\|_{M_{1\otimes(1/v)}^\infty} \|\varphi_1\|_{M_v^1} \|\varphi_2\|_{M_v^1}.$$

PROOF. We start with $a \in \Sigma_1^1(\mathbb{R}^{2d})$, $\varphi_1,\varphi_2 \in \Sigma_1^1(\mathbb{R}^d)$ and use the Plancherel formula and the Fourier transform of the pointwise product of two STFTs (2.4) to rewrite the weak definition (1.5) of $A_a^{\varphi_1,\varphi_2}$: for every $f,g \in \Sigma_1^1(\mathbb{R}^d)$,

$$(4.2) \qquad |\langle A_a^{\varphi_1,\varphi_2} f, g\rangle| = |\langle a, \overline{V_{\varphi_1} f V_{\varphi_2} g}\rangle| = |\langle \hat{a}, \widehat{\overline{V_{\varphi_1} f V_{\varphi_2} g}}\rangle|$$

Next, we use the duality of modulation spaces (3.12) and the action of the operators $\mathcal{F}$ and T_J explained in Lemmata 4.6 and 4.7:

$$|\langle A_a^{\varphi_1,\varphi_2} f, g\rangle| = \left| \int_{\mathbb{R}^{2d}} \hat{a}(y,\eta) T_J(V_f g \overline{V_{\varphi_1}\varphi_2})(y,\eta)\, dy d\eta \right|$$
$$\lesssim \|\hat{a}\|_{M_{(1/v)\otimes 1}^\infty} \|T_J(V_f g \overline{V_{\varphi_1}\varphi_2})\|_{M_{v\otimes 1}^1}$$
$$\asymp \|a\|_{M_{1\otimes(1/v)}^\infty} \|V_f g \overline{V_{\varphi_1}\varphi_2}\|_{M_{v\otimes 1}^1}.$$

Using (3.19), we can write $\|V_f g \overline{V_{\varphi_1}\varphi_2}\|_{M_{v\otimes 1}^1} \asymp \|V_f g \overline{V_{\varphi_1}\varphi_2}\|_{W(\mathcal{F}L^1, L_v^1)}$. Next, using Lemma 3.6, we derive:

(i)$V_{\varphi_1}\varphi_2 \in W(\mathcal{F}L^1, L_v^1)(\mathbb{R}^{2d})$, with $\|V_{\varphi_1}\varphi_2\|_{W(\mathcal{F}L^1, L_v^1)} \lesssim \|\varphi_1\|_{M_v^1}\|\varphi_2\|_{M_v^1}$.

(ii) $V_g f \in W(\mathcal{F}L^1, L^\infty)(\mathbb{R}^{2d})$, with $\|V_g f\|_{W(\mathcal{F}L^1, L^\infty)} \lesssim \|f\|_{M^{p,q}} \|g\|_{M^{p',q'}}$. The pointwise product of Wiener amalgam spaces (3.20) yields the following estimate:

$$\|V_f g \overline{V_{\varphi_1}\varphi_2}\|_{W(\mathcal{F}L^1, L^1_v)} \lesssim \|V_g f\|_{W(\mathcal{F}L^1, L^\infty)} \|V_{\varphi_1}\varphi_2\|_{W(\mathcal{F}L^1, L^1_v)}$$
$$\lesssim \|f\|_{M^{p,q}} \|g\|_{M^{p',q'}} \|\varphi_1\|_{M^1_v} \|\varphi_2\|_{M^1_v}.$$

To sum up,

$$|\langle A_a^{\varphi_1,\varphi_2} f, g\rangle| \lesssim \|a\|_{M^\infty_{1\otimes(1/v)}} \|f\|_{M^{p,q}} \|g\|_{M^{p',q'}} \|\varphi_1\|_{M^1_v} \|\varphi_2\|_{M^1_v},$$

and the standard density argument provides the desired result. ∎

REMARK 4.9. To compare Theorem 4.8 to existing results, we observe that the two most important conditions for $A_a^{\varphi_1,\varphi_2}$ to be bounded are: (Polynomial weights) $a \in M^\infty_{1/\tau_s}(\mathbb{R}^{2d})$, see [**4, 34**], (Exponential weights) $a \in M^\infty_{1/\mu_s}(\mathbb{R}^{2d})$, see [**8**]. Our case include both of them: take either $v(\zeta) = \tau_s(\zeta)$ or $v(\zeta) = \mu_s(\zeta)$. But our result goes far beyond the previous examples: e.g., every $v(\zeta) = e^{a|\zeta|^b}(1+|\zeta|)^s \log^r(e+|\zeta|)$, for parameters $a, r, s \geq 0$, $0 \leq b \leq 1$, are fitting weights for Theorem 4.8.

4.2. Necessary Conditions.
This section is devoted to show that the sufficient conditions obtained so far are optimal. We start with a simple observation that justifies the optimality of what follows. It concerns the behavior of symbols under frequency shifts M_ζ, $\zeta \in \mathbb{R}^{2d}$.

LEMMA 4.10. *Let $m \in \mathcal{M}_v$. If $a \in M^\infty_{1\otimes m}$ and $\zeta \in \mathbb{R}^{2d}$, then $M_\zeta a \in M^\infty_{1\otimes m}$ with*

(4.3) $$\|M_\zeta a\|_{M^\infty_{1\otimes m}} \lesssim v(\zeta) \|a\|_{M^\infty_{1\otimes m}}.$$

The proof is a straightforward rearrangement of [**4**, Thm. 11.3.5]. Consequently, under the assumption of Theorem 4.8, we obtain the boundedness of both $A_a^{\varphi_1,\varphi_2}$ and $A_{M_\zeta a}^{\varphi_1,\varphi_2}$, for every $\zeta \in \mathbb{R}^{2d}$, with

$$\|A_{M_\zeta a}^{\varphi_1,\varphi_2}\|_{B(M^{p,q})} \lesssim \|M_\zeta a\|_{M^\infty_{1\otimes(1/v)}} \|\varphi_1\|_{M^1_v} \|\varphi_2\|_{M^1_v} \lesssim v(\zeta) \|a\|_{M^\infty_{1\otimes(1/v)}} \|\varphi_1\|_{M^1_v} \|\varphi_2\|_{M^1_v}.$$

The optimality of the result in Theorem 4.8 is then proved by the subsequent conditions.

THEOREM 4.11. *Let v be a submultiplicative weight on $\mathbb{R}^{2d}$. Let $a \in (\Sigma_1^1)'(\mathbb{R}^{2d})$, $\varphi_1, \varphi_2 \in \Sigma_1^1$. If there exists a constant $C = C(a, \varphi_1, \varphi_2) > 0$ and some indices $1 \leq p, q \leq \infty$ such that*

(4.4) $$\|A_{M_\zeta a}^{\varphi_1,\varphi_2}\|_{B(M^{p,q})} \leq C v(\zeta), \qquad \forall \zeta \in \mathbb{R}^{2d},$$

then $a \in M^\infty_{1\otimes 1/v}$.

PROOF. As done in [**4**, Thm. 4.3], we first compute the time shift of $\overline{V_{\varphi_1} f} \cdot V_{\varphi_2} g$. For $z = (z_1, z_2) \in \mathbb{R}^{2d}$, Lemma 2.1(i) justifies the following computations

$$T_{(z_1, z_2)}(\overline{V_{\varphi_1} f} \cdot V_{\varphi_2} g)(x, \omega) = (\overline{V_{\varphi_1} f}\, V_{\varphi_2} g)(x - z_1, \omega - z_2)$$
$$= e^{-2\pi i(\omega - z_2)z_1} \overline{V_{\varphi_1}(M_{z_2} T_{z_1} f)}(x, \omega)\, e^{2\pi i(\omega - z_2)z_1} V_{\varphi_2}(M_{z_2} T_{z_1} g)(x, \omega)$$
$$= \overline{V_{\varphi_1}(M_{z_2} T_{z_1} f)}(x, \omega)\, V_{\varphi_2}(M_{z_2} T_{z_1} g)(x, \omega).$$

(4.5)

Take $f, g \in \Sigma_1^1(\mathbb{R}^d)$ and set $\Phi := \overline{V_{\varphi_1} f} \cdot V_{\varphi_2} g \in \Sigma_1^1(\mathbb{R}^{2d})$. We need to show that $V_\Phi a(1 \otimes v)$ is in $L^\infty(\mathbb{R}^{2d})$. At this point our arguments begin to divert slightly from those of [4, Thm. 4.3]. We shall use (4.5), the weak definition of $A_a^{\varphi_1, \varphi_2}$ and (4.4), to achieve our purpose:

$$
\begin{aligned}
|V_\Phi a(z, \zeta)| &= |\langle a, M_\zeta T_z \Phi \rangle| = |\langle M_{-\zeta} a, T_z \Phi \rangle| \\
&= |\langle M_{-\zeta} a, \overline{V_{\varphi_1}(M_{z_2} T_{z_1} f)} \, V_{\varphi_2}(M_{z_2} T_{z_1} g) \rangle| \\
&= |\langle A_{M_{-\zeta} a}^{\varphi_1, \varphi_2}(M_{z_2} T_{z_1} f), M_{z_2} T_{z_1} g \rangle| \\
&\leq C \, v(-\zeta) \|M_{z_2} T_{z_1} f\|_{M^{p,q}} \|M_{z_1} T_{z_2} g\|_{M^{p',q'}} \\
&= C \, v(\zeta) \|f\|_{M^{p,q}} \|g\|_{M^{p',q'}} < \infty,
\end{aligned}
$$

where in the last row we used the invariance of modulation spaces under time-frequency shifts and the even property of v.   ∎

REMARK 4.12. The compactness and Schatten class properties of localization operators with symbols in modulation spaces can also been attained but are beyond the scope of this paper. We refer the interested reader to the literature [5, 8, 19]. In these papers sufficient conditions for the operators property are attained via Weyl connection. We point out that techniques similar to the ones we have employed are fitting for the study of the previous properties as well, but will be the object of a forthcoming paper.

5. On the Weyl-Wick Connection

From then onwards, we consider the Gaussian window functions $\varphi_1(t) = \varphi_2(t) = \varphi(t) = 2^{d/4} e^{-\pi t^2}$, $t \in \mathbb{R}^d$. In this case, the Wigner distribution of the Gaussian φ is a Gaussian as well. Precisely, we have

$$
(5.1) \qquad\qquad W(\varphi, \varphi)(z) = 2^d e^{-2\pi z^2}, \qquad z \in \mathbb{R}^{2d}.
$$

Every tempered distribution $a \in \mathcal{S}'$ defines an anti-Wick operator $A_a := A_a^{\varphi, \varphi}$, acting from $\mathcal{S}(\mathbb{R}^d)$ into $\mathcal{S}'(\mathbb{R}^d)$ and the same holds true when replacing $\mathcal{S}$ by Σ_1^1 and related dual spaces. The Weyl connection immediately follows by (2.11). The Weyl symbol σ has here the expression

$$
(5.2) \qquad\qquad \sigma = a * 2^d e^{-2\pi z^2}.
$$

Clearly, the Weyl symbol σ is a smooth, analytic function. Thereby, not every pseudodifferential operator can be written in a Wick form. This problem is studied in in [30]. Here, to overcome this difficulty, a remainder term is introduced. To start with, recall the Shubin classes Γ_ρ^m, where $m \in \mathbb{R}$ and $0 < \rho \leq 1$, which consist of functions $a \in \mathcal{C}^\infty(\mathbb{R}^{2d})$ satisfying

$$
|\partial^\alpha a(z)| \leq C_\alpha \langle z \rangle^{m - \rho |\alpha|}, \qquad z \in \mathbb{R}^{2d}.
$$

Then, a pseudodifferential operator T_b with symbols $b \in \Gamma_\rho^m$ can be seen as [30, Theorem 24.2]

$$
T_b = A_a + R,
$$

where A_a is a Wick operator with symbol $a \in \Gamma_\rho^m$ and R is a *regularazing* remainder operator, that is, its Schwartz kernel lies in $\mathcal{S}(\mathbb{R}^{2d})$.

Our approach is different. In fact, we deal with the problem of finding an exact Weyl-Wick connection. Of course, this implies to restrict our consideration to very nice pseudodifferential operators, having analytic symbols. We focus on the Weyl form, but the Kohn-Nirenberg form is attainable similarly.

THEOREM 5.1. *Let* $m \in \mathcal{M}_v(\mathbb{R}^{2d})$. *If the Weyl symbol* σ *is an analytic function on* $\mathbb{R}^{2d}$ *such that*

$$(5.3) \qquad \hat{\sigma} \in \left\{ e^{-\pi \zeta^2/2} f(\zeta), \quad f \in M^\infty_{m \otimes 1}(\mathbb{R}^{2d}) \right\}.$$

Then, there exists a Wick symbol $a \in M^\infty_{1 \otimes m}(\mathbb{R}^{2d})$, *with*

$$(5.4) \qquad \hat{a}(\zeta) = \hat{\sigma}(\zeta) e^{\pi \zeta^2/2}, \quad \zeta \in \mathbb{R}^{2d},$$

and such that $L_\sigma = A_a$, *as bounded operators on every* $M^{p,q}$ *(hence on* L^2 *).*

Viceversa, if $a \in M^\infty_{1 \otimes m}(\mathbb{R}^{2d})$, *we have* $A_a = L_\sigma$, *with* σ *satisfying relations* (5.3) *and* (5.4).

PROOF. *Sufficient condition.* We use assumption (5.3) that guarantees the existence of $f \in M^\infty_{m \otimes 1}$ such that $\hat{\sigma}(\zeta) = e^{-\pi \zeta^2/2} f$. We then set $\hat{a}(\zeta) := \hat{\sigma}(\zeta) e^{\pi \zeta^2/2}$ and by taking the inverse Fourier transform of both sides of the previous equality we get the Weyl-Wick connection in (5.2). We next justify the previous calculations in terms of symbol function spaces and related operator properties. We shall show $a \in M^\infty_{1 \otimes m}$. Since

$$\hat{a} = e^{-\pi \zeta^2/2} f e^{\pi \zeta^2/2} = f,$$

we have $a = \mathcal{F}^{-1}(f)$. Fix a non-zero window $g \in \Sigma^1_1(\mathbb{R}^{2d})$, we shall use Lemma 4.6 together with the property: if f is in $M^\infty_{m \otimes 1}$, then $f^*(x) := f(-x)$ is in $M^\infty_{m \otimes 1}$ with equivalent norms (see [**22**, Proposition 12.1.3]). Namely, for every even weight m we can write

$$\|a\|_{M^\infty_{1 \otimes m}} = \|\mathcal{F}^{-1}(f)\|_{M^\infty_{1 \otimes m}} = \|\mathcal{F}(f^*)\|_{M^\infty_{1 \otimes m}} \asymp \|f^*\|_{M^\infty_{m \otimes 1}} \asymp \|f\|_{M^\infty_{m \otimes 1}} < \infty.$$

The operator A_a fulfills the assumptions of the Theorem 4.8 and, therefore, A_a is a bounded operator on every $M^{p,q}$. Finally, the Wick-Weyl connection of (2.10) gives $L_\sigma = A_a$ as bounded operators on every $M^{p,q}$.

Necessary condition. If $a \in M^\infty_{1 \otimes m}$, we know by Theorem 4.8 that A_a is bounded on every $M^{p,q}$ and the goal is achieved by using relation (5.2). ∎

The previous result underlines how much our approach differs from the classical one of [**30**, Theorem 24.2]. In fact, not only we can exactly write the Weyl operator in the Wick form, but also we allow anti-Wick symbols to be very rough, as in the following example.

EXAMPLE 5.2. Fix $r = (r_1, \ldots, r_{2d}) > 1$. Consider the Weyl function on $\mathbb{R}^{2d}$ defined by

$$(5.5) \qquad \sigma = 2^d e^{-2\pi z^2} * \sum_{\mu \in \mathbb{N}^{2d}} \frac{1}{(\mu!)^r} \partial^\mu \delta, \quad z \in \mathbb{R}^{2d}.$$

Then, $L_\sigma = A_a$, where the Wick symbol a is in $\mathcal{E}'_t$, $(t > 1)$, the space of ultra-distributions with compact support and is given by

$$a = \sum_{\mu \in \mathbb{N}^{2d}} \frac{1}{(\mu!)^r} \partial^\mu \delta.$$

PROOF. Since $a = \sum_{\mu \in \mathbb{N}^{2d}} \frac{1}{(\mu!)^r} \partial^\mu \delta \in M^{1,\infty}_{1/\mu_\epsilon} \hookrightarrow M^\infty_{1/\mu_\epsilon}$, for some $\epsilon > 0$ and with the weight μ_ϵ defined in (3.11) (see [8, Corollary 4.3]), we have $\hat{a} = f \in M^\infty_{e^{-\epsilon|x|} \otimes 1}$, so that $\hat{\sigma}$ lies in the set defined in (5.3). $\blacksquare$

In what follows we study the case of Wick-Weyl connections having both Weyl and Wick symbol smooth and regular. Precisely, σ and a are time-frequency shifts of Gaussian functions (hence they live in $\mathcal{S}^{1/2}_{1/2}(\mathbb{R}^{2d})$).

EXAMPLE 5.3. For every $0 < c < 2$, and $u, v \in \mathbb{R}^{2d}$, consider the Weyl symbol on $\mathbb{R}^{2d}$ defined by

$$(5.6) \qquad \sigma(z) := \sigma_{c,u,v}(z) = M_v T_u e^{-\pi c z^2}, \quad z \in \mathbb{R}^{2d}.$$

Then, $L_\sigma = A_a$, where the Wick symbol a is given by

$$(5.7) \qquad a(z) = a_{c,u,v}(z) = \left(\frac{2}{2-c}\right)^d e^{-2\pi \frac{c}{2-c} iuv} e^{\frac{\pi}{2-c} v^2} M_{\frac{2v}{2-c}} T_u e^{-\pi \frac{2c}{2-c} z^2}.$$

PROOF. The Fourier transform of the Weyl symbol σ is given by $\hat{\sigma}(\zeta) = \frac{1}{c^d} T_v M_{-u} e^{-\frac{\pi}{c} \zeta^2}$. The assumptions of Theorem 5.1 are fulfilled and $\hat{a}$ is given by (5.4). Namely,

$$
\begin{aligned}
\hat{a}(\zeta) &= \hat{\sigma}(\zeta) e^{\frac{\pi}{2} \zeta^2} = \frac{1}{c^d} e^{2\pi iuv} M_{-u} T_v e^{-\frac{\pi}{c} \zeta^2} \\
&= \frac{1}{c^d} e^{2\pi iuv} e^{-\frac{\pi}{c} v^2} M_{-u} e^{-\pi \frac{2-c}{2c} \zeta^2 + \pi \frac{2v}{c} \zeta} \\
&= \frac{1}{c^d} e^{2\pi iuv} e^{-\frac{\pi}{c} v^2} M_{-u} e^{-\pi \left(\sqrt{\frac{2-c}{2c}} \zeta - \sqrt{\frac{2}{c(2-c)}} v\right)^2} \\
&= \frac{1}{c^d} e^{2\pi iuv} e^{\frac{\pi}{2-c} v^2} M_{-u} T_{\frac{2v}{2-c}} e^{-\pi \frac{2-c}{2c} \zeta^2}.
\end{aligned}
$$

Finally, the inverse Fourier transform of the symbol $\hat{a}(\zeta)$ yields to the expression in (5.7). $\blacksquare$

The preceding example can be generalized as follows:

EXAMPLE 5.4. Fix $m \in \mathbb{N}$ and choose $0 < c_k < 2$, $\alpha_k \in \mathbb{R}$, $u_k, v_k \in \mathbb{R}^{2d}$, $k = 1, \ldots, m$. For the Weyl symbol σ on $\mathbb{R}^{2d}$ defined by

$$(5.8) \qquad \sigma(z) = \sum_{k=1}^m \alpha_k M_{v_k} T_{u_k} e^{-\pi c_k z^2}, \quad z \in \mathbb{R}^{2d},$$

we have $L_\sigma = A_a$ if and only if the Wick symbol a is given by

$$(5.9) \qquad a(z) = \sum_{k=1}^m \alpha_k \left(\frac{2}{2-c_k}\right)^d e^{-2\pi \frac{c_k}{2-c_k} iuv} e^{\frac{\pi}{2-c_k} v^2} M_{\frac{2v_k}{2-c_k}} T_{u_k} e^{-\pi \frac{2c_k}{2-c_k} z^2}.$$

It is clear from these examples that every time-frequency shift of Gaussian is a good candidate for a Wick symbol, whereas, for the Weyl case, the Gaussian cannot decay too fast at infinity, that is $\sigma(z) = O(e^{-\pi c z^2})$, with $0 < c < 2$. This Weyl property is, alternatively, the consequence of the semigroup property of Gaussian functions (see, e.g., [20]):

$$(e^{-\pi \alpha (\cdot)^2} * e^{-\pi \beta (\cdot)^2})(z) = (\alpha + \beta)^{-d} e^{-\pi \frac{\alpha \beta}{\alpha + \beta} z^2}, \quad \alpha, \beta > 0, \ z \in \mathbb{R}^{2d}.$$

For $a(z) = e^{-\pi \alpha z^2}$, $\beta = 2$ relation (5.2) gives the Weyl symbol $\sigma(z) = \left(\frac{2}{\alpha+2}\right)^d e^{-\pi \frac{2\alpha}{2+\alpha} z^2}$; observe that $0 < \frac{2\alpha}{2+\alpha} < 2$. If we introduce the linear space:

$$\mathcal{S}_c := \text{span}\{e^{-\pi c z^2}, z \in \mathbb{R}^{2d}, c > 0\},$$

what just said can be summarized as follows:

COROLLARY 5.5. For a Weyl symbol $\sigma \in \mathcal{S}_c$, $c < 2$, we have $L_\sigma = A_a$ where the Wick symbol a satisfies $a \in \mathcal{S}_{\frac{2c}{2-c}}$, and viceversa.

Other examples of Weyl symbols $\sigma \in \mathcal{S}_{1/2}^{1/2}$ that still provide Wick symbols in the same space are given by the Hermite functions [20, Chapter 1.7]. For $\alpha \in \mathbb{N}_0^d$, the αth (normalized, d-dimensional) *Hermite function* $h_\alpha(z)$ is given by

$$(5.10) \qquad h_\alpha(z) = \frac{2^{d/4}}{\sqrt{\alpha!}} \left(\frac{-1}{2\sqrt{\pi}}\right)^{|\alpha|} e^{\pi z^2} \partial^\alpha (e^{-2\pi z^2}).$$

The function $H_\alpha(z) = e^{\pi z^2} h_\alpha(z)$ is a polynomial of degree $|\alpha|$, called the αth *Hermite polynomial*. We have

$$(5.11) \qquad H_\alpha(z) = 2^{(d/4)+|\alpha|} \sqrt{\frac{\pi^{|\alpha|}}{\alpha!}} z^\alpha + (\text{terms of degree} < |\alpha|).$$

Recall that h_α is an eigenfunction of the Fourier transform $\mathcal{F}$:

$$(5.12) \qquad \mathcal{F} h_\alpha = (-1)^{|\alpha|} h_\alpha.$$

EXAMPLE 5.6. For $z \in \mathbb{R}^{2d}$, $\alpha \in \mathbb{N}_0^{2d}$, we choose the Weyl symbol $\sigma(z) = h_\alpha(z)$ (2d-dimensional Hermite function). Then, $L_\sigma = A_a$, where the Wick symbol a is given by

$$a(z) = (-1)^{|\alpha|} \mathcal{F}^{-1}\left(H_\alpha(\zeta) e^{(-\pi/2)\zeta^2}\right)(z).$$

The proof is a straightforward consequence of relations (5.4) and (5.12). Hence, in this case the Wick symbol is given by $a(z) = \sum_{\beta \leq \alpha} c_\beta z^\beta e^{-2\pi z^2}$, where $c_\beta \in \mathbb{C}$. More generally, since every polynomial $P_k(z)$ of degree k on $\mathbb{R}^{2d}$ can be written as $P_k(z) = \sum_{|\alpha| \leq k} c_\alpha H_\alpha(z)$, $c_\alpha \in \mathbb{C}$, the Weyl symbol $\sigma(z) = P_k(z) e^{-\pi z^2} \in \mathcal{S}_{1/2}^{1/2}$ corresponds to the Wick one $a(z) = Q_k(z) e^{-2\pi z^2} \in \mathcal{S}_{1/2}^{1/2}$, where $Q_k(z)$ is a suitable polynomial of degree k. Further extensions are Weyl symbols like $\sigma(z) = P_k(z) e^{-\pi c z^2}$, with $c < 2$ and $P_k(z)$ any polynomial of degree $k \in \mathbb{N}$, which provide Wick ones given by $a(z) = Q_k(z) e^{-\pi \frac{2c}{2-c} z^2}$. Again, what really matters in the Weyl-Wick correspondence is the Gaussian decay at infinity, independently of the choice of the polynomial P_k.

References

[1] F. A. Berezin, *Wick and anti-Wick symbols of operators.*, Mat. Sb. (N.S.), **86**(128) (1971), 578–610.

[2] P. Boggiatto and E. Cordero, *Anti-Wick quantization with symbols in L^p spaces*, Proc. Amer. Math. Soc., **130**(9) (2002), 2679–2685.

[3] P. Boggiatto, E. Cordero, and K. Gröchenig, *Generalized Anti-Wick operators with symbols in distributional Sobolev spaces*, Integral Equations and Operator Theory, **48** (2004), 427–442.

[4] E. Cordero and K. Gröchenig, *Time-frequency analysis of Localization operators*, J. Funct. Anal., **205**(1) (2003),107–131.

[5] E. Cordero and K. Gröchenig, *Necessary conditions for Schatten class localization operators*, Proc. Amer. Math. Soc., **133** (2005), 3573–3579.

[6] E. Cordero and K. Gröchenig, *Symbolic calculus and Fredholm property for localization operators*, J. Fourier Anal. Appl., **12**(4) (2006), 371–392.

[7] E. Cordero and L. Rodino, *Wick calculus: a time-frequency approach*, Osaka J. Math., **42**(1) (2005), 43–63.

[8] E. Cordero, S. Pilipović, L. Rodino and N. Teofanov, *Localization operators and exponential weights for modulation spaces*, Mediterr. J. Math., **2** (2005), 381–394.

[9] A. Córdoba and C. Fefferman, *Wave packets and Fourier integral operators*, Comm. Partial Differential Equations, **3**(11) (1978),979–1005.

[10] I. Daubechies, *Time-frequency localization operators: a geometric phase space approach*, IEEE Trans. Inform. Theory, **34**(4) (1988), 605–612.

[11] F. De Mari, H. G. Feichtinger and K. Nowak, *Uniform eigenvalue estimates for time-frequency localization operators*, J. London Math. Soc. (2), **65**(3) (2002), 720–732.

[12] F. De Mari and K. Nowak, *Localization type Berezin-Toeplitz operators on bounded symmetric domains*, J. Geom. Anal., **12** (2002), 9–27.

[13] H. G. Feichtinger, *Modulation spaces on locally compact abelian groups*, Wavelets and Their Applications, M. Krishna, R. Radha, S. Thangavelu, editors, Allied Publishers, 2003, pp. 99–140.

[14] H. G. Feichtinger, *Generalized amalgams, with applications to Fourier transform*, Canad. J. Math. **42** (3) (1990), 395–409.

[15] H. G. Feichtinger and K. Gröchenig, *Banach spaces related to integrable group representations and their atomic decompositions I*, J. Funct. Anal., **86**(2) (1989), 307–340.

[16] H. G. Feichtinger, *On a new Segal algebra*, Monatsh. Math., **92**(4) (1981), 269–289.

[17] H. G. Feichtinger and K. H. Gröchenig, *Banach spaces related to integrable group representations and their atomic decompositions II*, Monatsh. f. Math., **108** (1989), 129–148.

[18] H. G. Feichtinger and K. Nowak, *A First Survey of Gabor Multipliers*, In H. G. Feichtinger and T. Strohmer, editors, *Advances in Gabor Analysis*. Birkhäuser, Boston, 2002.

[19] C. Fernández and A. Galbis, *Compactness of time-frequency localization operators on* $L^2(\mathbb{R}^d)$, J. Funct. Anal., **233**(2) (2006), 399–419.

[20] G. B. Folland, *Harmonic Analysis in Phase Space*, Princeton Univ. Press, Princeton, NJ, 1989.

[21] I. M. Gelfand and G. E. Shilov, *Generalized Functions II*, Academic Press, 1968.

[22] K. Gröchenig, *Foundations of Time-Frequency Analysis*, Birkhäuser, Boston, 2001.

[23] K. Gröchenig and G. Zimmermann, *Hardy's theorem and the short-time Fourier transform of Schwartz functions*, J. London Math. Soc., **63** (2001), 205–214.

[24] K. Gröchenig, G. Zimmermann, *Spaces of test functions via the STFT*, Journal of Function Spaces and Applications, **2**(1) (2004) 25–53.

[25] C. Heil, *An introduction to weighted Wiener amalgams*, In M. Krishna, R. Radha, and S. Thangavelu, editors, *Wavelets and their Applications*, 2003, pp. 183–216.

[26] S. Pilipović, *Tempered ultradistributions*, Boll. Un. Mat. Ital., **7**(2-B) (1988) 235–251.

[27] S. Pilipović and N. Teofanov, *Pseudodifferential operators on ultra-modulation spaces*, J. Funct. Anal., **208** (2004), 194–228.

[28] L. Rodino, *Linear Partial Differential Operators in Gevrey Spaces*, World Scientific, 1993.

[29] J. Ramanathan and P. Topiwala, *Time-frequency localization via the Weyl correspondence*, SIAM J. Math. Anal., **24**(5) (1993), 1378–1393.

[30] M. A. Shubin, *Pseudodifferential Operators and Spectral Theory*, Springer-Verlag, Berlin, second edition, 2001.

[31] D. Tataru, *On the Fefferman-Phong inequality and related problems*, Comm. Partial Differential Equations, **27**(11-12) (2002), 2101–2138.

[32] N. Teofanov, *Ultradistributions and time-frequency analysis*, in Pseudo-differential Operators and Related Topics, Operator Theory: Advances and Applications, P. Boggiatto, L. Rodino, J. Toft, M.W. Wong, editors, vol. 164, Birkhäuser, 2006, pp. 173–191.

[33] J. Toft, *Continuity properties for modulation spaces, with applications to pseudo-differential calculus. I*, J. Funct. Anal., **207**(2) (2004), 399–429.

[34] J. Toft, *Continuity properties for modulation spaces, with applications to pseudo-differential calculus. II*, Ann. Global Anal. Geom., **26**(1) (2004), 73–106.

[35] M. W. Wong, *Wavelets Transforms and Localization Operators*, Operator Theory Advances and Applications, vol. 136, Birkhauser, 2002.

DEPARTMENT OF MATHEMATICS, UNIVERSITY OF TORINO, ITALY
E-mail address: elena.cordero@unito.it

DEPARTMENT OF MATHEMATICS, UNIVERSITY OF TORINO, ITALY
E-mail address: luigi.rodino@unito.it

Contemporary Mathematics
Volume **451**, 2008

Fractal wavelets of Dutkay-Jorgensen type for the Sierpinski gasket space

Jonas D'Andrea, Kathy D. Merrill, and Judith Packer

ABSTRACT. Several years ago, D. Dutkay and P. Jorgensen developed the concept of wavelets defined on a σ-finite fractal measure space, developed from an iterated affine system. They worked out in detail the wavelet and filter functions corresponding to the ordinary Cantor fractal subset of $\mathbb{R}$. In this paper we examine the construction of Dutkay and Jorgensen as applied to the fractal measure space corresponding to the Sierpinski gasket fractal. We develop a variety of high-pass filters, and as an application use the various families of wavelets to analyze digital photos.

1. Introduction

Two years ago, D. Dutkay and P. Jorgensen introduced the notion of multi-resolution analysis bases on σ-finite measure spaces built from dilations and translations on a fractal arising from an iterated affine function system [**DJ**]. Although their construction works in a very general setting, the details were mainly worked out in the one-dimensional setting, in particular for the ordinary Cantor set and its variants. In the case of the ordinary Cantor fractal, they used Hutchinson measure $\mathcal{H}$ on the inflated fractal measure space $\mathcal{R}$ and considered a multi-resolution $L^2(\mathcal{R}, \mathcal{H})$ constructed from dilation by 3 and integer translation. The self-similarity of the Cantor set under dilation by 3 gave a polynomial variant of a low-pass filter, and using "gap-filling" and "detail" high-pass filters allowed them to construct the wavelet. In further work on the Cantor fractal case, D. Dutkay used the polynomial low-pass filter to construct a probability measure ν on the solenoid Σ_3 and a mock Fourier transform $\mathcal{F} \colon L^2(\mathcal{R}, \mathcal{H}) \to L^2(\Sigma_3, \nu)$, such that Fourier-transformed version of the dilation operator corresponded to the shift automorphism on Σ_3, and the translation operator on $L^2(\mathcal{R}, \mathcal{H})$ corresponded to multiplication operators on $L^2(\Sigma_3, \nu)$ [**Dut**].

In this paper our aim is to study this construction in the case of the right triangle Sierpinski gasket. Accordingly, we let $\mathcal{S}_0$ be the points inside and on the right triangle with vertices $(0,0)$, $(0,1)$, and $(1,0)$ in $\mathbb{R}^2$. Consider the diagonal dilation matrix $A = \begin{pmatrix} 2 & 0 \\ 0 & 2 \end{pmatrix}$. Let

$$\mathcal{S}_1 = [A^{-1}(\mathcal{S}_0 + \tau_0)] \cup [A^{-1}(\mathcal{S}_0 + \tau_1)] \cup [A^{-1}(\mathcal{S}_0 + \tau_2)],$$

1991 *Mathematics Subject Classification.* Primary 54C40, 14E20; Secondary 46E25, 20C20.
Key words and phrases. Wavelet; Fractals; Frames; Fourier transform.

where $\tau_0 = (0,0)$, $\tau_1 = (1,0)$ and $\tau_2 = (0,1)$. Proceeding inductively, given $\mathcal{S}_n$, let

$$\mathcal{S}_{n+1} = [A^{-1}(\mathcal{S}_n + \tau_0)] \cup [A^{-1}(\mathcal{S}_n + \tau_1)] \cup [A^{-1}(\mathcal{S}_n + \tau_2)].$$

We thus have a nested sequence $\{\mathcal{S}_n\}_{n=0}^{\infty}$ of compact subsets of $\mathbb{R}^2$, and we define the Sierpinski gasket fractal by

$$\mathcal{S} = \cap_{n=0}^{\infty} \mathcal{S}_n.$$

The Sierpinski gasket $\mathcal{S}$ satisfies the self-similarity relation

$$A(\mathcal{S}) = \mathcal{S} \cup [\mathcal{S} + (1,0)] \cup [\mathcal{S} + (0,1)].$$

The Hausdorff dimension of $\mathcal{S}$ is known to be $s = \frac{\log 3}{\log 2}$. In the usual fashion one constructs the Hausdorff fractal measure corresponding to this dimension $\mathcal{H}^s$ on $\mathcal{S}$ ([**Hut**]), hereafter denoted by $\mathcal{H}$. Note that

$$\mathcal{H}(A^{-1}(\mathcal{S})) = \frac{1}{3}\mathcal{H}(\mathcal{S}) = \frac{1}{3},$$

and more generally, if E is a Borel subset of $\mathcal{S}$,

$$\mathcal{H}(A^{-1}(E)) = \frac{1}{3}\mathcal{H}(E).$$

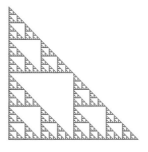

FIGURE 1. Sierpinski gasket.

Just as in the Dutkay and Jorgensen work on the Cantor set, we will construct our wavelets not on the Sierpinski gasket itself, but rather on an enlarged fractal that supports a structure closer to a standard multi-resolution analysis. In this respect, our work differs, for example, from Stricharz's wavelets for piecewise linear functions on triangulations of the Sierpinski gasket itself [**Str**]. In Section 2, we will describe the enlarged fractal for the Sierpinski gasket, and define the multi-resolution analysis structure supported on it. In Section 3, we will use this multi-resolution analysis to build a parametrized family of filters, and from them, wavelets on our space. We then go on to apply some of these wavelets to analyze digital photos in Section 4. We note that it is our use of the enlarged fractal, and thus the inclusion of "gap-filling" high-pass filters, that makes our wavelets reasonable candidates for describing images in $L^2(\mathbb{R}^2)$ rather than just on the fractal itself. The effects that using this type of wavelet has on image reconstruction will be explored further in Section 4.

2. A multi-resolution analysis corresponding to the Sierpinski gasket space

We now recall the σ-finite measure space from which we obtain the dilation and translation operators to build the promised multi-resolution analysis. Define the inflated fractal set $\mathcal{R}_\mathcal{S}$ associated to the Sierpinski gasket $\mathcal{S}$ by

$$\mathcal{R}_\mathcal{S} = \cup_{j=-\infty}^\infty \cup_{(m,n)\in\mathbb{Z}^2} [A^j(\mathcal{S} + (m,n))].$$

The probability measure $\mathcal{H}$ on $\mathcal{S}$ extends to a σ-finite measure that we also call $\mathcal{H}$ on $\mathcal{R}_\mathcal{S}$. This measure satisfies

$$\mathcal{H}(A^{-1}(E)) = \frac{1}{3}\mathcal{H}(E),$$

$$\mathcal{H}(E + (m,n)) = \mathcal{H}(E),$$

for every Borel subset E of $\mathcal{R}_\mathcal{S}$, and for every $(m,n) \in \mathbb{Z}^2$. Forming the Hilbert space $L^2(\mathcal{R}_\mathcal{S}, \mathcal{H})$, we can construct the unitary dilation operator D and a unitary representation of $\mathbb{Z}^2$ on $L^2(\mathcal{R}_\mathcal{S}, \mathcal{H})$ by

$$D(f)(s,t) = \sqrt{3}f(2s, 2t),$$

and

$$T_{(m,n)}f(s,t) = f(s - m, t - n).$$

These operators satisfy a standard commutation relation:

PROPOSITION 2.1. *Let D and $\{T_{(m,n)} : (m,n) \in \mathbb{Z}^2\}$ be the unitary operators on $L^2(\mathcal{R}_\mathcal{S}, \mathcal{H})$ defined above. Then*

$$T_{(m,n)}D = DT_{(2m,2n)}, \ \forall(m,n) \in \mathbb{Z}^2.$$

PROOF. This is an easy calculation. □

Thus, as in the paper by the third author, L.H. Lim, and K. Taylor, the operators $\{T_{(m,n)} : (m,n) \in \mathbb{Z}^2\}$ and $\{D^j : j \in \mathbb{Z}\}$ generate a representation of the generalized Baumslag-Solitar group $\mathbb{Q}_A \rtimes \mathbb{Z}$. Here the group of generalized A-adic rationals is defined by

$$\mathbb{Q}_A = \cup_{j=0}^\infty A^{-j}(\mathbb{Z}^2),$$

and the automorphism of $\mathbb{Z}$ on $\mathbb{Q}_A$ corresponds to the action of matrix multiplication by integer powers of A.

We now use the dilation operator D in the standard way to construct a multi-resolution analysis of the Hilbert space $L^2(\mathcal{R}_\mathcal{S}, \mathcal{H})$. Define a closed subspace $V_0 \subset L^2(\mathcal{R}_\mathcal{S}, \mathcal{H})$ by

$$V_0 = \overline{\text{span}}\{T_{(m,n)}(\chi_\mathcal{S}) : (m,n) \in \mathbb{Z}^2\}.$$

Here $\chi_\mathcal{S}$ is the characteristic function of the Sierpinski gasket triangle, and corresponds to the scaling function in the standard multi-resolution analysis set-up. The Sierpinski gasket $\mathcal{S}$ satisfies the self-similarity relation

(2.1) $$A(\mathcal{S}) = \mathcal{S} \cup [\mathcal{S} + (1,0)] \cup [\mathcal{S} + (0,1)],$$

and up to sets of measure 0, the above union is a disjoint union. It follows that its characteristic function $\chi_\mathcal{S}$ satisfies the dilation equation

(2.2) $$\chi_\mathcal{S}(A^{-1}(s,t)) = \chi_\mathcal{S}(s,t) + \chi_\mathcal{S}(s-1,t) + \chi_\mathcal{S}(s, t-1).$$

By construction, V_0 is invariant under the operators $\{T_{(m,n)} : (m,n) \in \mathbb{Z}^2\}$. For each $j \in \mathbb{Z}$, define

$$V_j = D^j(V_0).$$

We note that

$$\begin{aligned}
V_1 &= D(\overline{\text{span}}\{T_{(m,n)}(\chi_\mathcal{S}) : (m,n) \in \mathbb{Z}^2\}) \\
&= \overline{\text{span}}\{DT_{(m,n)}(\chi_\mathcal{S}) : (m,n) \in \mathbb{Z}^2\} \\
&= \overline{\text{span}}\{T_{(m/2,n/2)}D(\chi_\mathcal{S}) : (m,n) \in \mathbb{Z}^2\},
\end{aligned}$$

and consequently

$$V_0 \subseteq D(V_0) = V_1.$$

It follows that the closed subspaces $\{V_j\}_{j=-\infty}^{\infty}$ form an increasing nested sequence of closed spaces of $L^2(\mathcal{R}_\mathcal{S}, \mathcal{H})$.

The following result shows that the subspaces $\{V_j\}_{j=-\infty}^{\infty}$ form a multi-resolution analysis. The first three items are similar in nature to Proposition 2.8 of [DJ], although the proof of item (iii) is somewhat different.

PROPOSITION 2.2. *Let $\{V_j\}_{j=-\infty}^{\infty}$ be the subspaces of $L^2(\mathcal{R}_\mathcal{S}, \mathcal{H})$ constructed above, and let D, $\{T_{(m,n)} : (m,n) \in \mathbb{Z}^2\}$ be the unitary operators constructed in Proposition 2.1. Then*

(i) $D^{-1}(\chi_\mathcal{S}) = \frac{1}{\sqrt{3}}[\chi_\mathcal{S} + T_{(1,0)}(\chi_\mathcal{S}) + T_{(0,1)}(\chi_\mathcal{S})]$.

(ii) $\langle T_{(m,n)}(\chi_\mathcal{S}), \chi_\mathcal{S} \rangle = \delta_{(m,n),(0,0)}, \ (m,n) \in \mathbb{Z}^2$.

(iii) $\overline{\cup_{j=-\infty}^{\infty}[V_j]} = L^2(\mathcal{R}_\mathcal{S}, \mathcal{H})$.

(iv) $\cap_{j=-\infty}^{\infty}[V_j] = \{\vec{0}\}$.

PROOF. The proof of item (i) follows directly from the self-similarity relation of the Sierpinski gasket $\mathcal{S}$ and corresponding dilation equation outlined in Equations 2.1 and 2.2. Item (ii) is done by a similarly easy calculation, noting that translates of the Sierpinski gasket $\mathcal{S}$ by non-zero vectors in $\mathbb{Z}^2$ intersect $\mathcal{S}$ in sets of $\mathcal{H}$ measure 0.

As for item (iii), it will suffice to show that any Hausdorff measurable subset $E \subset \mathcal{R}_\mathcal{S}$ with $\mathcal{H}(E) < \infty$ has the property that χ_E is in the closure of the span of $\{D^j T_{(m,n)}(\chi_\mathcal{S}) \mid j, m, n \in \mathbb{Z}\}$. Since $\mathcal{R}_\mathcal{S} = \cup_{j=-\infty}^{\infty} \cup_{(m,n) \in \mathbb{Z}^2} [A^j(\mathcal{S} + (m,n))]$, we can write such a set $E = \cup E_{(j,m,n)}$, where $E_{(j,m,n)} = E \cap [A^j(\mathcal{S} + (m,n))]$. It is enough to show that the characteristic function of each set $E_{(j,m,n)}$ is in the closure of the span, so that by applying dilations and translations, we may without loss of generality assume that our arbitrary measurable set $E \subset \mathcal{S}$.

Let $\mathcal{V}$ be the collection of all the lower left vertices of subtriangles in S. By writing each $\{\vec{v}\} \in \mathcal{V}$ as a decreasing intersection: $\{\vec{v}\} = \cap_{n=1}^{\infty} T_n$, where each T_n is an nth dilate of a translate of $\mathcal{S}$, we see that $\mathcal{H}(\{\vec{v}\}) = \lim_{n \to \infty} \mathcal{H}(T_n) = \lim_{n \to \infty} \frac{1}{3^n} = 0$. By countable subadditivity of measures, we then have $\mathcal{H}(\mathcal{V}) = 0$. Now let $\mathcal{S}' = \mathcal{S} \sim \mathcal{V}$. Then $\mathcal{H}(\mathcal{S}') = \mathcal{H}(\mathcal{S}) = 1$. Since $\mathcal{S}$ is a metric space, so is $\mathcal{S}'$, although $\mathcal{S}'$ is no longer closed.

Thus, again without loss of generality, we assume that $E \subset \mathcal{S}'$. Let

$$\mathcal{T} = \{A^{-k}(\mathcal{S} + (i,j)) \cap \mathcal{S}' : \ k \in \mathbb{N}, \ (i,j) \in \{(0,0), (1,0), (0,1)\}\}.$$

Then $\mathcal{T}$, which consists of the "sub-Sierpinski gaskets" of $\mathcal{S}'$, is a semi-algebra of subsets of $\mathcal{S}'$; that is, finite disjoint unions of elements from $\mathcal{T}$ form an algebra of subsets of $\mathcal{S}'$. We denote this algebra by $\mathcal{A}$. Applying Hausdorff measure to the algebra $\mathcal{A}$, we obtain a set-valued function on $\mathcal{A}$, denoted by μ^*, which satisfies the conditions of the Carathéodory Extension Theorem. Therefore μ^* can be extended to an outer measure on all subsets of $\mathcal{S}'$, and μ^* agrees with Hausdorff measure $\mathcal{H}$ on the algebra $\mathcal{A}$ by construction. Thus μ^* determines a σ-algebra $\mathcal{M}$ of measurable sets, which contains the smallest σ-algebra $\mathcal{B}$ containing $\mathcal{A}$. If we denote the outer measure μ^* restricted to $\mathcal{M}$, now a measure, by μ, recall that $(X, \mathcal{M}, \mu)$ is a complete measure space. Moreover, since $\mu^*(\mathcal{S}') = \mathcal{H}(\mathcal{S}')$ is finite, the Carathéodory Extension Theorem tells us this extension of $\mathcal{H}$ on $\mathcal{A}$ to the smallest σ-algebra containing $\mathcal{A}$ is unique. It follows that $\mu = \mathcal{H}$ on $\mathcal{B}$, and also on $\mathcal{M}$. The question that remains is whether or not the σ-algebra that arises when one constructs the outer Hausdorff measure on subsets of $\mathcal{S}'$, is larger than the σ-algebra $\mathcal{M}$ arising using the Carathéodory Extension Theorem with the Hausdorff measure $\mu = \mathcal{H}$.

Note that the family of sets $\mathcal{T}$, in addition to being a semi-algebra of subsets of $\mathcal{S}'$, is a Vitali cover for any subset of $\mathcal{S}'$, that is, for each $x \in E$ and each $\delta > 0$, there is a subset $T \in \mathcal{T}$ with $x \in T$ and $0 < \mathcal{H}(T) \leq \delta$. By [**Ed2**], p.10, outer measures constructed using the Carathéodory Extension Theorem from Vitali covers are metric outer measures, and so satisfy $\mu^*(A \cup B) = \mu^*(A) + \mu^*(B)$ for any subsets A and B with $d(A, B) > 0$. Furthermore, if ν^* is a metric outer measure on a metric space X, then the σ-algebra of measurable sets with respect to this outer measure contains the σ-algebra of Borel sets of X([**Ed1**], 5.2.6). We apply this result to $(\mathcal{S}', \mathcal{M}, \mu)$ to deduce that μ^* is a metric outer measure and $\mathcal{M}$ contains the Borel sets of $\mathcal{S}'$. Thus, the σ-algebra of measurable sets of $\mathcal{S}'$ constructed from μ^* contains the Borel sets of $\mathcal{S}'$ and hence the open and closed sets of $\mathcal{S}'$. It follows that $\mathcal{H}$ agrees with μ on the Borel subsets of $\mathcal{S}$.

Finally, if one constructs a finite measure μ on a set X using an algebra $\mathcal{A}$ of sets using the Carathéodory Extension Theorem, then given any measurable set $G \subset X$, there exists a element R of $\mathcal{A}$ such that $\mu(G \Delta R) < \epsilon$ by Theorem 1.1.9 of [**Ed2**].

Applying this last result to our set-up, and the measurable subset E of $\mathcal{S}'$, we note that there exists a F_σ-set $F \subset E$ with $\mathcal{H}(E \Delta F) = \mathcal{H}(E \sim F) = 0$. We then deduce that there is a finite collection of finite sub-Sierpinski gasket triangles $T_i{}_{i=1}^n$ from $\mathcal{T}$ with $\mu(F \Delta \cup_{i=1}^n T_i) < \epsilon$. Since $\mu = \mathcal{H}$ on the σ-algebra of Borel sets of $\mathcal{S}'$, $\mathcal{H}(F \Delta \cup_{i=1}^n T_i) < \epsilon$. But since $\mathcal{H}(E \Delta F) = 0$, we see that $\mathcal{H}(E \Delta \cup_{i=1}^n T_i) < \epsilon$. Thus, χ_E is in the closure of the span of $\{D^j T_{(m,n)}(\chi_{\mathcal{S}}) \mid j, m, n \in \mathbb{Z}\}$.

It remains to establish item (iv). Note that if $(x, y) \in \text{support}(f)$ for $f \in \cap_{j=-\infty}^\infty V_j$, then for each $j \in \mathbb{Z}$, $(x, y) \in A^j(\mathcal{S} + (u_j, v_j))$ for some $(u_j, v_j) \in \mathbb{Z}^2$. For j large enough that $x^2 + y^2 < 2^{2j-1}$, this forces $(u_j, v_j) \in \{(0, 0), (0, -1), (-1, 0)\}$ and also forces (u_j, v_j) to be constant for these j. Thus, (x, y) must be in one of the nested unions $\cup_{j=-\infty}^\infty A^j \mathcal{S}$, $\cup_{j=-\infty}^\infty A^j(\mathcal{S} + (0, -1))$, or $\cup_{j=-\infty}^\infty A^j(\mathcal{S} + (-1, 0))$. Since $f \in V_{-j}$ must be constant on sets of the form $A^j(\mathcal{S} + (u, v))$, the fact that each union is nested means that f must be constant on each of these unions. As the measure of each union is infinite, these constants must all be 0.

$\square$

3. A parametrized family of high-pass filters for the Sierpinski gasket scaling function

Recall that in the multi-resolution analysis we have constructed for $L^2(\mathcal{R}_{\mathcal{S}}, \mathcal{H})$,

$$V_0 = \overline{\text{span}}\{T_{(m,n)}(\chi_{\mathcal{S}}) : (m, n) \in \mathbb{Z}^2\},$$

and the $\{T_{(m,n)}(\chi_S)\}$ form an orthonormal basis for V_0. Thus there is the standard isometric isomorphism $J : V_0 \rightarrow L^2(\mathbb{T}^2)$ given by

$$J(\sum_{(m,n)\in\mathbb{Z}^2} c_{m,n}T_{(m,n)}(\chi_S)) = \sum_{(m,n)\in\mathbb{Z}^2} c_{m,n}e_{(m,n)},$$

where $\{c_{m,n}\} \subset l^2(\mathbb{Z})$ and $\{e_{(m,n)}(z,w) = z^m \, w^n\}$ is the standard orthonormal basis for $L^2(\mathbb{T}^2)$.

Recall $D^{-1}(V_0) = V_{-1} \subset V_0$. It follows that $J(V_{-1}) \subset L^2(\mathbb{T}^2)$, and that

$$J(V_0) = J(V_{-1} \oplus W_{-1}) = J(V_{-1}) \oplus J(W_{-1}),$$

where $W_{-1} = V_{-1}^\perp \cap V_0$. Now we calculate

$$J(D^{-1}(\chi_S)) = J(\frac{1}{\sqrt{3}}[\chi_S + T_{(1,0)}(\chi_S) + T_{(0,1)}(\chi_S)]) = \frac{1}{\sqrt{3}}[e_{(0,0)} + e_{(1,0)} + e_{(0,1)}].$$

This function is our substitute for the low-pass filter, and we denote the above function by m_0, so that

$$m_0(z,w) = \frac{1}{\sqrt{3}}[e_{(0,0)}(z,w) + e_{(1,0)}(z,w) + e_{(0,1)}(z,w)].$$

Our aim is to find functions $\{\eta_l : l = 1,2,3\} \subset W_{-1}$ such that $\{T_{(2m,2n)}(\eta_l) : l = 1,2,3, (m,n) \in \mathbb{Z}^2\}$ form an orthonormal basis for W_{-1}. Note that $W_0 = D(W_{-1})$. Applying D and using the commutation relation $T_{(m,n)}D = DT_{(2m,2n)}$, the functions $\psi_k = D(\eta_k)$ will be our wavelet family for $L^2(\mathcal{R}_S, \mathcal{H})$, since it will then follow that

$$\cup_{l=1}^3 \{T_{(m,n)}(\psi_l) : (m,n) \in \mathbb{Z}^2\} = \cup_{l=1}^3 \{DT_{(2m,2n)}(\eta_l) : (m,n) \in \mathbb{Z}^2\}$$

will give an orthonormal basis for W_0.

Using the fact that J is a unitary isomorphism from the Hilbert space W_{-1} into $L^2(\mathbb{T})$, we let $m_l = J(\eta_l)$, $l = 1,2,3$. Part of the problem then comes down to computing when

$$\{z^{2m}w^{2n}m_l(z,w) : (m,n) \in \mathbb{Z}^2\}$$

is an orthonormal set in $L^2(\mathbb{T}^2)$.

LEMMA 3.1. *Let f be an element of $L^2(\mathbb{T}^2)$. Then the collection of functions $\{z^{2m}w^{2n}f : (m,n) \in \mathbb{Z}^2\}$ forms an orthonormal set in $L^2(\mathbb{T}^2)$ if and only if*

$$\sum_{j=0}^1 \sum_{k=0}^1 |f(ze^{\pi ij}, we^{\pi ik})|^2 = 4.$$

PROOF. The proof comes down to a simple calculation involving Fourier coefficients and the inner products $\langle f, z^{2m}w^{2n}f\rangle$, $(m,n) \in \mathbb{Z}^2$, which we leave to the reader. $\square$

LEMMA 3.2. *Suppose that f is as in Lemma 3.1. Then a function $g \in L^2(\mathbb{T}^2)$ is orthogonal to every function $z^{2m}w^{2n}f$ if and only if*

$$\sum_{j=0}^1 \sum_{k=0}^1 f(ze^{\pi ij}, we^{\pi ik})\overline{g(ze^{\pi ij}, we^{\pi ik})} = 0, \text{ a.e. on } \mathbb{T}^2.$$

PROOF. Again the proof involves calculations of Fourier coefficients and inner products defined by integrals, and is left to the reader. $\square$

This leads us to the following result, which summarizes a special case of results from [**BCM02**].

PROPOSITION 3.3. *(c.f.* [**BCM02**]*) Let* $m_0 = \frac{1}{\sqrt{3}}[e_{(0,0)} + e_{(1,0)} + e_{(0,1)}]$ *be the "mutant" low-pass filter on* $\mathbb{T}^2$ *defined earlier, and let* m_1, m_2, $m_3 \in L^2(\mathbb{T}^2)$. *Then* $\{\psi_l = D(J^{-1}(m_l) : l = 1, 2, 3\}$ *is a wavelet family for* $L^2(\mathcal{R}_\mathcal{S}, \mathcal{H})$ *if and only if the functions* $\{m_l\}$ *satisfy:*

$$(3.1) \qquad \sum_{j=0}^{1} \sum_{k=0}^{1} |m_l(ze^{\pi ij}, we^{\pi ik})|^2 = 4 \ a.e. \ on \ \mathbb{T}^2.$$

$$(3.2) \qquad \sum_{j=0}^{1} \sum_{k=0}^{1} m_l(ze^{\pi ij}, we^{\pi ik}) \overline{m_{l'}(ze^{\pi ij}, we^{\pi ik})} = 0, \ l \neq l',$$

and

$$(3.3) \qquad \sum_{j=0}^{1} \sum_{k=0}^{1} m_0(ze^{\pi ij}, we^{\pi ik}) \overline{m_l(ze^{\pi ij}, we^{\pi ik})} = 0, \ l = 1, 2, 3.$$

We now use a proposition based in linear algebra that has been used to create polynomial high-pass filters from polynomial low-pass filters as far back as 1992, by R. Gopinath and C. Burrus [**GB**].

THEOREM 3.4. *(c.f.* [**GB**]*) Let* $\vec{v}_0 = (\frac{1}{\sqrt{3}}, \frac{1}{\sqrt{3}}, \frac{1}{\sqrt{3}}, 0)$, $\vec{v}_1 = (a_{(0,0)}, a_{(1,0)}, a_{(0,1)}, a_{(1,1)})$, $\vec{v}_2 = (b_{(0,0)}, b_{(1,0)}, b_{(0,1)}, b_{(1,1)})$ *and* $\vec{v}_3 = (c_{(0,0)}, c_{(1,0)}, c_{(0,1)}, c_{(1,1)})$ *be vectors in* $\mathbb{C}^4$ *such that* $\{\vec{v}_0, \vec{v}_1, \vec{v}_2, \vec{v}_3\}$ *forms an orthonormal basis for* $\mathbb{C}^4$. *Then setting*

$$m_1 = \sum_{j=0}^{2} \sum_{k=0}^{2} a_{(j,k)} e_{(j,k)},$$

$$m_2 = \sum_{j=0}^{2} \sum_{k=0}^{2} b_{(j,k)} e_{(j,k)},$$

and

$$m_3 = \sum_{j=0}^{2} \sum_{k=0}^{2} c_{(j,k)} e_{(j,k)},$$

the functions m_1, m_2, *and* m_3 *satisfy Equations 3.1,3.2, and 3.3 with respect to* m_0, *so that* $\{\psi_l = D(J^{-1}(m_l)) : l = 1, 2, 3\}$ *is a wavelet family for* $L^2(\mathcal{R}_\mathcal{S}, \mathcal{H})$.

PROOF. Since the set $\{\vec{v}_0, \vec{v}_1, \vec{v}_2, \vec{v}_3\}$ forms an orthonormal basis for $\mathbb{C}^4$, the 4×4 matrix

$$M = \begin{pmatrix} 1/\sqrt{3} & 1/\sqrt{3} & 1/\sqrt{3} & 0 \\ a_{(0,0)} & a_{(1,0)} & a_{(0,1)} & a_{(1,1)} \\ b_{(0,0)} & b_{(1,0)} & b_{(0,1)} & b_{(1,1)} \\ c_{(0,0)} & c_{(1,0)} & c_{(0,1)} & c_{(1,1)} \end{pmatrix}$$

is unitary, as its rows form an orthonormal set. For $z = e^{2\pi is}$ and $w = e^{2\pi it}$, let $\vec{v}(z, w)$ denote the row vector consisting of the following functions from $C(\mathbb{T}^2)$: $(e_{(0,0)}, e_{(1,0)}, e_{(0,1)}, e_{(1,1)})$. Now note that $m_l(t) = \vec{v}_l \cdot \vec{v}(z, w)$, $l = 0, 1, 2, 3$. where

the "·" denotes dot product. We first verify that Equation 3.1 holds for m_1; the
proof for m_2 and m_3 will be identical.

$$\sum_{j=0}^{1}\sum_{k=0}^{1}|m_1(ze^{\pi ij}, we^{\pi ik})|^2 \; =$$

$$= \sum_{j=0}^{1}\sum_{k=0}^{1}(a_{(0,0)} + a_{(1,0)}ze^{\pi ij} + a_{(0,1)}we^{\pi ik} + a_{(1,1)}zwe^{\pi i(j+k)})$$

$$\overline{(a_{(0,0)} + a_{(1,0)}ze^{\pi ij} + a_{(0,1)}we^{\pi ik} + a_{(1,1)}zwe^{\pi i(j+k)})}$$

$$= \; 4|a_{(0,0)}|^2 + a_{(0,0)}[\overline{a_{(1,0)}}z\sum_{j=0}^{1}\sum_{k=0}^{1}e^{-\pi ij} + \overline{a_{(0,1)}}w\sum_{j=0}^{1}\sum_{k=0}^{1}e^{-\pi ik} + \overline{a_{(1,1)}}zw\sum_{j=0}^{1}\sum_{k=0}^{1}e^{-\pi i(j+k)}]$$

$$+ 4|a_{(1,0)}|^2 + a_{(1,0)}z[\overline{a_{(0,0)}}\sum_{j=0}^{1}\sum_{k=0}^{1}e^{\pi ij} + \overline{a_{(0,1)}}w\sum_{j=0}^{1}\sum_{k=0}^{1}e^{\pi i(j-k)} + \overline{a_{(1,1)}}zw\sum_{j=0}^{1}\sum_{k=0}^{1}e^{-\pi ik}]$$

$$+ 4|a_{(0,1)}|^2 + a_{(0,1)}w[\overline{a_{(0,0)}}\sum_{j=0}^{1}\sum_{k=0}^{1}e^{\pi ik} + \overline{a_{(1,0)}}z\sum_{j=0}^{1}\sum_{k=0}^{1}e^{\pi i(k-j)} + \overline{a_{(1,1)}}zw\sum_{j=0}^{1}\sum_{k=0}^{1}e^{-\pi ij}]$$

$$+ 4|a_{(1,1)}|^2 + a_{(1,1)}zw[\overline{a_{(0,0)}}\sum_{j=0}^{1}\sum_{k=0}^{1}e^{\pi i(j+k)} + \overline{a_{(1,0)}}z\sum_{j=0}^{1}\sum_{k=0}^{1}e^{-\pi ik} + \overline{a_{(0,1)}}w\sum_{j=0}^{1}\sum_{k=0}^{1}e^{-\pi ij}].$$

Repeatedly using the trivial equality

$$(3.4) \qquad\qquad\qquad \sum_{j=0}^{1} e^{\pm\pi ij} = 0,$$

we see that

$$\sum_{j=0}^{1}\sum_{k=0}^{1}|m_1(ze^{\pi ij}, we^{\pi ik})|^2 \; = \; 4|a_{(0,0)}|^2 + 4|a_{(1,0)}|^2 + 4|a_{(0,1)}|^2 + 4|a_{(1,1)}|^2 = 4.$$

With this and the identical calculations for m_2 and m_3, we have established Equation 3.1.

The proofs of the other two equations make similar use of 3.4. For example, to
establish the $l = 1$ case of Equation 3.3 we calculate that

$$\sum_{j=0}^{1}\sum_{k=0}^{1} m_0(ze^{\pi ij}, we^{\pi ik})\overline{m_1(ze^{\pi ij}, we^{\pi ik})}$$

$$= \sum_{j=0}^{1}\sum_{k=0}^{1}(\frac{1}{\sqrt{3}} + \frac{1}{\sqrt{3}}ze^{\pi ij} + \frac{1}{\sqrt{3}}we^{\pi ik})\overline{(a_{(0,0)} + a_{(1,0)}ze^{\pi ij} + a_{(0,1)}we^{\pi ik} + a_{(1,1)}zwe^{\pi i(j+k)})}$$

$$= 4\frac{1}{\sqrt{3}}\overline{a_{(0,0)}} + 4\frac{1}{\sqrt{3}}\overline{a_{(1,0)}} + 4\frac{1}{\sqrt{3}}\overline{a_{(1,0)}}$$

$$= 4\vec{v}_0 \cdot \overline{\vec{v}_1} \; = \; 0.$$

We leave the remainder of the details to the reader. $\square$

It follows from the above theorem that modulo permuting the wavelets, one can parametrize wavelet bases corresponding to the scaling function χ_S by ordered families of orthonormal bases for the orthogonal complement of $\{(\frac{1}{\sqrt{3}}, \frac{1}{\sqrt{3}}, \frac{1}{\sqrt{3}}, 0)\}$ in $\mathbb{R}^4$, or in $\mathbb{C}^4$ if we include complex linear combinations of dilates and translates of χ_S as our wavelets. These correspond to a parametrization by $O(3, \mathbb{R})$ or by $U(3, \mathbb{C})$, respectively. We thus have a concrete way to construct high-pass filters associated to m_0, and we now give some examples corresponding to the case where $\vec{v}_1 = (0, 0, 0, 1)$.

EXAMPLE 3.5. Let $\vec{v}_1 = (a_{(0,0)}, a_{(1,0)}, a_{(0,1)}, a_{(1,1)}) = (0, 0, 0, 1)$; then the associated filter function is

$$m_1(z, w) = zw.$$

This corresponds to the "gap-filling wavelet" in the discussion of the Cantor set wavelets given in [DJ].

For this choice of m_1, we now give a family of high-pass filters $\{m_2, m_3\}$ parametrized by a circle. We first need to find unit vectors $\vec{v}_2 = (b_{(0,0)}, b_{(1,0)}, b_{(0,1)}, 0)$ and $\vec{v}_3 = (c_{(0,0)}, c_{(1,0)}, c_{(0,1)}, 0)$ that are orthogonal to $\vec{v}_0 = (\frac{1}{\sqrt{3}}, \frac{1}{\sqrt{3}}, \frac{1}{\sqrt{3}}, 0)$ and to one another. If we just want real coefficients for our polynomials, this corresponds to finding (x, y, z) with $x^2 + y^2 + z^2 = 1$ and $\frac{1}{\sqrt{3}}x + \frac{1}{\sqrt{3}}y + \frac{1}{\sqrt{3}}z = 0$, i.e. $x + y + z = 0$. The plane $x + y + z = 0$ intersects the sphere $x^2 + y^2 + z^2 = 1$ in a circle. So suppose we have $(x, y, (-x - y))$ with $x^2 + y^2 + xy = \frac{1}{2}$. Let $\vec{v}_2 = (x, y, (-x - y), 0) = (b_{(0,0)}, b_{(1,0)}, b_{(0,1)}, 0)$. Then there will be a two choices of $\vec{v}_3 = (c_{(0,0)}, c_{(1,0)}, c_{(0,1)}, 0)$ lying on the desired circle (hence a unit vector that is perpendicular to $\vec{v}_1$ and is also perpendicular to $\vec{v}_2$. The vectors $\vec{v}_2$ and $\vec{v}_3$ correspond to the detail wavelets. We see these are parametrized by the Cartesian product of the circle formed from the intersection of the sphere $x^2 + y^2 + z^2 = 1$ and the plane $x + y + z = 0$ and the two point space $\{1, -1\}$.

We parametrize the family of possible $\{\vec{v}_2\}$ by using the angular variable θ : We can have

$$\vec{v}_2 = \frac{1}{\sqrt{2 + \sin 2\theta}}(\cos \theta, \sin \theta, -\sqrt{2}\sin(\theta + \frac{\pi}{4}), 0), \quad \theta \in [-\pi, \pi).$$

For each choice of $\vec{v}_2$, we have two possible choices of $\vec{v}_3$: one continuously parametrized choice is

$$\vec{v}_3 = \frac{1}{\sqrt{6 + 3\sin 2\theta}}(\sqrt{5}\sin(\theta + \alpha), -\sqrt{5}\cos(\theta - \alpha), \sqrt{2}\cos(\theta + \frac{\pi}{4}), 0),$$

for $\theta \in [-\pi, \pi)$, and where $\alpha = \arcsin \frac{1}{\sqrt{5}}$. Then

$$m_2(z, w) = \vec{v_2} \cdot (e_{(0,0)}, e_{(1,0)}, e_{(0,1)}, e_{(1,1)}) = \frac{\cos \theta + \sin \theta z - \sqrt{2}\sin(\theta + \frac{\pi}{4})w}{\sqrt{2 + \sin 2\theta}},$$

and

$$m_3(z, w) = c_{(0,0)} + c_{(1,0)}z + c_{(0,1)}w,$$

where $c_{(0,0)}, c_{(1,0)}$ and $c_{(0,1)}$ are the components of $\vec{v}_3$ parametrized above. For example, taking $\theta = -\frac{\pi}{4}$, we would have $m_2(z, w) = \frac{1}{\sqrt{2}} - \frac{1}{\sqrt{2}}z$ corresponding to the choice of vector $\vec{v}_2 = (\frac{1}{\sqrt{2}}, -\frac{1}{\sqrt{2}}, 0, 0)$. Then the possible choices of $\vec{v}_3$ are

$\vec{v}_3 = (\frac{-1}{\sqrt{6}}, \frac{-1}{\sqrt{6}}, \frac{2}{\sqrt{6}}, 0)$, (corresponding to the parametrization given by θ above) or $\vec{v}_3 = (\frac{1}{\sqrt{6}}, \frac{1}{\sqrt{6}}, \frac{-2}{\sqrt{6}}, 0)$. In the first case we have

$$m_3(z, w) = -\frac{1}{\sqrt{6}} - \frac{z}{\sqrt{6}} + \frac{2w}{\sqrt{6}}.$$

In the second case we have

$$m_3(z, w) = \frac{1}{\sqrt{6}} + \frac{z}{\sqrt{6}} - \frac{2w}{\sqrt{6}}.$$

Our wavelet family for the first choice of m_3 is given by $\psi_i = D(J^{-1}(m_i))$ so that

$$\psi_1 = D(T_{(1,1)}\chi_{\mathcal{S}}),$$

$$\psi_2 = \frac{1}{\sqrt{2}}D(\chi_{\mathcal{S}} - T_{(1,0)}\chi_{\mathcal{S}}),$$

$$\psi_3 = \frac{1}{\sqrt{6}}D(-\chi_{\mathcal{S}} - T_{(1,0)}\chi_{\mathcal{S}} + 2T_{(0,1)}\chi_{\mathcal{S}}).$$

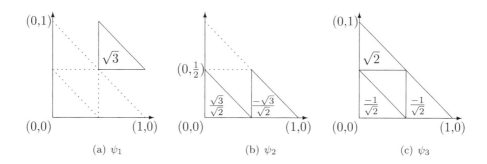

(a) ψ_1 (b) ψ_2 (c) ψ_3

FIGURE 2. Wavelets of Example 3.5.

The right triangles in the graphs of Figure 2 represent the sub-Sierpinski gaskets obtained by dilation of $\chi_{\mathcal{S}}$ and associated translates in the construction of ψ_i.

EXAMPLE 3.6. (The D'Andrea Code) We now do an example with complex coefficients. It has the benefit that the non-zero coefficients in the $\vec{v}_2$ and $\vec{v}_3$ have the same modulus, thus giving a certain symmetry to the detail wavelets. This cannot be achieved in the parametrization with real coefficients given above. We still take $m_1(z) = zw$ corresponding to the gap-filling wavelet but for the choice of detail wavelets we use complex coefficients for the $\vec{v}_i$, $i = 2, 3$. In particular, setting $\lambda = e^{\frac{2\pi i}{3}}$ we let

$$\vec{v}_2 = (\frac{1}{\sqrt{3}}, \frac{\lambda}{\sqrt{3}}, \frac{\lambda^2}{\sqrt{3}}, 0)$$

and

$$\vec{v}_3 = (\frac{1}{\sqrt{3}}, \frac{\lambda^2}{\sqrt{3}}, \frac{\lambda}{\sqrt{3}}, 0),$$

and check that the hypotheses of Theorem 3.4 hold.

In this case, we get

$$m_1(z, w) = zw, \ m_2(z, w) = \frac{1}{\sqrt{3}}[1 + \lambda z + \lambda^2 w],$$

and

$$m_3(z, w) = \frac{1}{\sqrt{3}}[1 + \lambda^2 z + \lambda w],$$

so that the wavelet family is given by

$$\psi_1 = D(T_{(1,1)}\chi_S),$$

$$\psi_2 = \frac{1}{\sqrt{3}}D(\chi_S + \lambda T_{(1,0)}\chi_S + \lambda^2 T_{(0,1)}\chi_S)$$

and

$$\psi_3 = \frac{1}{\sqrt{3}}D(\chi_S + \lambda^2 T_{(1,0)}\chi_S + \lambda T_{(0,1)}\chi_S).$$

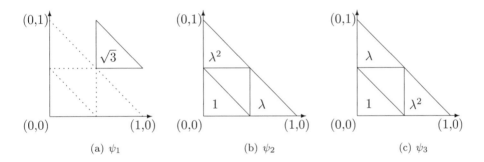

(a) ψ_1 (b) ψ_2 (c) ψ_3

FIGURE 3. Wavelets of Example 3.6.

4. Discrete Sierpinski gasket wavelet transform and image compression

In this section we will describe a discrete fractal wavelet transform based on the Sierpinski gasket space, using the filters of Examples 3.5 and 3.6. We will then apply this transform to images in $L^2(\mathbb{R}^2)$, and analyze how the results compare to those using the Haar wavelet, which is also discontinuous. For our Sierpinski gasket wavelet transform (DSGWT), we will use the same algorithm as the discrete Haar wavelet transform (DHWT), altering only the low and high-pass filters which are used. Thus, we begin by briefly describing the Haar wavelet and the DHWT.

The Haar scaling function and wavelet on $L^2(\mathbb{R}^2)$ are given by

$$\varphi = \chi_Q,$$

$$\psi_1 = \frac{1}{2}D(T_{(0,0)}\chi_Q + T_{(1,0)}\chi_Q - T_{(0,1)}\chi_Q - T_{(1,1)}\chi_Q),$$

$$\psi_2 = \frac{1}{2}D(T_{(0,0)}\chi_Q - T_{(1,0)}\chi_Q + T_{(0,1)}\chi_Q - T_{(1,1)}\chi_Q),$$

$$\psi_3 = \frac{1}{2}D(T_{(0,0)}\chi_Q - T_{(1,0)}\chi_Q - T_{(0,1)}\chi_Q + T_{(1,1)}\chi_Q),$$

where $\mathcal{Q} = [0,1) \times [0,1)$, and where

$$D(f)(s,t) = 2f(2s, 2t),$$

and

$$T_{(m,n)}f(s,t) = f(s-m, t-n)$$

are the unitary operators on $L^2(\mathbb{R}^2)$ analogous to the operators of the same name that were defined in Section 2 on $L^2(\mathcal{R}_S, \mathcal{H})$.

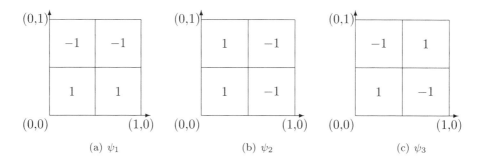

FIGURE 4. Haar wavelets on $L^2(\mathbb{R}^2)$.

The scaling function φ defines a multi-resolution analysis of $L^2(\mathbb{R}^2)$ in the usual way with

$$V_j = \overline{\mathrm{span}}\{D^j T_{(m,n)}(\varphi) : (m,n) \in \mathbb{Z}^2\}.$$

Also in the usual way, if we let W_j be defined by $V_{j+1} = V_j \oplus W_j$, then

$$\cup_{i=1}^{3}\{D^j T_{(m,n)}(\psi_i) : (m,n) \in \mathbb{Z}^2\}$$

forms an orthonormal basis of W_j.

The Haar scaling function and wavelet can be constructed from filters which can in turn be described by unit vectors in $\mathbb{R}^4$ in a similar fashion to the Sierpinski wavelets of the previous section. The Haar low-pass filter is described by the unit vector,

$$\vec{v}_0 = (\frac{1}{2}, \frac{1}{2}, \frac{1}{2}, \frac{1}{2}),$$

and high-pass filters by the unit vectors

$$\vec{v}_1 = (\frac{1}{2}, \frac{1}{2}, \frac{-1}{2}, \frac{-1}{2}),$$

$$\vec{v}_2 = (\frac{1}{2}, \frac{-1}{2}, \frac{1}{2}, \frac{-1}{2}),$$

and

$$\vec{v}_3 = (\frac{1}{2}, \frac{-1}{2}, \frac{-1}{2}, \frac{1}{2}).$$

We consider a $2^n \times 2^n$ pixel grayscale image, supported on the unit square, as a function in V_n, and represent it by a $2^n \times 2^n$ matrix B whose entries are intensity

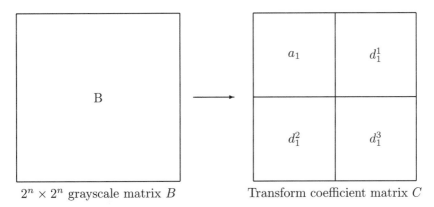

$2^n \times 2^n$ grayscale matrix B Transform coefficient matrix C

FIGURE 5. Representation of 2×2 sub-matrix decomposition.

values. The DHWT allows us to decompose B in the following manner. We take a 2×2 sub-matrix of B and write it as a column vector,

$$
\begin{pmatrix} b_{i,j} & b_{i,j+1} \\ b_{i+1,j} & b_{i+1,j+1} \end{pmatrix} \simeq \begin{pmatrix} b_{i+1,j} \\ b_{i+1,j+1} \\ b_{i,j} \\ b_{i,j+1} \end{pmatrix} = \vec{b}_{ij}.
$$

Multiplication by the matrix M, whose rows are the filters $\vec{v}_0, \vec{v}_1, \vec{v}_2$, and $\vec{v}_3$ (as in the proof of Theorem 3.4), yields the column vector,

$$
\vec{c}_{ij} = M\vec{b}_{ij} = \begin{pmatrix} \vec{v}_0 \\ \vec{v}_1 \\ \vec{v}_2 \\ \vec{v}_3 \end{pmatrix} \vec{b}_{ij} = \begin{pmatrix} a_{\frac{i+1}{2},\frac{j+1}{2}} \\ d^1_{\frac{i+1}{2},\frac{j+1}{2}} \\ d^2_{\frac{i+1}{2},\frac{j+1}{2}} \\ d^3_{\frac{i+1}{2},\frac{j+1}{2}} \end{pmatrix},
$$

whose entries are transform coefficients a, d^1, d^2, and d^3 which are stored in the matrix C. Since B was chosen to be a $2^n \times 2^n$ matrix, the indices (i,j) on $\vec{b}$ and $\vec{c}$ are always both odd. The indices on the entries of $\vec{c}$ are the positions of these values in the $2^{n-1} \times 2^{n-1}$ sub-matrices of C of the same name depicted in Figure 5. Hence, transform coefficients, corresponding to an original 2×2 block in B, have the same relative position in their respective sub-matrix of C, (either a_1, d^1_1, d^2_1, or d^3_1,) as the position of the original 2×2 block in B.

This describes a standard algorithm for computing the transform coefficients in a discrete setting based on the Haar wavelet family using nothing more than matrix multiplication. If we renormalize the unit square to a $2^n \times 2^n$ matrix of pixels and note that the matrix uses column information for the x shift and row information for the y shift, then the transform coefficients correspond to standard inner products in $L^2(\mathbb{R}^2)$ given by,

$$
a_{\frac{i+1}{2},\frac{j+1}{2}} = \langle f, D^{n-1} T_{(\frac{j-1}{2}, 2^{n-1} - \frac{i+1}{2})}(\varphi) \rangle,
$$

$$
d^1_{\frac{i+1}{2},\frac{j+1}{2}} = \langle f, D^{n-1} T_{(\frac{j-1}{2}, 2^{n-1} - \frac{i+1}{2})}(\psi_1) \rangle,
$$

$$
d^2_{\frac{i+1}{2},\frac{j+1}{2}} = \langle f, D^{n-1} T_{(\frac{j-1}{2}, 2^{n-1} - \frac{i+1}{2})}(\psi_2) \rangle,
$$

and

$$d^3_{\frac{i+1}{2},\frac{j+1}{2}} = \langle f, D^{n-1}T_{(\frac{j-1}{2},2^{n-1}-\frac{i+1}{2})}(\psi_3)\rangle.$$

Note that these are the coefficients of the basis functions in the decomposition of

$$f \in V_n = V_{n-1} \oplus W_{n-1}$$

on the support of f corresponding to the original 2×2 block in B. Hence,

$$f_{|E} = D^{n-1}T_{(\frac{j-1}{2},2^{n-1}-\frac{i+1}{2})}(a\varphi + d^1\psi_1 + d^2\psi_2 + d^3\psi_3),$$

where E is the support of the original 2×2 block in B, (we have suppressed the indices on the coefficients a, d^1, d^2, and d^3). In this manner we decompose B into

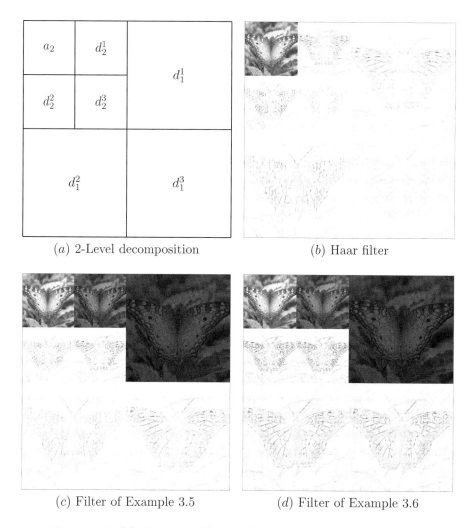

(a) 2-Level decomposition (b) Haar filter

(c) Filter of Example 3.5 (d) Filter of Example 3.6

FIGURE 6. (a) 2-level coefficient decomposition scheme; (b)-(d) decomposed butterfly using various filters. Note that detail submatrices with white backgrounds have been modified for display purposes and are scaled negative images of the actual sub-matrices.

C, which consists of the sub-matrices a_1, d_1^1, d_1^2, and d_1^3, as shown in Figure 5 (with the subscripts indicating that we have performed one level of decomposition).

We can continue decomposing f by transforming the sub-matrix a_1 of C, achieving a second level of decomposition shown in Figure 6. This is analogous again to the decomposition of V_n as seen by

$$V_n = V_{n-2} \oplus W_{n-2} \oplus W_{n-1}.$$

Note that in this case, the entries in a_2 are the coefficients of the basis elements of V_{n-2} for this representation of f. The entries of the detail matrices $d_2^k, k = 1, 2, 3$, are the coefficients of the basis elements of W_{n-2} corresponding to the appropriate translation and dilation of ψ_k respectively for this representation of f. Similarly, $d_1^k, k = 1, 2, 3$, correspond to W_{n-1}. Since $f \in V_n$, we can perform n such decompositions by successive iterations on V_j for $j = 1, ..., n$ yielding,

$$V_n = V_0 \oplus (\oplus_{j=0}^{n-1} W_j).$$

Note that the sub-matrix a_n consists of a single entry, which is the coefficient of φ. For further details regarding the Haar wavelet family and its discrete wavelet transform see [**Mal99**].

We have detailed the DHWT because of its familiarity and because the algorithm and theory just described are essentially the same for the DSGWT. We need only change the filters, $\vec{v}_i$ for $i = 0, 1, 2, 3$, to those described in Section 3. Recall that for Examples 3.5 and 3.6 this means that $\vec{v}_0 = (\frac{1}{\sqrt{3}}, \frac{1}{\sqrt{3}}, \frac{1}{\sqrt{3}}, 0)$ and $\vec{v}_1 = (0, 0, 0, 1)$. If we consider our matrix B as the discretization of a function $f \in L^2(\mathcal{R}_\mathcal{S}, \mathcal{H})$, then this approximation to f lives in V_n of the multi-resolution on $L^2(\mathcal{R}_\mathcal{S}, \mathcal{H})$ described in Section 2, and everything we just said about the DHWT still applies to the DSGWT substituting $L^2(\mathcal{R}_\mathcal{S}, \mathcal{H})$ in place of $L^2(\mathbb{R}^2)$. If however we keep B as the discretization of a function $f \in L^2(\mathbb{R}^2)$ there is nothing preventing us from using the DSGWT to decompose f. Essentially, we are then treating level n Sierpinski gaskets as pixels. Since the transform relies on the fact that M is unitary we are just reorganizing the information provided in B into a new coefficient matrix C. But now our coefficients no longer match up nicely to the standard inner product on $L^2(\mathbb{R}^2)$ of f with the appropriate basis elements since the measures on $\mathcal{R}_\mathcal{S}$ and $\mathbb{R}^2$ are different, i.e., if the unit square $\mathcal{Q} \in \mathbb{R}^2$ is represented by a 2×2 matrix then a single entry of this matrix corresponds to an area of $\frac{1}{4}$ under the $\mathbb{R}^2$ Lebesgue measure whereas a single entry of this matrix under the fractal measure $\mathcal{H}$ corresponds to an area of $\frac{1}{3}$ since it is one-third of the discretized unit Sierpinski gasket. This technicality does not affect the implementation of the DSGWT on a given matrix regardless of one's point of view.

Figure 6 shows two levels of decomposition of a 512×512 grayscale image (butterfly of Figure 7a [**UGr**]) into transform coefficient matrices, which have also been visualized as grayscale images. These transform matrices illustrate the desirability of using certain wavelets in image compression. Grayscale intensity values take on values between zero (black) and one (white). The detail matrices with white backgrounds in the transform matrices of figure 6 are scaled negative images. Hence, the large amounts of white or near white areas of these transform matrices are entries which are nearly zero. Transforming the image at different levels amounts to reorganizing the information of the image with respect to the basis functions of our multi-resolution analysis. Under this reorganization, energy is concentrated in the transform coefficients where the structure of the image is well-represented by the structure of the basis functions at a particular level of resolution. Thresholding or quantizing the transform matrices zeros out a certain number of entries. As long the entries being zeroed out are near zero, we are only throwing away a small amount of the information contained in the transform matrix, and a reasonable reconstruction of the original image can be obtained. Sparse matrices are easily compressed using various coding schemes which we do not discuss here. So, if we can represent an image efficiently by basis functions of a multi-resolution analysis, this representation will admit a sparse matrix under thresholding which contains most of the original information and can be compressed easily. Again, we are taking advantage of image structure being similar to basis function structure at different levels of resolution.

(a) Original   (b) Haar filter

(c) Filter of Example 3.5   (d) Filter of Example 3.6

FIGURE 7. (a) Original 512×512 grayscale image; (b)-(d) reconstruction using 3% of transform coefficients.

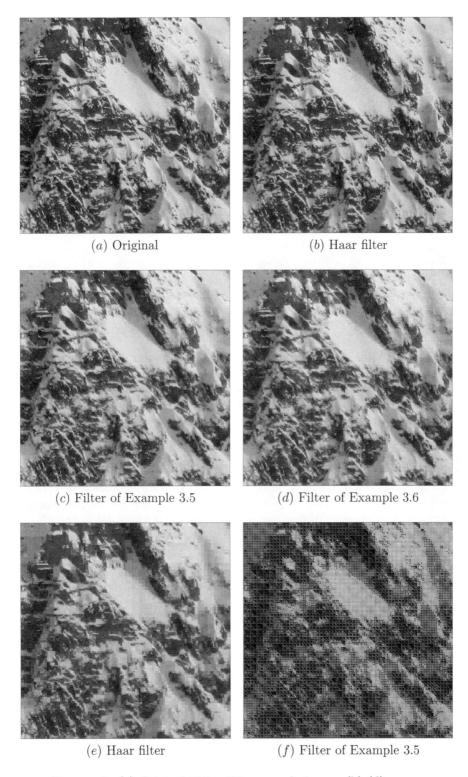

(a) Original

(b) Haar filter

(c) Filter of Example 3.5

(d) Filter of Example 3.6

(e) Haar filter

(f) Filter of Example 3.5

FIGURE 8. (a) Original 512×512 grayscale image; (b)-(d) reconstruction using 30% of transform coefficients; (e),(f) using 3% of transform coefficients.

Reconstruction of an image from its transform coefficients is accomplished via an inverse wavelet transform. For our examples this amounts to the matrix multiplication,

$$M^* \vec{c} \;=\; M^* M \vec{b} \;=\; \vec{b},$$

where M^* is the adjoint of the filter matrix M defined in the discrete wavelet transform and $\vec{c}$ is a transform coefficient vector of the form,

$$\vec{c} \;=\; \begin{pmatrix} a \\ d^1 \\ d^2 \\ d^3 \end{pmatrix}.$$

If $\vec{c}$ contains entries from coefficient sub-matrices at the jth level of decomposition (i.e. from $a_j, d_j^1, d_j^2,$ or d_j^3), then the resulting $\vec{b}$ obtained from the inverse transform will be part of the coefficient sub-matrix a_{j-1}. Image reconstruction is perfect when all of the transform coefficients are kept at all levels. Lossy compression schemes utilize a quantization or thresholding step to remove a certain percentage of transform coefficients as previously described. Figures 7 and 8 show 512×512 grayscale images and reconstructed images using 3% or 30% of the transform coefficients after 9 levels of decomposition.

Note, that for this lossy scheme, the DSGWT doesn't concentrate energy in the transform coefficients of Figure 7 as well as Haar because the structure of the butterfly image isn't well correlated to the structure of the Sierpinski gasket on any level. The result is that the Haar transform outperforms the gasket transform in this task. In particular, upon closer inspection we see that the 'gap' coefficients, those detail coefficients computed from the filter associated with the "gap-filling wavelet" ψ_1, produced by the DSGWT are merely being rearranged from B to d_1^1 in C and from a_j to d_{j+1}^1 at each level of the transform process. This is easily seen in Figures 6c and 6d by noticing that the coefficient sub-matrices d_1^1 and d_2^1 appear to be approximations of the original image. They are, in the sense that they are sampled directly from the upper-right corner of the 2×2 blocks tiling the original image or tiling a_1 respectively. In contrast, the approximation given by the sub-matrices a_1 or a_2 are 'averages' of more than one entry of those same 2×2 blocks. As a result, in the quantization step, when 'gap' coefficients are thrown away they become unrecoverable. This is readily seen in both Figures 7 and 8. Enough transform coefficient information has been lost due to thresholding that 'gap' coefficients have been thrown away and the reconstructed images have 'holes' on those 'gap' supports where there is nothing to reconstruct. The 'gap' information contained in an image matrix is approximately one-third the total image information since it constitutes one-fourth of our transform matrix C at each level of decomposition, thus following a geometric series. Both Figures 7 and 8 throw out more than two-thirds of the transformed information forcing the deletion of unrecoverable 'gap' information as already mentioned. Figure 8 shows a portion of a snow covered mountain with 'gasket' features. At 30%, both the DHWT and DS-GWT perform fairly well in this figure, although a bit of bandedness is detectable in the two DSGWT reconstructions. At 3%, weaknesses of both transforms are evident, blockiness in the DHWT reconstruction and noticeable bandedness in the DSGWT reconstruction caused by missing 'gap' coefficients as previously described.

Although the large scale structure of the mountain appears amenable to the Sierpinski gasket transform, the mountain does not have the gasket structure at all levels, thus bandedness is still a problem on smaller scales. It is interesting to note, that in the absence of information on the transform side, both the Haar and Sierpinski gasket transforms introduce their structure into the reconstructed images, as demonstrated by Figures 7 and 8. This is to be expected given our discussion of multi-resolutions and the fact that the corresponding basis functions of each transform are the building blocks in the reconstructed images.

5. Conclusion

Very recently, N. Larsen and I. Raeburn have developed an abstract theory relating multi-resolution analyses and directly limit systems of Hilbert spaces constructed from pure isometries [**LarRae**]. It is our belief that the fractal wavelets and multi-resolution analyses of Dutkay and Jorgensen can be fit very nicely into this framework, and in future work with Raeburn, the third author hopes to pursue this idea. This Sierpinski gasket case would be a perfect case to which one could apply the abstract theory of direct limits.

In particular, we conjecture that taking the partial isometry S defined on $L^2(\mathbb{T}^2)$ by

$$S(f)(z, w) = \frac{1}{\sqrt{3}}(1 + z + w)f(2z, 2w),$$

then S is a pure isometry on $L^2(\mathbb{T}^2)$, and forming the direct limit Hilbert space from S via the Larsen-Raeburn construction, one could obtain a multi-resolution analysis and wavelet families isomorphic to the wavelet families described here. It then would be possible to use this direct limit characterization to construct a Fourier transform from $L^2(\mathcal{R}_\mathcal{S}, \mathcal{H})$ to $L^2(\Sigma_2 \times \Sigma_2, \nu)$, corresponding to that constructed by Dutkay in [**Dut**], Corollary 5.8, in the Cantor set case. Here Σ_2 represents the solenoid viewed as the compact Pontryagin dual of the dyadic rationals, and ν is a probability measure on $\Sigma_2 \times \Sigma_2$ determined by the "low-pass" filter mentioned above. We leave this discussion to further research.

ACKNOWLEDGEMENTS. The authors gratefully acknowledge helpful conversations with Lawrence Baggett, Iain Raeburn, and Arlan Ramsay on the topic of this article.

References

[BCM02] L. W. Baggett, J. E. Courter, and K. D. Merrill, *The construction of wavelets from generalized conjugate mirror filters in $L^2(\mathbb{R}^n)$*, Appl. Comput. Harmon. Anal. **13** (2002), 201–223.

[BJMP] L.W. Baggett, P.E.T. Jorgensen, K.D. Merrill and J.A. Packer, *Construction of Parseval wavelets from redundant filter systems*, J. Math. Physics **46** (2005), 0–28.

[BLPR] L.W. Baggett, N. Larsen, J. Packer, and I. Raeburn, *Direct limits, MRA's, and Wavelets*, in preparation.

[BMM99] L. W. Baggett, H. A. Medina, and K. D. Merrill, *Generalized multi-resolution analyses and a construction procedure for all wavelet sets in $\mathbb{R}^n$*, J. Fourier Anal. Appl. **5** (1999), 563–573.

[BJ97] O. Bratteli and P.E.T. Jorgensen, *Isometries, shifts, Cuntz algebras and multiresolution wavelet analysis of scale N*, Integral Equations Operator Theory **28** (1997), 382–443.

[Dau] I. Daubechies, *Ten Lectures on Wavelets*, CBMS-NSF Regional Conference Series in Applied Math., vol. 61, SIAM, Philadelphia, 1992.

[Dut] D. Dutkay, *Low-pass filters and reprentations of the Baumslag Solitar group*, Trans. Amer. Math. Soc. **358** (2006), no. 12, 5271–5291.

[DJ] D. Dutkay and P.E.T. Jorgensen, *Wavelets on fractals*, Rev. Math. Iberoamericana **22** (2006), no. 1, 131–180.

[Ed1] G. A. Edgar, *Measure, Topology, and Fractal Geometry*, New York : Springer-Verlag, 1990.

[Ed2] G. A. Edgar, *Integral, Probability, and Fractal Measures*, New York : Springer, 1998.

[GB] R. A. Gopinath, C. S. Burrus, *Wavelet transforms and filter banks*, in "Wavelets: A Tutorial in Theory and Applications", pp. 603–654, C. K. Chui, Ed., Academic Press, Inc., San Diego, CA, 1992.

[Hut] J. E. Hutchinson, *Fractals and self-similarity*, Indiana Univ. Math. J. **30** (1981), 713-747.

[Jor03] P.E.T. Jorgensen, *Matrix factorization, algorithms, and wavelets*, Notices Amer. Math. Soc., to appear.

[LarRae] N. Larsen and I. Raeburn, *From filters to wavelets via direct limits*, Contemp. Math., to appear.

[LPT] L.H. Lim, J.A. Packer, and K. Taylor, *A direct integral decomposition of the wavelet representation*, Proc. Amer. Math. Soc. **129** (2001), 3057-3067.

[Mal99] S.G. Mallat, *A Wavelet Tour of Signal Processing*, 2nd ed., Academic Press, Orlando–San Diego, 1999.

[Mey93] Y. Meyer, *Wavelets and operators*, Different Perspectives on Wavelets (San Antonio, TX, 1993) (Ingrid Daubechies, ed.), Proc. Sympos. Appl. Math., vol. 47, American Mathematical Society, Providence, 1993, pp. 35–58.

[Roy] H. L. Royden, *Real Analysis*, Third Edition, New York: Macmillan Publishing Co., 1988.

[Str] R. S. Strichartz, *Piecewise linear wavelets on Sierpinski gasket type fractals*, J. Fourier Anal. Appl. **3** (1997), 387–416.

[UGr] Computer Vision Group, University of Granada, http://decsai.ugr.es/cvg/CG/base.htm.

DEPARTMENT OF MATHEMATICS, CAMPUS BOX 395, UNIVERSITY OF COLORADO, BOULDER, CO, 80309-0395
   E-mail address: dandrea@colorado.edu

DEPARTMENT OF MATHEMATICS, COLORADO COLLEGE, COLORADO SPRINGS, CO 80903-3294
   E-mail address: kmerrill@coloradocollege.edu

DEPARTMENT OF MATHEMATICS, CAMPUS BOX 395, UNIVERSITY OF COLORADO, BOULDER, CO, 80309-0395
   E-mail address: packer@euclid.colorado.edu

Contemporary Mathematics
Volume **451**, 2008

Texture Identification of Tissues Using Directional Wavelet, Ridgelet and Curvelet Transforms

Lucia Dettori and Ahmed I. Zayed

ABSTRACT. The wavelet transform on the real line is an effective tool for detecting point singularities. However, because of its lack of orientation, the multi-dimensional wavelet transform does not perform very well when the singularities are of higher dimensions.

In the last few years new geometric multiscale transforms, such as the directional wavelet, ridgelet and curvelet transforms, have been developed to, among other things, detect intermediate-dimensional singularities. The problem of detecting singularities is closely related to image processing because in two-dimensional images smooth regions are separated by edges which are typically smooth curves.

In this article we discuss an application of the directional wavelet, ridgelet and curvelet transforms in medical image processing. We will compare the performance of these transforms in texture identification of tissues in images obtained from computed tomography scans.

1. Introduction

Since the introduction of the wavelet transform in the early 1980s, several generalizations and hybrids of it have been developed either to enrich the mathematical foundations of the subject or to enlarge the scope of its applications. The wavelet transform has proved to be very useful in many applications; nevertheless, it has its natural limitations. One of these limitations is apparent in an application in image processing.

One of the salient features of the wavelet transform is its ability to detect point singularities effectively. This is a consequence of the fact that if the mother wavelet ψ has N vanishing moments and f is a smooth signal that behaves like a polynomial of degree at most N, then the wavelet transform of f vanishes. That is to say, the wavelet transform is blind to the regions where the signal is smooth and can only see regions where the signal is not smooth.

This nice feature of the wavelet transform also has its limitation. The wavelet transform neither performs well nor does it represent singularities effectively when

1991 *Mathematics Subject Classification.* Primary: 42C40; Secondary: 94A080.
Key words and phrases. Wavelet, directional wavelet, ridgelet, and curvelet transforms, texture classification.

the singularities are of higher dimensions. This shortfall of the wavelet transform is closely related to the question of how to extend the wavelet transform to higher dimensions. The extension of the continuous wavelet transform from one dimension to higher dimensions may be done in a number of different ways. Some of these extensions are more amenable to the geometry of the space than others and some allow more degrees of freedom either in preserving certain properties of the transform, such as invariance under translation and rotation, or in detecting certain properties of the transformed functions, such as singularities along certain directions.

One of the generalizations of the wavelet transform is the directional wavelet transform, which in addition to scaling and translation, it accounts for space rotation. This makes the directional wavelet transform better at detecting singularities of higher dimensions. For other generalizations of the continuous wavelet transforms to higher dimensions, see [1].

In the last 5 years, there have been some recent developments in the subject that led to the discovery of new multiscale transforms to remedy shortfalls of the wavelet transform. Chief among them are the ridgelet and curvelet transforms which are new geometric, multi-scale schemes to represent singularities of higher dimensions [3, 4, 5, 6, 7, 8, 12, 13, 26]. They outperform the continuous wavelet transform in taking into account the geometrical features of a signal and detecting its singularities.

The Ridgelet transform, which may be viewed as an application of a one-dimensional wavelet transform to a slice of the Radon transform of the signal, detects singularities along straight lines more efficiently than the wavelet transform, while the curvelet transform, which may be viewed as a directional parabolic dilation transform in which the parabolic dilation is applied in the frequency domain, can detect singularities along curves better than the ridgelet and wavelet transform.

The problem of detecting singularities is closely related to image processing because in two-dimensional images smooth regions are separated by edges which are typically smooth curves. That is two-dimensional images generally have one-dimensional singularities. This suggests that the wavelet transform may not be the best tool to use for some image processing applications.

The aim of this article is to discuss different generalizations and hybrids of the wavelet transform and to investigate one of their applications in medical imaging. We will compare the performance of these transforms in texture identification of tissues in images obtained from computed tomography scans. More precisely we are interested in identifying the following five organs: heart, liver, spleen, kidney, and backbone, from images of normal tissues taken from these organs.

The article is divided into two parts; the first part gives an overview of the mathematical foundations of the directional wavelet, ridgelet, and curvelet transforms. Although our applications are two dimensional, we shall introduce the theory in n-dimensions, with $n \geq 2$, since the extension from two to higher dimensions is straightforward. In the second part we present the results and compare the performance of these transforms in the problem of texture identification.

2. Preliminaries

Although our applications are two dimensional, we will introduce the some of concepts in n dimensions since the notation is the same. A point or a vector in

$I\!R^n$ will be denoted by $\underline{x} = \mathbf{x} = (x_1, \cdots, x_n)$, where $x_i \in I\!R$, $1 \le i \le n$, and for $\mathbf{x}, \boldsymbol{\omega} \in I\!R^n$ we denote their inner product by $\langle \mathbf{x}, \boldsymbol{\omega} \rangle = \mathbf{x}.\boldsymbol{\omega} = \sum_{i=1}^n x_i \omega_i$. Moreover, we set

$$dx = dx_1 \cdots dx_n, \quad f(\mathbf{x}) = f(x_1, ..., x_n),$$

and

$$\int_{I\!R^n} \cdots dx = \int_{I\!R} \cdots \int_{I\!R} dx_1 \cdots dx_n.$$

The n-dimensional Fourier transform ($n \ge 1$) of a function $f \in L^1(I\!R^n)$ is defined as follows

$$(2.1) \qquad \mathcal{F}(f)(\boldsymbol{\omega}) = \widehat{f}(\underline{\omega}) = \frac{1}{(2\pi)^{n/2}} \int_{I\!R^n} f(\mathbf{x}) e^{-i(\boldsymbol{\omega} \cdot \mathbf{x})} dx, \quad \mathbf{x}, \boldsymbol{\omega} \in I\!R^n,$$

so that

$$(2.2) \qquad f(\mathbf{x}) = \frac{1}{(2\pi)^{n/2}} \int_{I\!R^n} \widehat{f}(\boldsymbol{\omega}) e^{i(\boldsymbol{\omega} \cdot \mathbf{x})} d\omega,$$

whenever the integral exists.

With this definition of the Fourier transform, we have $F(1)(\boldsymbol{\omega}) = (2\pi)^{n/2} \delta(\boldsymbol{\omega})$ or

$$(2.3) \qquad \int_{I\!R^n} e^{-i(\boldsymbol{\omega} \cdot \mathbf{x})} dx = (2\pi)^n \, \delta(\boldsymbol{\omega}),$$

where δ stands for the delta function. Parseval's relation takes the symmetric form

$$(2.4) \qquad \left\langle \widehat{f}, \widehat{g} \right\rangle = \langle f, g \rangle.$$

Moreover, if we define the convolution of f and g as

$$(2.5) \qquad h(\mathbf{x}) = (f * g)(\mathbf{x}) = \frac{1}{(2\pi)^{n/2}} \int_{R^n} f(\mathbf{t}) g(\mathbf{x} - \mathbf{t}) dt,$$

then

$$(2.6) \qquad \widehat{h}(\boldsymbol{\omega}) = \widehat{f}(\boldsymbol{\omega}) \widehat{g}(\boldsymbol{\omega}).$$

We define the translation, dilation, parabolic dilation, modulation, and rotation operators as follows

    i) Translation: $(T_{\mathbf{b}} f)(\mathbf{x}) = f(\mathbf{x} - \mathbf{b})$, $\mathbf{b} \in I\!R^n$,

    ii) Dilation: $(D_a f)(\mathbf{x}) = a^{-n/2} f(a^{-1} \mathbf{x})$, $a > 0$,

    iii) Parabolic dilation: $(\mathbf{D}_a f)(x_1, x_2, \cdots, x_n) = f(\sqrt{a} x_1, a x_2, \cdots, a x_n)$, $a > 0$,

    iv) Modulation: $(E_{\mathbf{b}} f)(\mathbf{x}) = e^{i(\mathbf{b} \cdot \mathbf{x})} f(\mathbf{x})$, $\mathbf{b} \in I\!R^n$,

    v) Rotation: $(R_{\underline{\theta}} f)(\mathbf{x}) = f(r_{\underline{\theta}}(\mathbf{x}))$,

where $r_{\underline{\theta}}$ is an $n \times n$ rotation matrix which preserves the inner product. For $n = 2$, $r_{\underline{\theta}}$ takes the form

$$r_{\underline{\theta}} = r_{\theta} = \begin{pmatrix} \cos\theta & \sin\theta \\ -\sin\theta & \cos\theta \end{pmatrix}, \; 0 \le \theta \le 2\pi.$$

The operators $D_a R_{\underline{\theta}}$ and $\mathbf{D}_a R_{\underline{\theta}}$ are sometimes called directional dilation and directional parabolic dilation operators.

Parabolic dilations have been used in harmonic analysis for almost three decades, where they were first used by Fefferman [14], Seeger, Sogge, and Stein [20] to study boundedness of Fourier integral operators and more recently they were used by H.

Smith [**25**] to construct a Hardy space for Fourier integral operators. The parabolic surface $x_1^2 = c(x_2 + x_3 + \cdots + x_n)$ is invariant under parabolic dilations. Unlike the wavelet transform which is constructed by using uniform dilation on all coordinates in the spatial domain, the curvelet transform is based on parabolic dilations in the frequency domain.

3. The Wavelet, Directional Wavelet, and Ridgelet Transforms

First, let us recall that the definition of the wavelet transform and its inversion formula in one dimension. The wavelet transform of a function $f(t) \in L^2(I\!R)$ with respect to a mother wavelet ψ is given by

$$W_f(a, b) = <f, \psi_{a,b}> = \frac{1}{\sqrt{|a|}} \int_{I\!R} f(t)\overline{\psi}((t-b)/a)\, dt, \quad a, b \in I\!R, a \neq 0,$$

where $\psi_{a,b}(t) = \psi((t-b)/a)/\sqrt{|a|}$, ψ is the analyzing wavelet and $\overline{\psi}$ is the complex conjugate of ψ.

The wavelet ψ is assumed to satisfy the admissibility condition

$$C_\psi = 2\pi \int_{I\!R} \frac{|\hat{\psi}(\xi)|^2}{|\xi|} d\xi < \infty.$$

The synthesis formula or the inversion formula is given by

$$f(t) = \frac{1}{C_\psi} \int_{I\!R} \int_{I\!R} W_f(a, b)\psi_{a,b}(t)\, db \frac{da}{a^2}.$$

From now on we assume that the mother wavelet is real. Let $\mathbf{b} = \underline{b} \in I\!R^n, a \in I\!R, a > 0$, and r_θ be a rotation matrix. For a function $\psi(\mathbf{x})$ with $\mathbf{x} \in I\!R^n$, define

$$\psi_{\mathbf{b},a,\underline{\theta}}(\mathbf{x}) = T_{\mathbf{b}} D_a R_{\underline{\theta}} \psi(\mathbf{x}) = a^{-n/2} \psi\left(a^{-1} r_{\underline{\theta}}(\mathbf{x} - \mathbf{b})\right)$$

For $n = 2$, this takes the simple form

$$\psi_{\mathbf{b},a,\theta}(x) =$$
$$a^{-1}\psi\left(\frac{(x_1 - b_1)\cos\theta + (x_2 - b_2)\sin\theta}{a}, \frac{-(x_1 - b_1)\sin\theta + (x_2 - b_2)\cos\theta}{a}\right)$$

In view of the relations

$$\mathcal{F} D_a = D_{1/a}\mathcal{F}, \quad \mathcal{F} T_{\mathbf{b}} = E_{-\mathbf{b}}\mathcal{F}, \quad \mathcal{F} R_{\underline{\theta}} = R_{\underline{\theta}}\mathcal{F},$$

it is easy to see that

$$(3.1) \qquad \hat{\psi}_{\mathbf{b},a,\underline{\theta}}(\boldsymbol{\omega}) = a^{n/2} e^{-i(\boldsymbol{\omega} \cdot \mathbf{b})} \hat{\psi}(a r_{\underline{\theta}}(\boldsymbol{\omega})).$$

In the next definition, we give two generalizations of the wavelet transform to higher dimensions, one is straightforward and the other accounts of orientation.

DEFINITION 3.1. The n-dimensional wavelet transform of a function $f(\mathbf{x}) \in L^2(I\!R^n)$ with respect to an admissible wavelet $\psi(\mathbf{x})$ is defined as

$$(3.2) \qquad (W_\psi f)(\mathbf{b}, a) = \langle f, \psi_{\mathbf{b},a} \rangle = a^{-n/2} \int_{I\!R^n} f(\mathbf{x})\psi(a^{-1}(\mathbf{x} - \mathbf{b}))d\mathbf{x}.$$

More generally, we define the n-dimensional directional wavelet transform as

$$(W_\psi f)(\mathbf{b}, a, \underline{\theta}) = \langle f, \psi_{\mathbf{b},a,\underline{\theta}} \rangle$$
$$(3.3) \qquad\qquad = a^{-n/2} \int_{I\!R^n} f(\mathbf{x})\psi\left(a^{-1} r_{\underline{\theta}}(\mathbf{x} - \mathbf{b})\right) d\mathbf{x}.$$

A wavelet ψ is said to be admissible if

$$(3.4) \qquad \int_{\mathbb{R}^n} \frac{\left|\widehat{\psi}(\mathbf{k})\right|^2}{|\mathbf{k}|^n} d\mathbf{k} < \infty$$

By Parseval's relation for the Fourier transform (2.4) and (3.1) we have

$$(3.5) \qquad (W_\psi f)(\mathbf{b}, a, \underline{\theta}) = a^{n/2} \int_{\mathbb{R}^n} \widehat{f}(\boldsymbol{\omega}) e^{i(\boldsymbol{\omega} \cdot \mathbf{b})} \overline{\widehat{\psi}(a r_{\underline{\theta}}(\boldsymbol{\omega}))} d\boldsymbol{\omega}.$$

Let $\mathbb{R}^+$ be the set of all positive real numbers, $S^{n-1} = \{\mathbf{x} \in \mathbb{R}^n, \|\mathbf{x}\| = 1\}$ be the unit sphere in $\mathbb{R}^n$ and $d\sigma$ be the unit surface area on S^{n-1}. Let $\Gamma = \{(a, \mathbf{b}, \underline{\theta}) : a \in \mathbb{R}^+, \mathbf{b} \in \mathbb{R}^n, \underline{\theta} \in S^{n-1}\}$ and $d\mu = da\, d\mathbf{b}\, d\sigma / a^{n+1}$.

THEOREM 3.2. *The directional wavelet transform is a continuous transformation from $L^2(\mathbb{R}^n)$ into $L^2(\Gamma, d\mu)$ that preserves the inner product up to a constant. In fact,*

$$(3.6) \qquad \langle W_\psi f, W_\psi g \rangle_{L^2(\Gamma), d\mu} = \langle W_\psi f, W_\psi g \rangle_{d\mu} = c_\psi \langle f, g \rangle_{L^2(\mathbb{R}^n)},$$

where

$$c_\psi = (2\pi)^n \widetilde{C}_\psi,$$

and

$$\widetilde{C}_\psi = \int_{\mathbb{R}^n} \frac{\left|\widehat{\psi}(\mathbf{k})\right|^2}{|\mathbf{k}|^n} d\mathbf{k}$$

Before we introduce the ridgelet transform, let us recall the definition of the Radon transform.

Let $\boldsymbol{\xi} = (\xi_1, \cdots, \xi_n)$ be a unit vector in $\mathbb{R}^n$ or equivalently a point on the unit sphere S^{n-1} in $\mathbb{R}^n$, that is $\sum_{i=1}^n \xi_i^2 = 1$. The Radon transform of a function $f(\mathbf{x})$ is defined as

$$R[f](p, \boldsymbol{\xi}) = f^\dagger(p, \boldsymbol{\xi}) = \int_{\mathbf{x} \cdot \boldsymbol{\xi} = p} f(\mathbf{x})\, d\mathbf{x} = \int_{\mathbb{R}^n} f(\mathbf{x}) \delta(p - \boldsymbol{\xi} . \mathbf{x}) d\mathbf{x},$$

where $p = \mathbf{x} \cdot \boldsymbol{\xi} = x_1 \xi_1 + ... + x_n \xi_n$ is a hyperplane.

DEFINITION 3.3. Let $\psi \in L^2(\mathbb{R}^n)$ be a one-dimensional admissible wavelet, and $\gamma = (b, a, \mathbf{u})$ be the ridgelet parameter which runs through the set

$$\Gamma = \{(b, a, \mathbf{u}) : b, a \in \mathbb{R}, a > 0, \mathbf{u} \in S^{n-1}\}.$$

The ridgelet $\psi_\gamma(\mathbf{x})$ is defined as

$$\psi_\gamma(\mathbf{x}) = a^{-1/2} \psi\left(\frac{\mathbf{u} \cdot \mathbf{x} - b}{a}\right), \quad \mathbf{x} \in \mathbb{R}^n,$$

and the ridgelet transform of $f \in L^2(\mathbb{R}^n), n \geq 2$, is defined as

$$(R_\psi f)(\gamma) = \int_{\mathbb{R}^n} f(\mathbf{x}) \psi_\gamma(\mathbf{x}) d\mathbf{x} = \langle f, \psi_\gamma \rangle_{L^2(\mathbb{R}^n)}$$

In $\mathbb{R}^2$ the ridgelet transform takes the form $R_f(a, b, \theta) = \langle f, \psi_{a,b,\theta} \rangle$, with

$$\psi_{a,b,\theta}(\mathbf{x}) = a^{-1/2} \psi\left(\frac{u_\theta \cdot x - b}{a}\right), \quad a > 0, \mathbf{x} \in \mathbb{R}^2, b \in \mathbb{R}, \theta \in [0, 2\pi),$$

u_θ is a unit vector pointing in direction θ, and ψ is an admissible wavelet.

THEOREM 3.4. *Let* $d\mu = da\,db\,d\sigma/a^{n+1}$, *and* ψ *be a real-valued, one-dimensional admissible wavelet. Then, the ridgelet transform is a continuous transformation from* $L^2(\mathbb{R}^n)$ *into* $L^2(\Gamma, d\mu)$ *that preserves the inner product up to a constant. In fact,*

$$\langle R_\psi f, R_\psi g \rangle_{(L^2(\Gamma), d\mu)} = \langle R_\psi f, R_\psi g \rangle_{d\mu} = c_\psi \langle f, g \rangle,$$

where $c_\psi = 2A(2\pi)^n$ *and*

$$A = \int_0^\infty \frac{\left|\hat{\psi}(\zeta)\right|^2}{\zeta^n}\,d\zeta\,.$$

The proof of the theorem depends on the close relationship between the ridgelet transform and the Radon transforms, which can be outlined as follows. But first, let us denote the Radon transform of f by $(Rf)(p, \mathbf{u}) = f\dagger(p, \mathbf{u})$ and for a fixed $\mathbf{u}$, we use the notation

$$(R_\mathbf{u} f)(p) = f_\mathbf{u}^\dagger(p).$$

Then

$$\begin{aligned}
(R_\psi f)(\gamma) &= \int_{\mathbb{R}^n} f(\mathbf{x})\psi_\gamma(\mathbf{x})d\mathbf{x} = a^{-1/2}\int_{\mathbb{R}^n} f(\mathbf{x})\psi\left(\frac{\mathbf{u}\cdot\mathbf{x} - b}{a}\right)d\mathbf{x} \\
&= a^{-1/2}\int_{\mathbb{R}} \psi\left(\frac{p-b}{a}\right)dp\int_{\mathbf{x}\cdot\mathbf{u}=p} f(\mathbf{x})d\mathbf{x}
\end{aligned}$$

$$(3.7)\qquad = a^{-1/2}\int_{\mathbb{R}} \psi\left(\frac{p-b}{a}\right)(R_\mathbf{u} f)(p)dp = a^{-1/2}\int_{\mathbb{R}} \psi\left(\frac{p-b}{a}\right) f_\mathbf{u}^\dagger(p)dp.$$

Thus, the Ridgelet transform may be viewed as an application of a one-dimensional wavelet transform to a slice of the Radon transform.

4. The Curvelet Transform

The curvelet transform differs in some essential ways from both the wavelet and ridgelet transforms. Unlike the wavelet transform which uses translation, dilation,and rotation in the spatial domain, the curvelet transform employs rotation and translation in the spatial domain and some kind of parabolic polar dilation in the frequency domain. Because there are few versions of the curvelet transforms, we shall discuss in some details properties of the version that will be used in the sequel.

Curvelet transforms are transforms based on affine parabolic scaling. The following version of the curvelet transform is based on a transform introduced by Hart Smith [25]. Let $P_{a,\theta}$ be a parabolic directional dilation of $\mathbb{R}^2$ given in matrix form by

$$P_{a,\theta} = D_{1/a} R_{-\theta}, \quad a > 0$$

where

$$D_{1/a} = \begin{pmatrix} 1/\sqrt{a} & 0 \\ 0 & 1/a \end{pmatrix},$$

and R_θ is a rotation matrix. A curvelet transform can be defined as

$$\tilde{\Gamma} = \langle f, \phi_{a,b,\theta} \rangle, \quad a > 0, \mathbf{b} \in \mathbb{R}^2, \theta \in [-\pi, \pi),$$

where

$$\phi_{a,b,\theta} = \phi\left(P_{a,\theta}(\mathbf{x} - \mathbf{b})\right) \mathrm{Det}^{1/2}(P_{a,\theta}),$$

for some suitable function ϕ. This transform may be viewed as a directional wavelet transform associated with an affine parabolic dilation.

Another version of the curvelet transform, which was introduced by E. Candes and D. Donoho [5, 6, 7], is preferred because it yields a tight frame when discretized appropriately.

Consider a pair of positive and real-valued functions $W(r)$ and $V(t)$, which we will call the 'radial window' and 'angular window', respectively. Let W be supported on $r \in (1/2, 2)$ and V be supported on $t \in [-1, 1]$ and assume that they satisfy the *admissibility* conditions conditions:

$$(4.1) \qquad \int_0^\infty W^2(r)\frac{dr}{r} = 1$$

$$(4.2) \qquad \int_{-1}^1 V^2(t)dt = 1$$

These windows are used in the frequency domain to construct the basic element of the curvelet transform kernel. A family of complex -valued waveforms with three parameters: scale $a > 0$, location $\mathbf{b} \in I\!\!R^2$, and orientation $\theta \in [0, 2\pi)$ (or $(-\pi, \pi)$ according to convenience below). At scale a, the family is generated by translation and rotation of a basic element $\gamma_{a,0,0}$:

$$\gamma_{a,\mathbf{b},\theta}(\mathbf{x}) = \gamma_{a,0,0}(R_\theta(\mathbf{x} - \mathbf{b})),$$

The basic element $\gamma_{a,0,0}$ is defined by its Fourier transform

$$(4.3) \qquad \hat{\gamma}_{a,0,0}(r, \omega) = W(ar)V(\omega/\sqrt{a})a^{3/4}, \ 0 < a < a_0,$$

where, $a_0 < \pi^2$, is a fixed number - the coarsest scale of the transform. Notice that the Fourier transform of the basic curvelet element is defined by means of a parabolic dilation on its polar coordinates.

DEFINITION 4.1. The Continuous Curvelet Transform of a function f is defined as

$$(4.4) \qquad \Gamma_f(a, \mathbf{b}, \theta) = \langle f, \gamma_{a,\mathbf{b},\theta} \rangle, \ a < a_0, \ \mathbf{b} \in I\!\!R^2, \ \theta \in [-\pi, \pi).$$

Let $B_\sigma^2(I\!\!R^2)$, $\sigma > 0$, be a space of bandlimited functions

$$B_\sigma^2(I\!\!R^n) = \left\{ f \in L^2(I\!\!R^n) : \hat{f}(\xi) = 0, \text{ for } |\xi| \geq \sigma \right\}$$

and define $H_\sigma^2(I\!\!R^n)$ as

$$H_\sigma^2(I\!\!R^n) = \left\{ f \in L^2(I\!\!R^n) : \hat{f}(\xi) = 0, \text{ for } |\xi| \leq \sigma \right\},$$

and for $0 < \sigma_1 < \sigma_2$, define

$$B_{\sigma_1,\sigma_2}^2(I\!\!R^n) = \left\{ f \in L^2(I\!\!R^n) : \hat{f}(\xi) = 0, \text{ for } \sigma_1 \leq |\xi| \leq \sigma_2 \right\}.$$

The spaces B_σ^2 and H_σ^2 are also called spaces of low and high frequency signals. Any function $f \in L^2(I\!\!R^n)$ can be written as $f = f_L + f_H$, where $f_L \in B_\sigma^2$ and $f_H \in H_\sigma^2$.

In the next theorem we give some properties of the curvelet transform.

THEOREM 4.2. *Let $f, g \in H^2_\sigma(\mathbb{R}^2)$ where $\sigma = 2/a_0$. Let V and W obey the admissibility conditions (4.1) and (4.2). Then*

$$(4.5) \qquad \langle \Gamma_f(a, \mathbf{b}, \theta), \Gamma_g(a, \mathbf{b}, \theta) \rangle_{d\mu} = \langle f, g \rangle,$$

and

$$(4.6) \qquad f(\mathbf{x}) = \int \Gamma(a, \mathbf{b}, \theta) \gamma_{a,\mathbf{b},\theta}(\mathbf{x}) d\mu$$

Moreover, the following Parseval formula holds

$$(4.7) \qquad \|f\|^2_{L^2} = \int |\Gamma(a, b, \theta)|^2 \, d\mu,$$

where, $d\mu$ denotes the measure $d\mu = \frac{da}{a^3} \frac{d\mathbf{b}}{(2\pi)^2} \, d\theta$.

PROOF. From (4.4) we have

$$\langle \Gamma_f(a, \mathbf{b}, \theta), \Gamma_g(a, \mathbf{b}, \theta) \rangle_{d\mu} = \int d\mu \left(\int_{\mathbb{R}^2} f(\mathbf{x}) \overline{\gamma}_{a,\mathbf{b},\theta}(\mathbf{x}) d\mathbf{x} \right) \left(\int_{\mathbb{R}^2} \overline{g}(\mathbf{y}) \gamma_{a,\mathbf{b},\theta}(\mathbf{y}) d\mathbf{y} \right)$$

$$(4.8) \qquad = \int_{\mathbb{R}^2} \int_{\mathbb{R}^2} f(\mathbf{x}) \overline{g}(\mathbf{y}) d\mathbf{x} d\mathbf{y} \int \overline{\gamma}_{a,\mathbf{b},\theta}(\mathbf{x}) \gamma_{a,\mathbf{b},\theta}(\mathbf{y}) d\mu,$$

but

$$\int \overline{\gamma}_{a,\mathbf{b},\theta}(\mathbf{x}) \gamma_{a,\mathbf{b},\theta}(\mathbf{y}) d\mu = \int \overline{\gamma}_{a,0,\theta}(\mathbf{x} - \mathbf{b}) \gamma_{a,0,\theta}(\mathbf{y} - \mathbf{b}) d\mu$$

$$= \int \frac{da}{a^3} \, d\theta \int \overline{\gamma}_{a,0,\theta}(\mathbf{x} - \mathbf{y} + \mathbf{b}) \gamma_{a,0,\theta}(\mathbf{b}) \frac{d\mathbf{b}}{(2\pi)^2}$$

$$(4.9) \qquad = \frac{1}{2\pi} \int \frac{da}{a^3} \, d\theta \, (\gamma_{a,0,\theta} * \tilde{\gamma}_{a,0,\theta}) (\mathbf{y} - \mathbf{x}),$$

where $\tilde{\gamma}(z) = \overline{\gamma}(-z)$. Since $\hat{\tilde{\gamma}} = \overline{\hat{\gamma}}$ and

$$(A * B)(\mathbf{x}) = \frac{1}{2\pi} \int_{\mathbb{R}^2} \hat{A}(\xi) \hat{B}(\xi) e^{i(\mathbf{x}.\xi)} d\xi,$$

we have by combining (4.8) and (4.9)

$$\langle \Gamma_f(a, \mathbf{b}, \theta), \Gamma_g(a, \mathbf{b}, \theta) \rangle_{d\mu} = \frac{1}{(2\pi)^2} \int_{\mathbb{R}^2} \int_{\mathbb{R}^2} f(\mathbf{x}) \overline{g}(\mathbf{y}) d\mathbf{x} d\mathbf{y} \int \frac{da}{a^3} \, d\theta$$

$$\times \int_{\mathbb{R}^2} |\hat{\gamma}_{a,0,\theta}(\xi)|^2 e^{i\xi.(\mathbf{y}-\mathbf{x})} d\xi$$

$$(4.10) \qquad = \int \frac{da}{a^3} \, d\theta \int_{\mathbb{R}^2} |\hat{\gamma}_{a,0,\theta}(\xi)|^2 \hat{f}(\xi) \overline{\hat{g}}(\xi) d\xi,$$

$$(4.11) \qquad = \int_{\mathbb{R}^2} \left(\int \frac{da}{a^3} \, d\theta |\hat{\gamma}_{a,0,\theta}(\xi)|^2 \right) \hat{f}(\xi) \overline{\hat{g}}(\xi) d\xi,$$

which will complete the proof of (4.5) if we verify that

$$(4.12) \qquad 1 = \int |\hat{\gamma}_{a,0,\theta}(\xi)|^2 \, d\theta \frac{da}{a^3}, \text{ for all } \xi \text{ with } |\xi| > \sigma.$$

We will see that this follows from the admissibility conditions (4.1) and (4.2).

First, let us observe that

$$\hat{\gamma}_{a,0,\theta}(\xi) = \frac{1}{2\pi} \int_{\mathbb{R}^2} \gamma_{a,0,\theta}(\mathbf{x}) e^{-i(\mathbf{x}.\xi)} d\mathbf{x} = \frac{1}{2\pi} \int_{\mathbb{R}^2} \gamma_{a,0,0}(R_\theta \mathbf{x}) e^{-i(\mathbf{x}.\xi)} d\mathbf{x},$$

where

$$R_\theta = \begin{pmatrix} \cos\theta & \sin\theta \\ -\sin\theta & \cos\theta \end{pmatrix}, \quad -\pi \le \theta \le \pi.$$

If we set $\xi = re_\omega = r(\cos\omega, \sin\omega)$, and $\mathbf{y} = R_\theta \mathbf{x}$, we have

$$\hat{\gamma}_{a,0,\theta}(\xi) = \frac{1}{2\pi} \int_{I\!\!R^2} e^{-ir(e_\omega \cdot R_{-\theta}\mathbf{y})} \gamma_{a,0,0}(\mathbf{y}) d\mathbf{y}$$

$$= \frac{1}{2\pi} \int_{I\!\!R^2} e^{-ir(R_\theta e_\omega \cdot \mathbf{y})} \gamma_{a,0,0}(\mathbf{y}) d\mathbf{y}$$

(4.13)
$$= \hat{\gamma}_{a,0,0}(R_\theta \xi) = \hat{\gamma}_{a,0,0}(re_{\omega-\theta}).$$

Hence, from the condition (4.3), it follows that

$$\hat{\gamma}_{a,0,\theta}(\xi) = W(ar)V((w-\theta)/\sqrt{a})a^{3/4}.$$

Thus,

(4.14)
$$\int_0^{a_0} \int_{-\pi}^{\pi} |\hat{\gamma}_{a,0,\theta}(\xi)|^2 \, d\theta \frac{da}{a^3} = \int_0^{a_0} \int_{-\pi}^{\pi} W^2(ar)V^2\left(\frac{w-\theta}{\sqrt{a}}\right)a^{3/2} d\theta \frac{da}{a^3};$$

and using the admissibility condition of V, we have $\int_{-\pi}^{\pi} V^2((w-\theta)/\sqrt{a}) \, dw = a^{1/2}$, since $1 < \pi/\sqrt{a_0}$. Hence, (4.12) reduces to

$$1 = \int_0^{a_0} W^2(ar)\frac{da}{a} \quad \text{for all } r = |\xi| \ with \ = \xi \in \text{supp } \hat{f}.$$

Now, for $r = |\xi|$ with $\xi \in \text{supp } \hat{f}$, we have $r > 2/a_0$, and by a simple rescaling of variables and using the admissibility condition of W, we obtain

$$\int_0^{a_0} W(ar)^2 \frac{da}{a} = \int_0^{a_0 r} W(a)^2 \frac{da}{a} = \int_{1/2}^2 W(a)^2 \frac{da}{a} = 1,$$

which yields (4.12) and hence (4.5).

The proof of (4.7) follows from (4.5) by taking $f = g$.

To prove (4.6), we use (4.5) to obtain

$$\int_{I\!\!R^2} f(\mathbf{x})\overline{g}(\mathbf{x})d\mathbf{x} = \int \Gamma_f(a,\mathbf{b},\theta)\overline{\Gamma}_g(a,\mathbf{b},\theta)(\mathbf{x})d\mu$$

$$= \int \Gamma_f(a,\mathbf{b},\theta)d\mu \left(\int_{I\!\!R^2} \gamma_{a,\mathbf{b},\theta}(\mathbf{x})\overline{g}(\mathbf{x}) \, d\mathbf{x} \right)$$

(4.15)
$$= \int_{I\!\!R^2} \left(\int \Gamma_f(a,\mathbf{b},\theta)\gamma_{a,\mathbf{b},\theta}(\mathbf{x})d\mu \right) \overline{g}(\mathbf{x}) \, d\mathbf{x}.$$

But since (4.15) is valid for all $g \in H_\sigma^2$, (4.6) holds.

$\square$

We can extend the curvelet transform to low frequency signals and hence to finite energy signals. Let f be an $L^2(I\!\!R^2)$ function, and let

(4.16)
$$P_H(f)(\mathbf{x}) = \int_{a<a_0} \Gamma_f(a,\mathbf{b},\theta)\gamma_{a,\mathbf{b},\theta}(\mathbf{x})d\mu.$$

The integral on the right-hand side is understood to mean

$$\int_{a<a_0} \Gamma_f(a,\mathbf{b},\theta)\gamma_{a,\mathbf{b},\theta}(\mathbf{x})d\mu = \int_0^{a_0} \int_{-\pi}^{\pi} \int_{I\!\!R^2} \Gamma_f(a,\mathbf{b},\theta)\gamma_{a,\mathbf{b},\theta}(\mathbf{x}) \frac{da}{a^3} \, d\theta \, \frac{d\mathbf{b}}{(2\pi)^2}.$$

It is easy to see that $\hat{\gamma}_{a,\mathbf{b},\theta}(\xi) = e^{-i(\mathbf{b}\cdot\xi)}\hat{\gamma}_{a,0,\theta}(\xi)$. Hence,

$$P_H(f)(\mathbf{x}) = \int \left(\int_{\mathbb{R}^2} f(\mathbf{y})\overline{\gamma}_{a,0,\theta}(\mathbf{y}-\mathbf{b})d\mathbf{y} \right) \gamma_{a,0,\theta}(\mathbf{x}-\mathbf{b})d\mu$$

$$(4.17) \qquad = \int \frac{da}{a^3}\, d\theta \left(\int_{\mathbb{R}^2} f(\mathbf{y})d\mathbf{y} \right) \left(\int_{\mathbb{R}^2} \overline{\gamma}_{a,0,\theta}(\mathbf{y}-\mathbf{b})\gamma_{a,0,\theta}(\mathbf{x}-\mathbf{b})\frac{d\mathbf{b}}{(2\pi)^2} \right)$$

which, in view of (4.9) and (4.11), leads to

$$P_H(f)(\mathbf{x}) = \frac{1}{(2\pi)^2} \int \frac{da}{a^3}\, d\theta \left(\int_{\mathbb{R}^2} f(\mathbf{y})d\mathbf{y} \right) \int_{\mathbb{R}^2} |\hat{\gamma}_{a,0,\theta}(\xi)|^2 e^{i\xi\cdot(\mathbf{x}-\mathbf{y})}d\xi$$

$$(4.18) \qquad = \frac{1}{(2\pi)} \int_{\mathbb{R}^2} \left(\int \frac{da}{a^3}\, d\theta |\hat{\gamma}_{a,0,\theta}(\xi)|^2 \right) \hat{f}(\xi)e^{i\xi\cdot\mathbf{x}}d\xi.$$

Therefore,

$$(4.19) \qquad \widehat{P_H(f)}(\xi) = \hat{f}(\xi) \int \frac{da}{a^3}\, d\theta |\hat{\gamma}_{a,0,\theta}(\xi)|^2,$$

which, in view of (4.14), gives

$$(4.20) \qquad \widehat{P_H(f)}(\xi) = \hat{f}(\xi) \int_0^{a_0} W^2(ar)\frac{da}{a}, \qquad \text{where } r = |\xi|.$$

Let us set

$$\widehat{P_H(f)}(\xi) = \hat{f}(\xi) \left[\hat{\Psi}(\xi) \right]^2,$$

where $\hat{\Psi}(\xi) = \left(\int_0^{a_0|\xi|} |W(a)|^2\, da/a \right)^{1/2}$. The definition makes sense because the integrand is real and nonnegative.

It is easy to see that $0 \leq \hat{\Psi}^2(\xi) \leq 1$, $\hat{\Psi}^2(\xi) = 1$ if $r = |\xi| > \sigma = 2/a_0$, and $\hat{\Psi}^2(\xi) = 0$ if $r < \sigma/4$.

Let $\hat{\Phi}(\xi) = \sqrt{1 - \hat{\Psi}^2(\xi)}$. Then $0 \leq \hat{\Phi}^2(\xi) \leq 1$, and

$$(4.21) \qquad \hat{\Phi}^2(\xi) = \begin{cases} 0 & \text{if } |\xi| > \sigma, \\ 1 & \text{if } |\xi| < \sigma/4\,. \end{cases}$$

Define

$$(4.22) \qquad \widehat{P_L(f)} = \hat{f} - \widehat{P_H(f)} = \hat{\Phi}^2\hat{f}.$$

The next theorem extends the curvelet transform to functions in $L^2(\mathbb{R}^2)$.

THEOREM 4.3. *Let Φ be defined as above and set*

$$\Gamma_f^\Phi(\mathbf{b}) = \int_{\mathbb{R}^2} f(\mathbf{x})\overline{\Phi}_{\mathbf{b}}(\mathbf{x})d\mathbf{x},$$

where $\Phi_{\mathbf{b}}(\mathbf{x}) = \Phi(\mathbf{x} - \mathbf{b})$. Then for $f, g \in L^2(\mathbb{R}^2)$

(1)

$$(4.23) \qquad \langle \Gamma_f^\Phi(\mathbf{b}), \Gamma_g^\Phi(\mathbf{b}) \rangle = \langle \hat{\Phi}\hat{f}, \hat{\Phi}\hat{g} \rangle,$$

where the integral on the left-hand side is taken with respect to the measure $d\mathbf{b}/(2\pi)^2$.

(2)

$$(4.24) \qquad \frac{1}{(2\pi)^2} \int_{I\!\!R^2} \left| \Gamma_f^\Phi(\mathbf{b}) \right|^2 d\mathbf{b} = \| \Phi * f \|_2^2$$

(3)

$$(4.25) \qquad P_L(f)(\mathbf{x}) = \frac{1}{(2\pi)^2} \int_{I\!\!R^2} \Gamma_f^\Phi(\mathbf{b}) \Phi_\mathbf{b}(\mathbf{x}) d\mathbf{b}.$$

(4)

$$(4.26) \qquad \langle \Gamma_f(a, \mathbf{b}, \theta), \Gamma_g(a, \mathbf{b}, \theta) \rangle = \langle \hat{\Psi}\hat{f}, \hat{\Psi}\hat{g} \rangle.$$

(5)

$$(4.27) \qquad \int_{\{a<a_0\}} |\Gamma_f a, \mathbf{b}, \theta|^2 \, d\mu = \| \Psi * f \|_2^2.$$

(6)

$$(4.28) \qquad f(\mathbf{x}) = \frac{1}{(2\pi)^2} \int \langle f, \Phi_{a_o,\mathbf{b}}, \rangle \Phi_{a_o,\mathbf{b}}(\mathbf{x}) d\mathbf{b} + \int_{a<a_0} \langle f, \gamma_{a,\mathbf{b},\theta} \rangle \gamma_{a,\mathbf{b},\theta}(\mathbf{x}) d\mu$$

(7)

$$(4.29) \qquad \|f\|_2^2 = \frac{1}{(2\pi)^2} \int |\langle f, \Phi_{a_o,\mathbf{b}}, \rangle|^2 d\mathbf{b} + \int_{a<a_0} |\langle f, \gamma_{a,\mathbf{b},\theta} \rangle|^2 \, d\mu$$

PROOF. For (4.23) we start with

$$
\begin{aligned}
\langle \Gamma_f^\Phi(\mathbf{b}), \Gamma_g^\Phi(\mathbf{b}) \rangle &= \frac{1}{(2\pi)^2} \int_{I\!\!R^2} \left(\int_{I\!\!R^2} f(\mathbf{x}) \overline{\Phi}(\mathbf{x}-\mathbf{b}) d\mathbf{x} \right) \left(\int_{I\!\!R^2} \overline{g}(\mathbf{y}) \Phi(\mathbf{y}-\mathbf{b}) d\mathbf{y} \right) d\mathbf{b} \\
&= \frac{1}{(2\pi)^2} \int_{I\!\!R^2} \int_{I\!\!R^2} f(\mathbf{x}) \overline{g}(\mathbf{y}) d\mathbf{x} d\mathbf{y} \left(\int_{I\!\!R^2} \overline{\Phi}(\mathbf{x}-\mathbf{b}) \Phi(\mathbf{y}-\mathbf{b}) d\mathbf{b} \right) \\
&= \frac{1}{(2\pi)^2} \int_{I\!\!R^2} \int_{I\!\!R^2} f(\mathbf{x}) \overline{g}(\mathbf{y}) d\mathbf{x} d\mathbf{y} \left(\int_{I\!\!R^2} \left| \hat{\Phi}(\xi) \right|^2 e^{i\xi \cdot (\mathbf{y}-\mathbf{x})} d\xi \right) \\
(4.30) \qquad &= \int_{I\!\!R^2} \left| \hat{\Phi}(\xi) \right|^2 \hat{f}(\xi) \overline{\hat{g}}(\xi) d\xi = \langle \hat{\Phi}\hat{f}, \hat{\Phi}\hat{g} \rangle.
\end{aligned}
$$

Now setting $f = g$, we obtain

$$\frac{1}{(2\pi)^2} \int_{I\!\!R^2} |\Gamma_f^\Phi(\mathbf{b})|^2 d\mathbf{b} = \left\| \hat{\Phi}\hat{f} \right\|_2^2,$$

which by Parseval's equality is equivalent to (4.24).

To prove (4.25), let us denote the right hand-side by H. Thus,

$$
\begin{aligned}
H(\mathbf{x}) &= \frac{1}{(2\pi)^2} \int_{I\!\!R^2} \left(\int_{I\!\!R^2} f(\mathbf{y}) \overline{\Phi}(\mathbf{y}-\mathbf{b}) d\mathbf{y} \right) \Phi(\mathbf{x}-\mathbf{b}) d\mathbf{b} \\
(4.31) \qquad &= \frac{1}{(2\pi)^2} \int_{I\!\!R^2} f(\mathbf{y}) d\mathbf{y} \left(\int_{I\!\!R^2} \overline{\Phi}(\mathbf{y}-\mathbf{b}) \Phi(\mathbf{x}-\mathbf{b}) d\mathbf{b} \right),
\end{aligned}
$$

and as in the proof of (4.9) and in view of the fact that $\hat{\Phi}$ is positive and real, we obtain

$$H(\mathbf{x}) = \frac{1}{(2\pi)^2} \int_{I\!\!R^2} f(\mathbf{y}) d\mathbf{y} \left(\int_{I\!\!R^2} \hat{\Phi}^2(\xi) e^{i\xi \cdot (\mathbf{x}-\mathbf{y})} d\xi \right)$$

$$(4.32) \qquad = \frac{1}{(2\pi)} \int_{I\!\!R^2} \hat{f}(\xi) \hat{\Phi}^2(\xi) e^{i\xi \cdot \mathbf{x}} d\xi,$$

which implies that $H(\mathbf{x}) = P_L(f)(\mathbf{x})$.

As in (4.11), we obtain

$$\langle \Gamma_f(a, \mathbf{b}, \theta), \Gamma_g(a, \mathbf{b}, \theta) \rangle = \int_{I\!\!R^2} \left| \hat{\Psi}(\xi) \right|^2 \hat{f}(\xi) \overline{\hat{g}}(\xi) d\xi,$$

which is equivalent to (4.26). Now Equation (4.27) follows from (4.26) and (4.28) is equivalent to (4.22).

Finally, (4.24) and (4.28) imply that

$$\frac{1}{(2\pi)^2} \int_{I\!\!R^2} \left| \Gamma_f^{\Phi}(\mathbf{b}) \right|^2 d\mathbf{b} = \int_{I\!\!R^2} \left| \hat{\Phi}(\xi) \right|^2 \left| \hat{f}(\xi) \right|^2 d\xi,$$

and

$$\int_{a<a_0} \left| \Gamma_f(a, \mathbf{b}, \theta) \right|^2 d\mu = \int_{I\!\!R^2} \left| \hat{\Psi}(\xi) \right|^2 \left| \hat{f}(\xi) \right|^2 d\xi,$$

and by adding these two equations we obtain (4.29). $\square$

We can think of the 'full continuous curvelet transform' as consisting of curvelets at fine scales and isotropic father wavelets at course scales. For many purposes, it is only the behavior of the fine-scale elements that matters. Yet, for functions in the space $B_{\sigma/4, \sigma}^2(I\!\!R^2)$, the last theorem takes the following nicer form.

COROLLARY 4.4. Let $f, g \in B_{\sigma/4, \sigma}^2(I\!\!R^2)$. Then

$$(4.33) \qquad \langle \Gamma_f^{\Phi}(\mathbf{b}), \Gamma_g^{\Phi}(\mathbf{b}) \rangle = \int_{|\xi| \le \sigma/4} \hat{f}(\xi) \overline{\hat{g}}(\xi) d\xi,$$

and

$$(4.34) \qquad \langle \Gamma_f(a, \mathbf{b}, \theta), \Gamma_g(a, \mathbf{b}, \theta) \rangle = \int_{\sigma \le |\xi|} \hat{f}(\xi) \overline{\hat{g}}(\xi) d\xi,$$

and hence

$$(4.35) \qquad \langle \Gamma_f^{\Phi}(\mathbf{b}), \Gamma_g^{\Phi}(\mathbf{b}) \rangle + \langle \Gamma_f(a, \mathbf{b}, \theta), \Gamma_g(a, \mathbf{b}, \theta) \rangle = \langle f, g \rangle.$$

In the following sections we discuss applications of the wavelet, ridgelet, and curvelet transforms in medical imaging, in particular, in texture classification of normal tissues in medical images obtained by Computed Tomography (CT) scans. For digital implementation, discrete versions of these transforms are needed. We followed [27] and [28] for the wavele transform, [11] for the ridgelet and [9] for the curvelet transform.

5. Texture Classification of Normal Tissues in Computed Tomography

This research is part of an ongoing project aimed at developing an automated imaging system for classification of tissues in medical images obtained by Computed Tomography (CT) scans [**18, 21, 22, 23, 24, 10, 29, 30**]. Classification of human organs in CT scans using shape or gray level information is particularly challenging due to the changing shape of organs in a stack of slices in 3D medical images and the gray level intensity overlap in soft tissues. However, healthy organs are expected to have a consistent texture within tissues across multiple slices. The research presented here focuses on using texture analysis for the classification of tissues.

Texture is a commonly used feature in the analysis and interpretation of images. One way of characterizing texture is by calculating a set of local statistical properties of the pixel gray level intensity, measuring variations in a surface such as smoothness, coarseness and regularity. Traditionally, texture features have been calculated using a variety of image processing techniques including run-length statistics, co-occurrence matrices, statistical moments, fractal dimensions, and Gabor filtering (see for example [**16, 29, 30**]. In this section we present multi-resolutions texture classification algorithms based on texture features extracted from the wavelet, ridglet and curvelet transforms of the CT images.

By decomposing the image into a series of high-pass and low-pass bands, the wavelet transform extracts directional details that capture horizontal, vertical and diagonal variations. However, these three linear directions are limiting and might not capture enough directional information in noisy images, such as medical CT scans which do not have strong horizontal, vertical, and diagonal directional elements.

Ridgelets, like wavelets, provide multi-resolution texture information; however they capture structural information of an image based on multiple radial directions in the frequency domain. Semler and Dettori in [**22, 24**] showed that the multi-directional capabilities of the ridgelet transform provide better texture discrimination than its wavelet counterpart.

One of the limitations of this approach is the fact that ridgelets are most effective in detecting linear radial structures, which are also not dominant in medical images. In the next section we will show that these limitations of the wavelet and ridgelet transforms can be rectified by using the curvelet transform. Preliminary tests show that using texture features extracted from the curvelet transform of the CT images significantly improve the texture classification algorithm. For more details see also [**22, 23**], and [**10**].

5.1. Methodology and data set. The texture classification algorithm consists of three main steps: segmentation of regions of interest from CT scans, extraction of the most discriminative wavelet, ridglet, and curvelet-based texture features, and creation of a classifier that generates classification rules that automatically identify the various tissues. The general algorithm is illustrated in Figure 1.

The algorithms were tested on 3D data extracted from two normal chest and abdomen CT studies from Northwestern Memorial Hospital. The 3D DICOM image data set consists of consecutive 2D coronal slices, each being 512 by 512 pixels in size and having 16-bit gray level resolution. The two studies were conducted on two different patients and resulted in a total of 344 images. Two sample images from the original data set are shown in Fig. 2 and Fig. 3. It should be noted that the printed image is not capable of rendering the full spectrum of gray levels contained in the original 16-bit image; however, all calculations were done using the original data contained in the image without loss of gray-level information.

We are interested in identifying normal tissues in the following organs: backbone, heart, liver, kidney, and spleen. Consequently, a necessary first step was to segment the five organs from the initial data set. Ideally this step would be carried out with the assistance of expert radiologists. However, lacking such resource, an Active Contour Model ("Snake") algorithm was used instead; see [**17, 31**]. A snake is a function that approximates a boundary given a user-defined set of initial points and parameters determining the boundary's smoothness. The initial curve evolves to match the nearest internal boundary, depending heavily on gradient intensity measures. The boundary curve can then be used to separate the object from the background. Using this segmentation algorithm, the following slices were generated: 140 Backbone, 52 Heart, 58 Liver, 54 Kidney, and 40 Spleen.

Wavelets, ridgelets, and curvelets are extremely sensitive to contrast in gray-level intensity. Therefore, it was necessary to eliminate all background pixels to avoid mistaking the edge between the artificial background and the organ as a texture feature. Each slice was therefore further cropped, and only square subimages fully contained in the interior of the segmented area were generated. Since the background pixels are clearly identifiable by an artificially set gray-level, the cropping of the images was accomplished using a simple algorithm that would generate as many subimages of the desired size with an overlap of at most 50%. Figure 5 shows an example of a CT scan, the corresponding segmented heart image and four cropped interior slices of the heart.

The following additional constraints were dictated by the specific versions of the discrete Ridgelet and Curvelet transforms used: images should be square and have sides of 2^n pixels for Curvelets (and Wavelets); they should have sides containing a prime number of pixels for Ridgelets. Given the size of the original segmented images and in order to be able to have two comparable sets of data for Curvelets and Ridgelets, 32 by 32 pixels images were chosen for Curvelets and 31 by 31 pixels for Ridgelets. The final data set contained 2090 slices of "pure" single-organ tissue (363 Backbone, 446 Heart, 506 Liver, 411 Kidney, 364 Spleen). Fig. 4 shows a sample image for each of the organ tissues.

Table 1 presents a summary of the data set: how many distinct images per organ per patients were segmented and the total number of 32 by 32 "pure tissue" subimages generated for each organ.

TABLE 1. Data set summary

Segmented Organs	Patient 1	Patient 2	Cropped Images
Backbone	68	72	363
Heart	27	25	446
Liver	29	29	506
Kidney (L and R)	27	27	411
Spleen	20	20	364
Total	171	173	2090

The classification step was carried out using a decision tree classifier based on the Classification and Regression Tree (C&RT). The decision trees used in the experiments listed below were generate using the Decision Tree software package included in SPSS.

A decision tree predicts the class of an object from values of predictor variables. Two techniques can be used to grow a C&RT tree: the training&testing method and the cross-validation method. In the training&testing approach, the original data set is randomly divided into two groups: a training set and a testing set. These sets are designated by choosing a random percentage split. The training set (in this case designated to be 67%) of images is used to train the decision tree and derive the classification rules. The testing set (the remaining 33%) is used to estimate the accuracy of the tree when classifying new images.

The second technique used to grow a C&RT tree is the N-fold cross-validation method. The training data is randomly divided into N mutually exclusive folds, each with similar class proportions and approximately the same size. Training and testing of the tree are performed N times. For each iteration of training the tree, a subset of each fold is reserved to test the classifier accuracy. Using the default of 10-fold cross-validation proved to be slightly better than using the training and testing technique.

The classifier identifies the most relevant texture descriptors for each specific organ, and based on those selected descriptors, generates a set of decision rules. These sets of rules are then used for the classification of each region and a misclassification matrix is generated for each region (organ).

A misclassification matrix is a table that lists each organ and its true positives, true negatives, false positives and false negatives. For example let us consider the misclassification matrix for the heart. The number of true positives is the number of organs that are correctly classified as heart. The number of false positives is the number of organs that are incorrectly classified as heart. The number of true negatives is the number of non heart organs that are correctly classified as not heart. The number of false negatives is the number of hearts that are incorrectly classified as non heart organs.

To evaluate the performance of each classifier, specificity, sensitivity, precision, and accuracy rates were calculated from each of the misclassification matrices; see Table 2.

Both the misclassification metrix and the four performance measures are automatically calculated by the software package used to generate the decision tree, according to the formulae in Table 2.

TABLE 2. Performance measures

Measure	Definition
Sensitivity	True Positives / Total Positives
Specificity	True Negatives / Total Negatives
Precision	True Positives / (True Positives + False Positives)
Accuracy	(True Positives + True Negatives) / Total Sample

Sensitivity measures the accuracy among negative instances, and is calculated by dividing the number of true positives by the number of total positives which is the total number of that specific organ slices. Specificity measures the accuracy among positive instances, and is calculated by dividing the number of true negatives by the number of total negatives which is the number of all other organ slices. Precision measures how consistently the results can be reproduced and is calculated by dividing the number of true positives by the sum of the numbers of true positive and false positive. Accuracy reflects the overall correctness of the classifier, and is calculated by adding the true positives and negatives together and dividing by the entire number of organ slices.

In the medical domain, the most important performance measures are both specificity and sensitivity. Optimally, both specificity and sensitivity should be high, however, in practice, these two measures have a negative correlation. Since accuracy reflects both the sensitivity and specificity in relation to each other, this descriptor was examined to determine the overall correctness of the classifier.

5.2. Wavelet and Ridgelet-based Texture Classification. In this section we discuss an application of three wavelet transforms (Harr, Daubechies, and Coiflet) and ridgelet transforms to the classification of normal tissues in CT images.

The first step of texture features extraction consists of calculating the discretized wavelet transform of the 2D image. Discrete versions of the wavelet transform are widely available, see for example [**27**].

Once the medical images are preprocessed as described in the previous section, the Haar, Daubechies, and Coiflet wavelet filters are applied to each 32 by 32 image at two levels of resolution. For details of the filters used, see Fig. 6 and Fig. 7. At each resolution level three detail coefficient matrices are calculated, resulting in three matrices representing the vertical, horizontal and diagonal structures of the image (see Fig 8).

For the purpose of defining texture features the matrix containing the averaged coefficients (C region) is ignored and first order and second order statistics are calculated on each detail coefficient matrix (W region). One set of descriptors was derived from the histogram of the various W regions, and the other from a set of co-occurrence matrices calculated on the W regions.

A histogram was initially calculated from each wavelet detail matrix, measuring the frequency distribution of wavelet detail coefficients. Mean and Standard Deviation texture descriptors were then extracted from the histograms.

Co-occurrence matrices were also calculated for each detail sub-band and level of resolution. Following standard procedures for computational efficiency when using co-occurrence matrices, the values in the detail sub-bands were first reduced to 16 gray-level bins.

A co-occurrence matrix applied to the wavelet details captures the spatial dependence of wavelet detail coefficients, depending on different directions and distances specified. A co-occurrence matrix C is an n by n matrix, where n is the number of gray levels within the image. The matrix $C(i, j)$ counts the number of pixel pairs at set distances and directions having the intensities i and j. Four co-occurrence matrices were calculated for each wavelet detail matrix at each resolution level for four directions: 0, 45, 90, and 135 degrees at a distance of one. The following nine Haralick texture descriptors were then extracted from each co-occurrence matrix: energy, entropy, contrast, homogeneity, sum-mean, variance, maximum probability, inverse difference moment, and cluster tendency Haralick. See Fig. 9 for details on such texture features. Reduction of the feature vector is imperative to avoid over-fitting and to obtain a manageable feature space. Extensive testing showed the optimal feature vector for the wavelet-based algorithm to be: mean, standard deviation, energy, entropy, contrast, homogeneity, sum-mean, variance, cluster tendency, inverse difference moment, and maximum probability, calculated from wavelet details for two levels of resolution and averaged over both co-occurrence directions and wavelet details, resulting in 22 descriptor.

Table (3) summarizes the performance of the three wavelets for each organ tissue.

TABLE 3. Percentage of performance measures for wavelet-based texture features

Organ	Descriptor	Sensitivity	Specificity	Precision	Accuracy
Backbone	Haar	82.6	96.1	82.6	93.7
	Daubechies	83.5	97.3	91.6	93.6
	Coiflet	85.9	96.8	90.7	93.1
Heart	Haar	59.0	92.1	67.0	85.0
	Daubechies	49.1	91.8	57.4	84.0
	Coiflet	67.1	89.4	58.9	85.3
Kidney	Haar	77.7	91.4	69.9	88.6
	Daubechies	63.2	92.0	55.7	88.0
	Coiflet	87.3	94.4	82.6	92.8
Liver	Haar	87.3	94.4	82.6	92.8
	Daubechies	85.4	82.9	64.9	83.6
	Coiflet	81.6	87.4	64.3	85.8
Spleen	Haar	65.5	94.3	69.7	89.5
	Daubechies	40.2	96.2	64.3	88.2
	Coiflet	35.2	97.6	70.8	88.6
Average	Haar	74.4	93.7	74.4	89.9
	Daubechies	64.2	92.0	66.8	87.5
	Coiflet	66.8	92.7	69.5	88.2

In these preliminary tests, accuracy performance rates are in the 83.6%-93.7% range, showing that Wavelet-based feature vectors have the potential to do well in classifying normal tissues in CT scans. It is interesting to note that texture descriptors based on the Haar wavelet consistently outperform descriptors based on the Daubechies and Coiflet wavelets for most organs and performance measures. The only exception is the backbone, for which the Daubechies and Coiflet wavelets produce slightly better results. This is possibly due to the fact that since the backbone has high contrasting elements, the larger windows of Daubechies and Coiflets are able to capture more discriminating features. A complete analysis of all classification performance measures can be found in [21].

While wavelet-based texture descriptors showed promising results in applications to texture classification in CT scans, their limitation lies in the fact that the details coefficients matrices are most sensitive to vertical, horizontal, and diagonal features. The texture of organ tissues in medical images does not normally present strong vertical, horizontal or diagonal features.

The ridgelet transform corresponds to applying a multi-resolution wavelet to the Radon transform of the image. The Radon transform is able to provide multi-directional information in the frequency domain. Thus, the ridgelet transform can capture multiple directions, in addition to the horizontal, vertical and diagonal offered by the wavelet. The ridgelet transform gives rotation invariant structural information on multiple directions and scales and is therefore better suited for texture classification of medical images.

Similar to the wavelet approach, a discretized ridgelet transform, and Radon transform are needed. We use the Finite Ridgelet Transform (FRIT) recently proposed in [**11**]. Here we summarize the approach introduced in [**11**], and refer the reader to the original article for more details of the algorithm. The implementation of the FRIT used in this article is a direct adaptation of the Matlab code associated with [**11**] (see `http://www.ifp.uiuc.edu/ minhdo/software/`). The FRIT consists of first calculating a Finite Radon Transform (FRAT) and then applying a one-dimensional discrete wavelet transform. The Finite Radon Transform is defined as the summation of image pixels over a set of "lines" which are the adaptation to a finite matrix of the lines used in the continuous Radom transform; see Section 4 above.

Let p be a prime number, then $Z_p = 0, 1, ..., p-1$ is a finite field with modulo p operations. The FRAT of a real function f on the finite grid Z_p^2 (the image) is the summation over the lines $L_{k,l} = \{(i, j) : j = ki + l \pmod p), i \in Z_p\}, 0 \le k \le p$, and $L_{p,l} = \{(l, j) : j \in Z_p\}$. A reordering of such lines eliminates the wrap around effect and is key factor in proving an modified version of the Discrete projection-slice theorem stating that a Discrete Fourier Transform of the FRAT projection is equivalent to a 2-D Discrete Fourier Transform along those discrete lines. Consequently, the FRIT can be computed in three steps:

> **Step 1:** Calculate the 2D Fast Fourier Transform of the image;
>
> **Step 2:** Apply a 1D Inverse Fourier Transform on each radial direction of the Radon Projection
>
> **Step 3:** Apply a 1D Discrete Haar Wavelet Transform to each radial direction. This steps generates three details matrices for each direction. As it was the case for the 2D wavelet transform, the matrix containing the averaged coefficients is not used in the calculation of the texture features. This step is repeated to the desired number of levels of Wavelet resolutions.

In our case, $p = 31$, generating 32 radial directions on which to calculate the 1D Discrete Wavelet Transform. Since the Haar wavelet perfomed the best for texture classification, it was chosen for the last step of the Ridgelet discretization. Fig. 10 illustrates the steps of the Discrete Ridgelet Transform.

FRIT-based texture features are extracted on the 1D wavelet coefficients from the last step of the FRIT. Given the size of our images, there are 32 radial directions, therefore there are 32 Wavelet transforms each generating 1 matrix of interest, half the size of the original image, at each resolution level.

The limited literature on ridgelet-based descriptors suggested the use of a combination of mean, standard deviation, and energy signals (see for example [**19**]). Each of these statistics was calculated for each radial direction and resolution level of the wavelet details matrix. One of the goals and challenges of this research was to identify the optimal combination of descriptors. The following four feature vectors were investigated:

EE: Energy and Entropy signatures averaged over radial directions

EEMS: Energy, Entropy, Mean, and Standard Deviation signatures averaged over radial directions

Eng: Energy signatures

Ent: Entropy signatures

Each of these feature vectors was computed for three levels of resolution yielding: 6 descriptors, 12 descriptors, and 96 descriptors respectively. Table 4 shows performance rates (expressed in percentage) for all ridgelet-based texture feature vectors. The accuracy of Entropy signatures are in the 91%-97% range, and clearly outperform all other feature vectors.

TABLE 4. Performance measures for ridgelet-based texture feature

Organ	Descriptor	Sensitivity	Specificity	Precision	Accuracy
	Ent	90.9	98.7	93.5	97.3
Backbone	Eng	88.4	96.0	82.3	94.7
	EE	73.3	96.6	82.1	92.6
	EEMS	87.6	94.0	75.4	92.9
	Ent	77.8	97.9	90.8	93.6
Heart	Eng	56.3	95.2	76.1	86.9
	EE	44.6	87.4	49.1	78.3
	EEMS	33.2	92.6	55.0	80.0
	Ent	94.2	92.3	79.4	92.7
Kidney	Eng	93.2	88.6	72.2	89.7
	EE	72.2	86.5	56.3	83.4
	EEMS	91.7	82.7	62.7	84.9
	Ent	72.5	97.7	88.5	92.7
Liver	Eng	68.6	94.9	76.8	89.7
	EE	89.5	84.9	65.3	86.0
	EEMS	47.9	92.1	56.3	86.5
	Ent	83.8	93.4	72.9	91.7
Spleen	Eng	62.5	92.9	65.0	87.6
	EE	25.5	97.2	65.5	84.6
	EEMS	48.0	92.1	56.3	84.4
	Ent	83.8	96.0	85.0	93.6
Average	Eng	73.8	93.5	74.4	89.7
	EE	61.0	90.5	63.7	85.0
	EEMS	63.8	91.0	63.6	85.7

When considering all ridgelet-based texture vectors, accuracy rates are in the 78.3%-97.3% range. However, the Entropy feature vector alone appears to outperformed all other descriptors, including all wavelet-based descriptors, with accuracy rates in the 91.7%-97.3%. A further analysis of the performance of these feature vectors using different combinations of resolution levels also showed that the best results are obtained when considering only the first two level of resolutions instead of all three for the discrete wavelet transform in the FRIT. Accuracy rates for the FRIT Entropy descriptor with two resolutions levels are in the 94%-98% range as shown in Table 6.

5.3. Curvelet-based Texture Classification. A recent extension of the wavelet and ridgelet transforms is the curvelet transform. The curvelet transform is proven to be particularly effective at detecting image activity along curves instead of radial directions; see Section 5 above. Curvelets also capture structural information along multiple scales, locations, and orientations. Instead of capturing structural information along radial lines, the curvelet transform captures this structural activity along radial 'wedges' in the frequency domain (see Fig. 11). Given the complex texture structure of CT images, it was reasonable to predict that using curvelet-based texture descriptor would improve our texture classification algorithm. This conjecture was proved true for the right combination of curvelet-based texture features.

In this application, the Discrete Curvelet Transform introduced by Candes and Donoho in [**11**] was used. This is a discretization of their continuous curvelet transform [**5**], which uses a "wrapping" algorithm; see also Section 5 above. This approach uses a spatial grid to translate curvelets at each scale and angle, assuming a regular rectangular grid defining 'Cartesian' curvelets. This method applies a 2D Fast Fourier transform to the image for each scale and angle. The result is then wrapped around the origin. The 2D Inverse Fast Fourier transform is then applied, resulting in discrete curvelet coefficients. The transform consists of four steps:

> **Step 1:** Application of a 2-dimensional fast Fourier transform of the image
> **Step 2:** Formation of a product of scale and angle windows
> **Step 3:** Wrapping of the resulting product image around the origin
> **Step 4:** Application of a 2D Inverse Fast Fourier transform.

The implementation of the Discrete Curvelet Transform used in this article is a direct adaptation of the Matlab code associated with [**9**]. The code is made available from the authors at `http://www.curvelet.org`. As mentioned above, the Finite Ridgelet transform restricted our images to a prime number side. Given the overall size of the segmented images, the cropped images were 31 by 31. The Discrete Curvelet Transform constrains the images to be of size 2^n by 2^n. To make the comparison between the two classification algorithms meaningfull, the data set for the Curvelet-based algorithm consists of 32 by 32 images (each containing one additional row and column compared to the Ridgelet data set).

The discrete curvelet transform can be calculated to various resolutions or scales and angles. Two parameters are involved in the digital implementation of the curvelet transform: number of resolutions and number of angles at the coarsest level. The parameters were bound by the following two constraints: the maximum number of resolutions depends on the original image size, and the number of angles at the second coarsest level must be at least 8 and a multiple of four. Since our images were 32 by 32 pixels, the maximum possible resolution extraction was three levels of resolution.

The following angles were explored: 12, 16, and 20. Using these parameters resulted in 14, 18, and 22 matrices containing structural information at three levels of resolution. Tests showed that the best results were achieved when using 16 angles. While this version of Discrete Curvelet Tansform is completely reversible and isometric, the original authors do not make claims of it being invariant under rotation and this feature has not been thoroughly tested in our case.

Very limited literature ([**23, 21, 15**]) exists on which curvelet-based texture features are optimal for classification. So, motivated by the encouraging results with

Ridgelets we chose to extract the following texture features from the images transformed using the Discrete Curvelet Transform described above: energy, entropy, mean, and standard deviation. Each of these texture features were computed for three levels of resolution and each radial 'wedge'(16 angles) yielding 18, 36, and 72 descriptors respectively. The following four feature vectors were investigated:

Eng: Energy signatures
Ent: Entropy signatures
EE: Energy and Entropy signatures
EEMS: Energy, Entropy, Mean, and Standard Deviation signatures.

Our analysis indicates that the best feature vector is the one containing all four curvelet-based texture descriptors (EEMS). It also indicates that using such feature vector significantly improves the wavelet-based and ridgelet-based classification algorithm.

Table 5 compares performance rates for all curvelet-based texture feature vectors.The EEMS signatures are in the 87.1%-98.9% range, and clearly outperform all other feature vectors.

TABLE 5. Performance comparison of curvelet-based texture feature vectors

Organ	Descriptor	Sensitivity	Specificity	Precision	Accuracy
	EEMS	99.4	98.8	95.3	98.9
Backbone	EE	94.2	99.1	96.3	98.2
	Ent	92.7	98.7	94.4	97.5
	EEMS	89.7	99.0	95.5	87.1
Heart	EE	75.55	95.9	82.0	91.9
	Ent	74.0	96.2	82.8	91.8
	EEMS	96.0	98.1	93.5	97.6
Kidney	EE	92.3	92.1	77.1	92.2
	Ent	89.1	92.5	77.4	91.7
	EEMS	95.9	98.5	94.3	98.0
Liver	EE	81.1	97.8	90.6	94.4
	Ent	89.2	97.3	88.6	94.2
	EEMS	91.8	98.9	94.9	97.6
Spleen	EE	78.4	95.6	79.7	92.5
	Ent	80.0	94.9	77.7	94.2
	EEMS	94.6	98.7	94.7	97.9
Average	EE	84.3	96.1	85.1	93.8
	Ent	83.6	95.9	84.2	93.5

TABLE 6. Performance comparison of the best wavelet (Haar), ridgelet (Entropy), and curvelet (EEMS) texture feature

Organ	Descriptor	Sensitivity	Specificity	Precision	Accuracy
Backbone	Wavelet	82.6	96.1	82.6	93.7
	Ridgelet	91.5	99.3	96.8	98.0
	Curvelet	99.4	98.8	95.3	98.9
Heart	Wavelet	59.0	92.1	67.0	85.0
	Ridgelet	82.5	97.5	88.5	94.6
	Curvelet	89.7	99.0	95.5	97.1
Kidney	Wavelet	77.7	91.4	69.9	88.6
	Ridgelet	95.4	93.3	82.0	93.8
	Curvelet	96.0	98.1	93.5	97.6
Liver	Wavelet	87.3	94.4	82.6	92.8
	Ridgelet	86.9	95.9	84.4	94.0
	Curvelet	95.9	98.5	94.3	98.0
Spleen	Wavelet	65.5	94.3	69.7	89.5
	Ridgelet	76.9	97.6	88.0	93.8
	Curvelet	91.8	98.9	94.9	97.6
Average	Wavelet	74.4	93.7	74.4	89.9
	Ridgelet	86.6	96.7	88.0	94.8
	Curvelet	94.6	98.7	94.7	97.9

Finally, Table 6 compares performance rates for the best texture feature vector for curvelets (EEMS), ridgelets (Entropy)[24], and Wavelets (Haar-based vector)[23]. It is clear that Curvelet-based descriptors perform better than both the wavelet and ridgelet, with accuracy rates approximately 5% - 12% and 1% - 4% higher respectively.

In conclusion, our preliminary tests show that the algorithms presented in this article are able to classify normal tissues in CT scans with high accuracy rates. The Curvelet based algorithm appears to have a slight advantage over Wavelet and Ridgelet based techniques.

This was expected since the curvelet transform is able to capture multi-directional features in wedges, as opposed to lines or points as it is the case for the Ridgelet or Wavelet transform. The multi-directional features in Curvelets prove to be very effective in the texture classification of the more organic textures in medical images. Although, theoretically curvelets are more complex, there are no significant differences in the time performance. The robustness of the alghoritm will be further tested on images from additional patients and obtained on different CT scans. Therefore, the authors intend to explore the use of 3D curvelet descriptors as well as using curvelet-based descriptors to classify anomalies in the various tissues found in CT scans.

References

[1] J. P. Antoine, R. Murenzi, P. Vandergheynst, and S. T. Ali, *Two-Dimensional Wavelets and Their Relatives,* Cambridge University Press, Cambridge, united kingdom (2004).

[2] E. Bolker, The finite Radon transform,in Integral Geometry (Contemp. Math. Vol 63), S. Helgason, R. Bryant, V. Guillemin, and R. Wells Jr., Eds., (1987), pp. 27-50.

[3] E. Candés, Ridgelets and representations of mutilated Sobolev functions, SIAM J. Math. Anal., Vol.33 (2001), pp. 347-368.

[4] E. Candés, Harmonic analysis of Neural Networks, Appl. Comput. Harm. Analy., Vol. 6 (1999), pp. 197-218.

[5] E. Candes and D. Donoho, Continuous curvelet transform I: Resolution of the wavefront set, *Appl. Comput. Harmon. Anal.,* 19 (2005), 162-197.

[6] E. Candes and D. Donoho, Continuous curvelet transform II: Discretization and frames, *Appl. Comput. Harmon. Anal.,* 19 (2005), 198-222.

[7] E. Candes and D. Donoho, *Curvelets, Multi-resolution Representation, and Scaling Laws, Wavelet Applications in Signal and Image Processing VIII,* SPIE 4119, (2000).

[8] E. Candés and D. Donoho, Ridgelets: A key to higher-dimensional intermittency, *Phil. Trans. Roy. Soc.Lond. A*, Vol. 357 (1999), pp. 2495 - 2509.

[9] E. Candes, L. Demanet, D. Donoho, and L. Ying, *Fast Discrete Curvelet Transforms,*SIAM Multiscale Modeling and Simulation. (2005).

[10] L. Dettori and L. Semler, *A comparison of wavelet, ridgelet, and curvelet-based texture classification algorithms in computed tomography,* Computers in Biology and Medicine, Vol. 37, (2007) Issue 4, Pages 486-498.

[11] M. Do, and M. Vetterli, "The Finite Ridgelet Transform for Image Representation." IEEE Transactions on Image Processing, Vol. 12 (2003), p. 16 - 28.

[12] D. Donoho, Emergin Applications of Geometric Multiscale Analysis, Inter. Congress of Math., Vol. 1, (2002), pp. 209-233.

[13] D. Donoho and M. Duncan, Digital curvelet transform: Strategy, implementation and experiments, in Proc. Aerosense 2000, Wavelet Applications, Vol II, SPIE 2002, Vol 4056, pp. 12-29.

[14] C. Fefferman, A note on spherical summation multipliers, *Israel J. Math,* 15 (1973), 44-52

[15] R. Haralick, L. Shapiro, *Computer and Robot Vision,* Addison-Wesley Publishing Co., (1992).

[16] B. Kara and N. Watsuji, *Using Wavelets for Texture Classification,* IJCI Proceedings of International Conference on Signal Processing, ISN 1304-2386, (2003).

[17] Kass, Witkin, and Terzopolous, *Snakes Active Contours Models,* International Journal of Computer Vision,(1998).

[18] A. Kurani, D. Xu, J. Furst, and D. Raicu, *Co-occurrence matrices for volumetric data,* The 7th IASTED International Conference on Computer Graphics and Imaging - CGIM 2004, Kauai, Hawaii, USA, (2004).

[19] H. LeBorgne, and N. O'Connor. Ridgelet-based Signature for Natural Image Classification, CORIA'05 - 2nd Conference on Information Retrieval and its Applications, Grenoble, France,March 2005.

[20] A. Seeger, C. Sogge, and E. Stein, Regularity properties of Fourier integral operators, ann. Math., 134 (1993), 231-251.

[21] L. Semler, L. Dettori, J. Furst, Wavelet-Based Texture Classification of Tissues in Computed Tomography, *Proceedings of the 18th IEEE International Symposium on Computer-Based Medical Systems,* (CBMS'05), Dublin, Ireland, June 2005, 265-270. (2005)

[22] L. Semler, L. Dettori, B. Kerr, Ridgelet-based texture classification of tissues in computed tomography, submitted to SPIE Symposium on Medical Imaging, San Diego, CA, June 2005.

[23] L. Semler and L. Dettori, *Curvelet-Based Texture Classification of Tissues in Computed Tomography,* Proceedings of International Conference on Computer Vision Theory and Applications, (2006).

[24] L. Semler and L. Dettori, *A Comparison of Wavelet-Based and Ridgelet-Based Texture Classification of Tissues in Computed Tomography,* Proceedings of International Conference on Computer Vision Theory and Applications, (2006).

[25] H. Smith, A Hardy space for Fourier integral operators, *J. Geom. Anal.,* 8 (1998), 629-653.

[26] J. Starck, D. Donoho, E. Candes, Astronomical Image Representation by the Curvelet Transform. *Astronomy & Astrophysics,* 398,(1999), 785-800.

[27] E. Stollnitz, T. Derose, and D. Salesin, Wavelets for Computer Graphics: Theory and Applications, Kaufman (1996).

[28] G. Van de Wouwer, P. Scheunders, and D. Van Dyck, Statistical Texture Characterization from Discrete Wavelet Representations, IEEE Transactions on Image Processing, Vol. 8, (1999), 592-598.

[29] D. Xu, J. Lee, D. Raicu,J. Furst, and D. Channin, *Texture Classification of Normal Tissues in Computed Tomography,* The 2005 Annual Meeting of the Society for Computer Applications in Radiology, (2005).

[30] D. H. Xu, A. Kurani, J. D. Furst, and D. S. Raicu, *Run-length encoding for volumetric texture,* The 4th IASTED International Conference on Visualization, Imaging, and Image Processing - VIIP 2004, Marbella, Spain, (2004).

[31] Xu, Yezzi, and Prince, *On the Relationship between Parametric and geometric Active Contours,* Asilomar Conference on Signals, Systems and Computers,(2000).

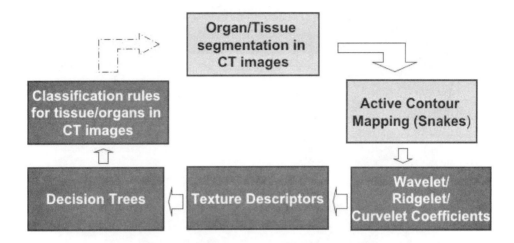

FIGURE 1. Overview of the texture classification algorithm

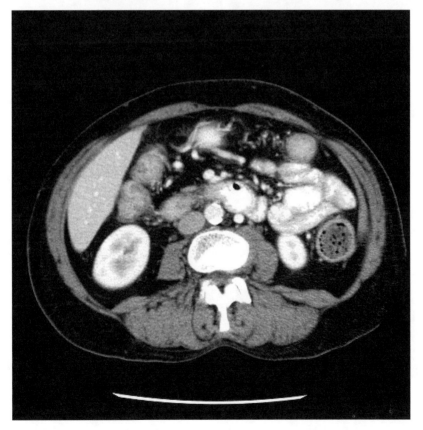

FIGURE 2. Example of abdominal DICOM Image

SCHOOL OF COMPUTER SCIENCE, TELECOMMUNICATIONS, INFORMATION SYSTEMS, DEPAUL UNIVERSITY CHICAGO, IL 60614
  E-mail address: ldettori@cs.depaul.edu

DEPARTMENT OF MATHEMATICAL SCIENCES, DEPAUL UNIVERSITY CHICAGO, IL 60614
  E-mail address: azayed@condor.depaul.edu

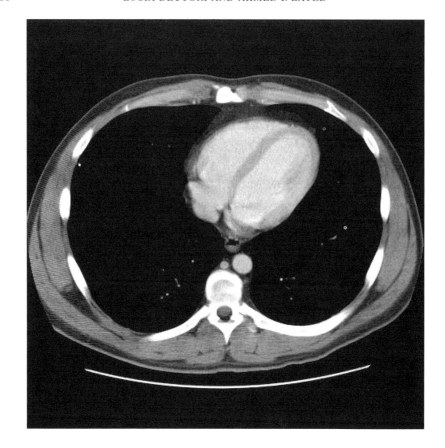

FIGURE 3. Example of chest DICOM Image

FIGURE 4. Examples of cropped tissues images: Backbone, Heart, Liver, Kidney, and Spleen

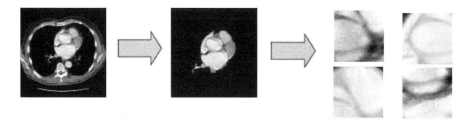

FIGURE 5. Segmentation and cropping: from 1 slice to multiple cropped 32 by 32 slices of Heart

$$\begin{bmatrix} h_2 & h_3 & h_4 & h_5 & 0 & 0 & 0 & 0 & h_0 & h_1 \\ g_2 & g_3 & g_4 & g_5 & 0 & 0 & 0 & 0 & g_0 & g_1 \\ h_0 & h_1 & h_2 & h_3 & h_4 & h_5 & 0 & 0 & 0 & 0 \\ g_0 & g_1 & g_2 & g_3 & g_4 & g_5 & 0 & 0 & 0 & 0 \\ 0 & 0 & h_0 & h_1 & h_2 & h_3 & h_4 & h_5 & 0 & 0 \\ 0 & 0 & g_0 & g_1 & g_2 & g_3 & g_4 & g_5 & 0 & 0 \\ 0 & 0 & 0 & 0 & h_0 & h_1 & h_2 & h_3 & h_4 & h_5 \\ 0 & 0 & 0 & 0 & g_0 & g_1 & g_2 & g_3 & g_4 & g_5 \\ h_4 & h_5 & 0 & 0 & 0 & 0 & h_0 & h_1 & h_2 & h_3 \\ g_4 & g_5 & 0 & 0 & 0 & 0 & g_0 & g_1 & g_2 & g_3 \end{bmatrix}$$

FIGURE 6. General form of the filter used for the Discrete Wavelet Transform

Scaling	h_0	h_1	h_2	h_3	h_4	h_5
Haar	$\frac{1}{\sqrt{2}}$	$\frac{1}{\sqrt{2}}$				
Daubechies	$\frac{1+\sqrt{3}}{4*\sqrt{2}}$	$\frac{3+\sqrt{3}}{4*\sqrt{2}}$	$\frac{3-\sqrt{3}}{4*\sqrt{2}}$	$\frac{1-\sqrt{3}}{4*\sqrt{2}}$		
Coiflet	$\frac{1-\sqrt{7}}{16*\sqrt{2}}$	$\frac{5+\sqrt{7}}{16*\sqrt{2}}$	$\frac{14+2\sqrt{7}}{16*\sqrt{2}}$	$\frac{14-2\sqrt{7}}{16*\sqrt{2}}$	$\frac{1-\sqrt{7}}{16*\sqrt{2}}$	$\frac{-3+\sqrt{7}}{16*\sqrt{2}}$
Wavelet	g_0	g_1	g_2	g_3	g_4	g_5
Haar	$\frac{1}{\sqrt{2}}$	$-\frac{1}{\sqrt{2}}$				
Daubechies	$\frac{1-\sqrt{3}}{4*\sqrt{2}}$	$\frac{-3-\sqrt{3}}{4*\sqrt{2}}$	$\frac{3+\sqrt{3}}{4*\sqrt{2}}$	$\frac{-1-\sqrt{3}}{4*\sqrt{2}}$		
Coiflet	$\frac{-3+\sqrt{7}}{16*\sqrt{2}}$	$\frac{-1+\sqrt{7}}{16*\sqrt{2}}$	$\frac{14-2\sqrt{7}}{16*\sqrt{2}}$	$\frac{-14-2\sqrt{7}}{16*\sqrt{2}}$	$\frac{5+\sqrt{7}}{16*\sqrt{2}}$	$\frac{-1+\sqrt{7}}{16*\sqrt{2}}$

FIGURE 7. Coefficients for the Haar, Daubeschies and Coiflet Discrete Wavelet Transforms

FIGURE 8. Multiresolution Wavelet decomposition of an image

Feature	Formula	What is measured?
Entropy	$-\sum_i^M \sum_j^N P[i,j] \log P[i,j]$	Measures the randomness of a gray-level distribution. The Entropy is expected to be high if the gray levels are distributed randomly through out the image.
Energy(Angular Second Moment)	$\sum_i^M \sum_j^N P^2[i,j]$	Measures the number of repeated pairs. The Energy is expected to be high if the occurrence of repeated pixel pairs is high.
Contrast	$\sum_i^M \sum_j^N (i-j)^2 P[i,j]$	Measures the local contrast of an image. The Contrast is expected to be low if the gray levels of each pixel pair are similar.
Homogeneity	$\sum_i^M \sum_j^N \dfrac{P[i,j]}{1+|i-j|}$	Measures the local homogeneity of a pixel pair. The Homogeneity is expected to be large if the gray levels of each pixel pair are similar
SumMean (Mean)	$\dfrac{1}{2}\sum_i^M \sum_j^N (iP[i,j]+jP[i,j])$	Provides the mean of the gray levels in the image. The SumMean is expected to be large if the sum of the gray levels of the image is high.
Variance	$\dfrac{1}{2}\sum_i^M \sum_j^N ((i-\mu)^2 P[i,j]+(j-\mu)^2 P[i,j])$	Variance tells us how spread out the distribution of gray-levels is. The Variance is expected to be large if the gray levels of the image are spread out greatly.
Maximum Probability (MP)	$\underset{i,j}{Max}\, P[i,j]$	Results in the pixel pair that is most predominant in the image. The MP is expected to be high if the occurrence of the most predominant pixel pair is high.
Inverse Difference Moment (IDM)	$\sum_i^M \sum_j^N \dfrac{P[i,j]}{|i-j|^k}\quad i\neq j$	Inverse Difference Moment tells us about the smoothness of the image, like homogeneity. The IDM is expected to be high if the gray levels of the pixel pairs are similar.
Cluster Tendency	$\sum_i^M \sum_j^N (i+j-2\mu)^k P[i,j]$	Measures the grouping of pixels that have similar gray-level values.

FIGURE 9. Haralic descriptors used for Wavelet texture features

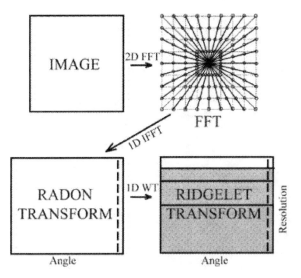

FIGURE 10. Haralic descriptors used for Wavelet texture features

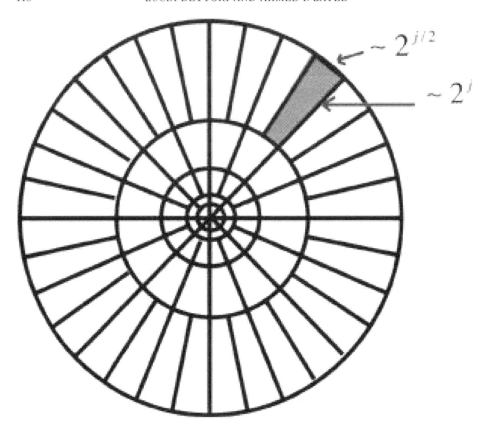

FIGURE 11. Haralic descriptors used for Wavelet texture features

Contemporary Mathematics
Volume **451**, 2008

Coxeter Groups, Wavelets, Multiresolution and Sampling

Mihaela Dobrescu and Gestur Ólafsson

ABSTRACT. In this short note we discuss the interplay between finite Coxeter groups and construction of wavelet sets, generalized multiresolution analysis and sampling.

Introduction

Finite reflection groups are examples of finite Coxeter groups. Those groups show up in a natural way in geometry as symmetry groups of geometric objects, fractal geometry in the classification of simple Lie algebras, in representation theory, theory of special functions and other places in analysis, cf. [**10, 15, 18, 17, 20, 21**] and the reference therein for few examples. On the other hand, the only example we know of, where those groups have shown up in the connection with wavelet theory and multiresolution analysis is in the book by P. Massopust [**15**] and related construction by him and his coworkers.

Let $\mathcal{D} \subseteq GL(n, \mathbb{R})$ and $\mathcal{T} \subseteq \mathbb{R}^n$ countable sets. Recall, that a $(\mathcal{D}, \mathcal{T})$-*wavelet* is a square integrable function ψ with the property that the set

$$(0.1) \qquad \{|\det d|^{\frac{1}{2}} \varphi(dx + t) \mid d \in \mathcal{D}, t \in \mathcal{T}\}$$

forms an orthonormal basis for $L^2(\mathbb{R}^n)$. A special class of wavelets are the ones corresponding to *wavelet sets*. Those are functions ψ such that $\mathcal{F}(\psi) = \chi_\Omega$ for a measurable subset Ω of $\mathbb{R}^n$. The set $\mathcal{D}$ is then called *the dilation set* and the set $\mathcal{T}$ is called *the translation set*. Quite often, one assume that the dilation set is a group and that the translation set is a latices, i.e., a discrete subgroup $\Gamma \subset \mathbb{R}^n$ such that $\mathbb{R}^n/\Gamma$ is compact. In particular, the simplest example is the group generated by one element, $\mathcal{D} = \{a^k \mid k \in \mathbb{Z}\}$. In [**16, 19**] more general sets of dilations were considered, and in general those dilations do not form a group. Even more general constructions can be found in [**1**].

In this article we consider the case where $\mathcal{D}$ is of the form $\mathcal{D} = \{a^k \mathrm{id} \mid k \in \mathbb{Z}\}W$, where a is an expansive matrix and W is a finite Coxeter group, see Section 2 for the definition. We use results from our previous article [**6**]. Those results are discussed

2000 *Mathematics Subject Classification.* 42C40,43A85.

Key words and phrases. Wavelet sets, spectral sets, tiling sets, subspace wavelets, wavelet transform.

M. Dobrescu was partially supported by DMS-0139783. The research of G. Ólafsson was supported by NSF grants DMS-0139783 and DMS-0402068.

in Section 1. We would like to remark, that this construction is more general than needed here. In particular it is not needed that $\mathcal{D}$ is a group.

Most of the examples of wavelet sets tend to be fractal like and symmetric around 0. Our aim is to construct wavelet functions that some directional properties in the frequency domain. The finite Coxeter group is then used to rotate the frequency domain to cover all of $\mathbb{R}^n$. Some two-dimensional examples are discussed at the end of Section 1 to motivate the construction in later sections. The construction is still fractal, but it is not symmetric around 0 anymore. The generalization of a rotation group for dimensions higher than two is a Coxeter group. The construction is generalized to higher dimensions in Section 2 and Section 5.

A natural question one asks when working with wavelets is if they are related with any multiresolution analysis. It is well known that one needs multiwavelets when working in higher dimensions which means that the wavelets coming from the wavelets sets described above are not associated with any multiresolution analysis.

In Section 3 and Section 4 we discuss the construction of scaling sets and the associated multiresolution analysis and multiwavelets for this situation. In particular, our wavelet is still directional in the frequency domain. In fact the support of the Fourier transform is supported in cones which are fundamental domains for the action of a Coxeter group on the Euclidian space. We are also illustrating our results with a few examples.

In the final section we discuss the relation between these results and sampling theory. Any square integrable function can be written as a sum of its projections on subspaces, where each subspace contains only signals supported in the frequency domain in the cones mentioned above. Each projection can then be sampled using a version of the Whittaker-Shannon-Kotel'nikov sampling theorem [**22**, **24**] for spectral sets stated in Theorem 6.1.

1. Existence of subspace wavelet sets

In this section we recall some general results from [**6**] which are the basic for the construction later in this work. Most of the literature deals with dilations groups generated by one element. The main idea here is to consider dilation sets that can be factorized as a product of finitely many groups (or more general set) and then use inductive construction to reduce the general case to the simple one. Let us remark that the statements in this section hold for more general settings, i.e., one could replace $\mathbb{R}^n$ by a measure space M and $\mathrm{GL}(n, \mathbb{R})$ by a group of automorphisms of M.

For $\mathcal{A}, \mathcal{B} \subset \mathrm{GL}(n, \mathbb{R})$ we say that the product $\mathcal{A}\mathcal{B} = \{ab \mid a \in \mathcal{A}, b \in \mathcal{B}\}$ is *direct* if $a_1 b_1 = a_2 b_2$, $a_1, a_2 \in \mathcal{A}$, $b_1, b_2 \in \mathcal{B}$, implies that $a_1 = a_2$ and $b_1 = b_2$. For the proof of the following statements see [**6**].

DEFINITION 1.1. A measurable tiling of a measure space (M, μ) is a countable collection of subsets $\{\Omega_j\}$ of M, such that

$$\mu(\Omega_i \cap \Omega_j) = 0 \,,$$

for $i \neq j$, and

$$\mu(M \backslash \bigcup_j \Omega_j) = 0 \,.$$

If $\Omega \subset \mathbb{R}^n$ is measurable and $\mathcal{A}$ a set of diffeomorphism of $\mathbb{R}^n$, then Ω is a $\mathcal{A}$-tile if $\{d(\Omega) \mid d \in \mathcal{A}\}$ is a measurable tiling of $\mathbb{R}^n$.

LEMMA 1.2. *Let $M \subseteq \mathbb{R}^n$ be measurable. Let $\mathcal{A}, \mathcal{B} \subset \mathrm{GL}(n, \mathbb{R})$ be two non-empty sets, such that the product $\mathcal{A}\mathcal{B}$ is direct. Let $\mathcal{D} = \mathcal{A}\mathcal{B} = \{ab \mid a \in \mathcal{A}, \ b \in \mathcal{B}\}$. Then there exists a $\mathcal{D}$-tile Ω for M if and only if there exists a measurable set $N \subseteq \mathbb{R}^n$, such that $\mathcal{A}N$ is a measurable tiling of M, and a $\mathcal{B}$-tile Ω for N.*

REMARK 1.3. We would like to remark at this point, that we do not assume that $\Omega \subseteq M$, nor that $N \subseteq M$. But this will in fact be the case in most applications because $\mathcal{D}$ will contain the identity matrix.

THEOREM 1.4 ((Construction of wavelet sets by steps, I)). *Let $\mathcal{M}, \mathcal{N} \subset \mathrm{GL}(n, \mathbb{R})$ be two non-empty subsets such that the product $\mathcal{M}\mathcal{N}$ is direct. Let $\mathcal{L} = \mathcal{M}\mathcal{N}$. Assume that $M \subseteq \mathbb{R}^n$ with $|M| > 0$, is measurable. Let $\mathcal{T} \subset \mathbb{R}^n$ be discrete. Then there exists a $(\mathcal{L}, \mathcal{T})$-wavelet set $\Omega \subset M$ for M if and only if there exists a $\mathcal{N}^T$-tiling set $N \subset M$ and a $(\mathcal{M}, \mathcal{T})$-wavelet set Ω_1 for N.*

Recall that if $\mathcal{D} \subseteq \mathrm{GL}(n, \mathbb{R})$ and $\mathcal{G} \subset \mathrm{GL}(n, \mathbb{R})$ is a group that acts on $\mathcal{D}$ from the right, then there exists a subset $\mathcal{D}_1 \subseteq \mathcal{D}$, such that $\mathcal{D} = \mathcal{D}_1 \mathcal{G}$ and the product is direct. Note that we do not assume that $\mathcal{G} \subset \mathcal{D}$.

THEOREM 1.5 ((Construction of wavelet sets by steps, II)). *Let $\mathcal{D} \subset \mathrm{GL}(n, \mathbb{R})$ and $M \subseteq \mathbb{R}^n$ measurable with $|M| > 0$. Let $\mathcal{T} \subset \mathbb{R}^n$ be discrete. Assume that $\mathcal{G} \subset \mathrm{GL}(n, \mathbb{R})$ is a group that acts on $\mathcal{D}$ form the right. Let $\mathcal{D}_1 \subseteq \mathcal{D}$ be such that $\mathcal{D} = \mathcal{D}_1 \mathcal{G}$ as a direct product. Then there exists a $(\mathcal{D}, \mathcal{T})$-wavelet set Ω for M if only only if there exists a $\mathcal{G}^T$-tiling set N for M and a $(\mathcal{D}_1, \mathcal{T})$-wavelet set Ω_1 for N.*

For a measurable set $M \subseteq \mathbb{R}^n$ such that $\overline{M} \setminus M$ has measure zero, let

$$L_M^2(\mathbb{R}^n) = \{f \in L^2(\mathbb{R}^n) \mid \mathrm{supp}(\mathcal{F}(f)) \subseteq M\}.$$

The question is then, how to obtain a wavelet set for the starting subset N. The following result gives one way to do that.

THEOREM 1.6 ((Existence of subspace wavelet sets)). *Let $M \subseteq \mathbb{R}^n$ be a measurable set, $|M| > 0$. Let $a \in \mathrm{GL}(n, \mathbb{R})$ be an expansive matrix and $\emptyset \neq \mathcal{D} \subset \mathrm{GL}(n, \mathbb{R})$. Assume that $\mathcal{D}^T$ is a multiplicative tiling of M, $a\mathcal{D} = \mathcal{D}$ and $a^T M = M$. If $\mathcal{T}$ is a lattice, then there exists a measurable set $\Omega \subseteq M$ such that $\Omega + \mathcal{T}$ is a measurable tiling of $\mathbb{R}^n$ and $\mathcal{D}^T \Omega$ is a measurable tiling of M. In particular, Ω is a $L_M^2(\mathbb{R}^n)$-subspace $(\mathcal{D}, \mathcal{T})$-wavelet set.*

EXAMPLE 1.7. We would like to note here that in general, for a given set $\mathcal{D}$, there are several ways to decompose it as a direct product. As an example take the set - which in fact is a group -

$$\mathcal{D} = \mathcal{D}_{a,m} = \{a^k R_{2\pi j/m} \mid k \in \mathbb{Z}, j = 0, \ldots m - 1\}.$$

Here $a > 1$ and R_θ stands for the rotation

$$R_\theta = \begin{pmatrix} \cos\theta & -\sin\theta \\ \sin\theta & \cos\theta \end{pmatrix}.$$

We can take $\mathcal{G} = \{R_{2\pi j/m} \mid j = 0, \ldots, m - 1\}$ and $\mathcal{D}_1 = \{a^k \mathrm{id} \mid k \in \mathbb{Z}\}$. Then we can take

$$
\begin{aligned}
N = \mathbb{R}_{2\pi/m}^2 \ &= \ \{r(\cos\theta, \sin\theta)^T \mid 0 \le \theta \le 2\pi/m, \ r > 0\} \\
&= \ \{(x, y) \in \mathbb{R}^2 \mid 0 \le x, y \quad \text{and} \quad y \le x \tan(2\pi/m)\}.
\end{aligned}
$$

But we could also take $\mathcal{G}$ as above and

$$\mathcal{D}_1 = \{(aR_{2\pi j/m})^k \mid k \in \mathbb{Z}\}$$

for some $0 \leq j \leq m - 1$. In this case we would take

$$N = \bigcup_{k \in \mathbb{Z}} (aR_{2\pi j/m})^k \{r(\cos\theta, \sin\theta)^T \mid a \leq r \leq a^2 \quad \text{and} \quad 0 \leq \theta \leq 2\pi/m\}$$

whose interior is not connected. Note that if we take $a = \sqrt{2}$ and $m = 4$, so $R_{2\pi/m} = R_{\pi/4}$, then

$$aR_{\pi/4} = \begin{pmatrix} 1 & -1 \\ 1 & 1 \end{pmatrix}$$

which often shows up in examples. As $a > 1$ it follows that $a\mathrm{id}$ and $aR_{2\pi j/m}$ are expansive matrices, Theorem 1.6 implies that in both cases and for any full rank lattice Γ, the set N contains a wavelet set Ω, which will be quite different for the two cases.

Let $\mathcal{T} = \mathbb{Z}^2$, $\mathcal{D} = \mathcal{D}_{a,m} = \{a^k R_{2\pi j/m} \mid k \in \mathbb{Z}, j = 0, \ldots m - 1\}$, and let

$$
\begin{aligned}
E &= [0,1] \times [0, \tan(2\pi/m)] \;\; if \;\; m \neq 2,4 \\
E &= [0,1]^2 \;\; if \;\; m = 4 \\
E &= [-1,1] \times [0,1] \;\; if \;\; m = 2 \\
F &= \{(x,y) \in \mathbb{R}^2_{2\pi/m} | 1 < x < a\}.
\end{aligned}
$$

The wavelet set Ω has the form

$$\Omega = \bigcup_{i=1}^{2} \bigcup_{j=1}^{\infty} \Omega_{i,j},$$

see figure 1. The description of the $\Omega_{i,j}$ is as follows

$$
\begin{aligned}
\Omega_{1,1} &= (E \setminus a^{-1}E) + (1,0) \\
\Omega_{2,1} &= a^{-2}\Big(F \setminus \big(E + (0,1)\big)\Big) \\
\Omega_{1,2} &= [(a^{-1}E \setminus a^{-2}E) \setminus \Omega_{2,1}] + (1,0) \\
\Omega_{2,2} &= a^{-3}[\Omega_{2,1} + (0,1)].
\end{aligned}
$$

For $j \geq 3$, we have the following formulas

$$\Omega_{1,j} = [(a^{-n+1}E \setminus a^{-n}E) \setminus \Omega_{2,j-1}] + (1,0)$$

and

$$\Omega_{2,j} = a^{-n-1}[\Omega_{2,j-1} + (0,1)].$$

From the construction, it is clear that Ω and E are $\mathcal{T}$-translation congruent, and Ω and F are $\mathcal{D}$-dilation congruent. On the other hand, F is a $\mathcal{D}_{a,m}$-multiplicative tile and $\{E, \mathcal{T}\}$ is a spectral pair. It follows that Ω is a $\mathcal{D}_{a,m}$-multiplicative tile and $\{\Omega, \mathcal{T}\}$ is a spectral pair. Thus Ω is a $(\mathcal{D}_{a,m}, \mathcal{T})$ wavelet set.

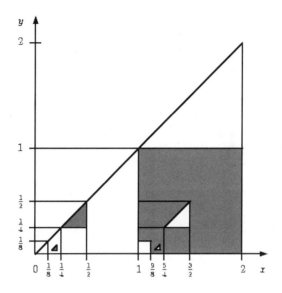

FIGURE 1. A $(\mathcal{D}_{2,4}, \mathcal{T})$ wavelet set.

2. Coxeter Groups

In the last example the group $\mathcal{G}$ was a special case of a Coxeter group, which we will introduce in this section. The generalization to higher dimensions of Example 1.7 are the sets of the form $\mathcal{D} = \{a^k w \mid k \in \mathbb{Z}, w \in \mathcal{W}(\Delta)\}$. The set N is then a convex cone which is a fundamental domain for W. Note that this decomposition of $\mathcal{D}$ is not the only one and we could also choose other sets for N. We could also replace a by any expansive matrix a such that $aN = N$. We start by collecting some well known facts on root systems and Coxeter groups. We use [13] as standard reference, but would also like to point out [5, 12, 15].

A finitely generated group W defined by the relations $(r_i r_j)^{m_{ij}} = 1$, where $m_{ij} \in \mathbb{N}$, $m_{ii} = 1$ and $m_{ij} = m_{ji}$, is called a *Coxeter group*. The finite Coxeter groups can be realized as finite reflexion groups in $\mathrm{O}(n)$. Let $\alpha \in \mathbb{R}^n$, $\alpha \neq 0$. A *reflection* along α is a linear map $r : \mathbb{R}^n \to \mathbb{R}^n$ such that

- $r(\alpha) = -\alpha$;
- The space $\{\lambda \in \mathbb{R}^n \mid r(\lambda) = \lambda\}$ is a hyperplane in $\mathbb{R}^n$.

Note that a reflection r is a non-trivial element of $\mathrm{O}(n)$ of order 2. The above reflection is given by

$$\lambda \mapsto \lambda - \frac{2(\lambda, \alpha)}{(\alpha, \alpha)} \alpha =: r_\alpha(\lambda) \,.$$

Furthermore,

$$\{\lambda \in \mathbb{R}^n \mid r(\lambda) = \lambda\} = H_\alpha := \{\lambda \in \mathbb{R}^n \mid (\alpha, \lambda) = 0\} \,.$$

A *finite reflexion group* is a finite subgroup $W \subset \mathrm{O}(n)$ generated by reflexions.

Let Δ be a finite set of nonzero vectors in $\mathbb{R}^n$. Then Δ is called *a root system* (in $\mathbb{R}^n$) (and its elements are called *roots*) if it satisfies the following three conditions:

(1) Δ generates $\mathbb{R}^n$;
(2) If $\alpha \in \Delta$, then $\Delta \cap \mathbb{R}\alpha = \{\alpha, -\alpha\}$;
(3) If $\alpha, \beta \in \Delta$, then $r_\alpha(\beta) \in \Delta$.

Note that sometimes it is not required that Δ generates $\mathbb{R}^n$. The reason is that this allows one to consider subsets $\Sigma \subset \Delta$ such that Σ is a root system in $\mathbb{R}\Sigma$ simply as root system in $\mathbb{R}^n$, see [**17**] and the reference therein for applications in analysis.

From now on Δ always stands for a root system in $\mathbb{R}^n$ and by $W = W(\Delta)$ we denote the reflection group generated by the reflections r_α, $\alpha \in \Delta$. Then W is a finite Coxeter group. Conversely, if W is a finite Coxeter group then there exists a $n \in \mathbb{N}$, and a root system $\Delta \subset \mathbb{R}^n$, such that $W \simeq W(\Delta)$, cf.[**5**], Chapter 1, pp 14 and 17.

Recall that a *total ordering* on a real vector space V is a transitive relation on V (denoted $<$) satisfying the following axioms:

(1) For each pair $\mu, \nu \in V$, exactly one of $\mu < \nu, \mu = \nu, \nu < \mu$ holds.
(2) Let $\mu, \nu, \eta \in V$. If $\mu < \nu$, then $\mu + \eta < \nu + \eta$.
(3) If $\mu < \nu$ and c is a nonzero real number, then $c\mu < c\nu$ if $c > 0$ and $c\nu < c\mu$ if $c < 0$.

We write $\mu > \nu$ if $\nu < \mu$. Given such a total ordering, we say that $\nu \in V$ is *positive* if $0 < \nu$ and negative if $\nu < 0$. Given a total ordering $<$ on $\mathbb{R}^n$ and a set of roots $\Delta \subset \mathbb{R}^n$ we set $\Delta^+ = \{\nu \in \Delta | 0 < \nu\}$. The elements in Δ^+ are the *positive roots*. We note that by (1) above and the fact that $0 \notin \Delta$ it follows that $\Delta = \Delta^+ \dot\cup (-\Delta^+)$, where $\dot\cup$ stands for disjoint union.

A subset Π of Δ is a *simple system* if Π is a vector space basis for $\mathbb{R}^n$, and each element of Δ is a linear combination of elements of Π with all coefficients having the same sign. It is easy to see that if Π is a simple system, then $w\Pi$ is also a simple system, for any $w \in W$.

THEOREM 2.1. *Every positive system contains a unique simple system Π. Furthermore, the Coxeter group $W(\Delta)$ is generated by the reflections r_α, $\alpha \in \Pi$.*

We now describe the construction of a fundamental domain for the action of the Coxeter group W on $\mathbb{R}^n$.

DEFINITION 2.2. Let G be a discrete group acting on $\mathbb{R}^n$. A closed subset D of $\mathbb{R}^n$ is called a fundamental domain for G, if D is a G-tile, i.e.,

$$\mathbb{R}^n = \bigcup_{g \in G} gD,$$

and $gD \cap hD$ has measure zero for all $g, h \in G$, $g \neq h$.

DEFINITION 2.3. A subset C of a vector space V is a cone if $\lambda C \subseteq C$, for any real $\lambda > 0$.

DEFINITION 2.4. A subset C of a vector space V is convex if for any vectors $u, v \in C$, the vector $(1 - t)u + tv$ is also in C for all $t \in [0, 1]$.

THEOREM 2.5. *Let $W = W(\Delta)$ and $\Pi \subset \Delta^+$ a simple system of roots. Then the convex cone*

$$C(\Pi) = \{\lambda \in \mathbb{R}^n \mid (\lambda, \alpha) \geq 0, \ \forall \alpha \in \Pi\} = \{\lambda \in \mathbb{R}^n \mid (\lambda, \alpha) \geq 0, , \ \forall \alpha \in \Delta^+\}$$

is a fundamental domain for the action of W on $\mathbb{R}^n$.

Note that if we replace Π by $w\Pi$, with $w \in W$, then the corresponding cone is wC, i.e., $C(w\Pi) = wC(\Pi)$. The open convex cones $C(\Pi)^o$ are called *chambers* and they are the connected components of the complement of $\bigcup_{\alpha \in \Pi} H_\alpha$ in $\mathbb{R}^n$. Given a chamber C associated with a simple system Π, its *walls* are defined to be the hyperplanes H_α, $\alpha \in \Pi$. The angle between any two walls is an angle of the form π/k, for some positive integer $k > 1$.

We apply now the results of Section 1 to this situation. Let $\Delta \subset \mathbb{R}^n$ be a root system and $\Pi = \{\alpha_1, \ldots, \alpha_n\}$ be a system of simple roots. Let $\Pi^* = \{\alpha_1^*, \ldots, \alpha_n^*\}$ be the corresponding dual basis, i.e., $(\alpha_i, \alpha_j^*) = \delta_{ij}$ for $i, j = 1, \ldots, n$. Then

$$C(\Pi) = \{t_1\alpha_1^* + \ldots + t_n\alpha_n^* \mid t_j \geq 0, \, j = 1, \ldots, n\}.$$

Let $a_j > 1$, $j = 1, \ldots, n$ and let A be a matrix such that $A(\alpha_j^*) = a_j\alpha_j^*$, $B = A^T$, $\mathcal{D}_1 = \{B^k \mid k \in \mathbb{Z}\}$, and $\mathcal{D} = \mathcal{D}_1 W$. Then $\mathcal{D}_1(C(\Pi)) = C(\Pi)$. Let Γ be a full rank lattice in $\mathbb{R}^n$ of the form $G\mathbb{Z}$ with $\det G = 1$. We finish this section with the following Theorem. The construction will be discussed in some more details in Section 5

THEOREM 2.6. *With the notation above there exists a* $(\mathcal{D}, \Gamma)$ *wavelet set* $\Omega \subset C(\Pi)$.

PROOF. This follows from Theorem 1.4 and Theorem 1.6. $\square$

We give now two example of Coxeter groups and we will come back to them in Section 3 when we will be able to describe the construction of multivavelets associated with MRA's.

EXAMPLE 2.7. Let $\mathbb{R}^2$ be the Euclidian plane, and let $\mathcal{D}_m$ be the dihedral group of order $2m$, consisting of the orthogonal transformations which preserve a regular m-sided polygon centered at the origin.

$\mathcal{D}_m$ contains m rotations through multiples of $2\pi/m$, and m reflections about the diagonals of the polygon. By 'diagonal', we mean a line joining two vertices or the midpoints of opposite sides if m is even, or joining a vertex to the midpoint of the opposite side if m is odd.

The group $\mathcal{D}_m$ is actually generated by reflections, since a rotation through $2\pi/m$ is a product of two reflections relative to a pair of adjacent diagonals which meet at an angle of $\theta = \pi/m$, see Figure 2.

The three dimensional case is more interesting. Let a, b, c be three linearly independent vectors such that the corresponding reflections lie in a finite group. That is only possible if $\sphericalangle(a, b)$, $\sphericalangle(a, c)$, $\sphericalangle(b, c)$ are rational multiple of π. This can be obtain by choosing $\sphericalangle(a, b)$ to be an arbitrary multiple of π and then choosing c such that $\sphericalangle(a, c) = \sphericalangle(b, c) = \pi/2$. In that case, the group generated by r_a and r_b, $< r_a, r_b >$, is a dihedral group and $< r_a, r_b, r_c >$ is the direct product of the dihedral group $< r_a, r_b >$ and the cyclic group of order 2 generated by r_c.

Except these direct products, there are only three 3-dimensional Euclidian reflection groups, the groups of symmetries of a regular tetrahedron, a cube, and a regular dodecahedron.

EXAMPLE 2.8. For each tetrahedron centered, there is a 'dual' tetrahedron which is congruent to the given one and has the property that each edge of the given tetrahedron is perpendicularly bisected by an edge of the dual. Together, the vertices of the two tetrahedra give the vertices of a cube. Let a and c be the

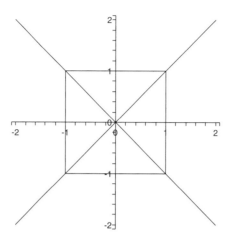

FIGURE 2. The dihedral group $\mathcal{D}_4$

position vectors of the midpoints of a pair of parallel but not opposite edges e_1 and e_2. Let b be the position vector of the midpoint of one of the edges on the opposite face to that determined by e_1 and e_2, which are not parallel to e_1 and e_2. Then we have the following:

$$\sphericalangle(a, b) = 2\pi/3, \quad r_a r_b \text{ has order 3;}$$
$$\sphericalangle(a, c) = \pi/2, \quad r_a r_c \text{ has order 2;}$$
$$\sphericalangle(b, c) = 2\pi/3, \quad r_b r_c \text{ has order 3;}$$

r_a, r_b, r_c are symmetries of the tetrahedron, and so these three reflections generate the group of all symmetries of the tetrahedron, which is just Sym(4).

3. Multiwavelets Associated with Multiresolution Analysis

This section and the next one contain our main results. We start by constructing scaling sets and by taking translates we obtain MRA multi-wavelets which are completely supported in a fundamental domain for the action of a Coxeter group on $\mathbb{R}^n$. Let A be an expansive matrix and $\mathcal{T}$ be a full rank lattice.

DEFINITION 3.1. A multiresolution analysis on $\mathbb{R}^n$ is a sequence of subspaces $\{V_j\}_{j \in \mathbb{Z}}$ of functions in $L^2(\mathbb{R}^n)$ satisfying the following properties:

i)   For all $j \in \mathbb{Z}$, $V_j \subseteq V_{j+1}$

ii) If $f(\cdot) \in V_j$, then $f(A\cdot) \in V_{j+1}$

iii) $\bigcap_{j \in \mathbb{Z}} V_j = \{0\}$

iv) $\overline{\bigcup_{j \in \mathbb{Z}} V_j} = L^2(\mathbb{R}^n)$

v)   There exists a function $\phi \in L^2(\mathbb{R}^n)$ such that $\{\phi(\cdot + t) | t \in \mathcal{T}\}$ is an orthonormal basis for V_0.

The function ϕ is called a *scaling function*. One can allow more than one scaling function, say m, and then the MRA has multiplicity m.

If we change condition *iv)* in the definition of the multiresolution analysis into

iv') $\overline{\bigcup_{j \in \mathcal{T}} V_j} = L_M^2(\mathbb{R}^n)$, for some subset $M \subseteq \mathbb{R}^n$,

then we get a *subspace multiresolution analysis*, SMRA.

We explain now how to construct a wavelet or multi-wavelets from a MRA. Let W_0 be the orthogonal complement of V_0 in V_1, that is, $V_1 = V_0 \oplus W_0$. In general, let $W_i = V_{i+1} \ominus V_i$, for each $j \in \mathbb{Z}$,

$$V_j = \bigoplus_{l=-\infty}^{j} W_l$$

and so

$$L^2(\mathbb{R}^n) = \bigoplus_{l=-\infty}^{\infty} W_l.$$

If there exists a function $\psi \in W_0$ such that $\{\psi(\cdot + t) | t \in \mathcal{T}\}$ is an orthonormal basis for W_0, then $\{\psi_{j,t} | t \in \mathcal{T}\}$ is an orthonormal basis for W_j, and $\{\psi_{j,t} | t \in \mathcal{T}, j \in \mathbb{Z}\}$ is an orthonormal basis for $L^2(\mathbb{R}^n)$, which means that ψ is an orthonormal wavelet associated with the given MRA..

Recall that $\mathcal{T}$ is a full rank lattice and A an expansive matrix. Set $B = A^T$. Suppose $B\mathcal{T} \subseteq \mathcal{T}$. Let $\mathcal{T}/B\mathcal{T}$ be the quotient group, where we identify its elements with their representative vectors in $\mathbb{R}^n$, $v_0, v_1, \dots v_{q-1}$, where $q = |\det(B)|$.

LEMMA 3.2. *Let K be a $\mathcal{T}$-tile such that $B^{-1}K \subset K$. Let*

$$K_i = (B^{-1}K + B^{-1}v_i + \mathcal{T}) \bigcap K.$$

Then

i) $K = \bigcup_{i=0}^{q-1} K_i$ up to measure zero and $K_i \bigcap K_j = 0$ up to measure zero for $i \neq j$.

ii) $BK_i \sim_{\mathcal{T}} K$.

PROOF. Let $x \in K_i \cap K_j$. Then there exist $u_1, u_2 \in K$, $v_i, v_j \in \mathcal{T}/B\mathcal{T}$ and $t_1, t_2 \in \mathcal{T}$, such that

$$x = B^{-1}u_1 + B^{-1}v_i + t_1 = B^{-1}u_2 + B^{-1}v_j + t_2,$$

and so $u_1 - u_2 \in \mathcal{T}$. But K is a $\mathcal{T}$ tile, so $u_1 = u_2$, and then $v_i - v_j = B(t_1 - t_2)$. Thus $v_i = v_j$ and so $i = j$. Let the notations be as before and let

$$K_{i,t} = B^{-1}K \bigcap (K - B^{-1}v_i - t).$$

Then $\bigcup_{t \in \mathcal{T}} (K_{i,t} + B^{-1}v_i + t) = K_i$.

Since K is a $\mathcal{T}$ tile, it follows that $K - B^{-1}v_i$ is also a $\mathcal{T}$ tile. Thus $K_{i,t}$ are measurewise disjoint and $\bigcup_{t \in \mathcal{T}} K_{i,t} = B^{-1}K$.

By definition, $K_i \subseteq K$, and

$$|K_i| = \Sigma_{t \in \mathcal{T}} |K_{i,t}| = |B^{-1}K| = q^{-1}|K|$$

for $i = 0, \dots, q-1$, and $|K_i \bigcap K_j| = 0$ for $i \neq j$. Therefore

$$K = \bigcup_{i=0}^{q-1} K_i$$

up to measure zero. Moreover,

$$B(K_i) = \bigcup_{t \in \mathcal{T}} B(K_{i,t} + B^{-1}v_i + t) = K + v_i + Bt$$

and thus

$$B(K_i) \sim_{\mathcal{T}} K.$$

$\square$

Let $K \subset \mathbb{R}^n$ be a measurable set. Set $V_0 = L^2_K(\mathbb{R}^n)$ and $V_j = \{f(A^j \cdot) | f(\cdot) \in V_0\}$.

DEFINITION 3.3. A set $K \subset \mathbb{R}^n$, $|K| = 1$, is a scaling set, if the sequence $\{V_j\}$ described above is a multiresolution analysis with scaling function $\phi = \mathcal{F}^{-1}\chi_K$.

THEOREM 3.4. A subset $K \subset \mathbb{R}^n$ is a scaling set if and only if $B^{-1}K \subseteq K$ and K is a $\mathcal{T} - tile$.

PROOF. Suppose first that $K \subset \mathbb{R}^n$ is a scaling set. Then $\phi = \mathcal{F}^{-1}\chi_K$ is a scaling function and so $\{\phi_{0,t}\}_{t \in \mathcal{T}}$ is an orthonormal basis for V_0. This implies that $\{\hat{\phi}_{0,t}\}_{t \in \mathcal{T}}$ is an orthonormal basis for $\hat{V}_0$. For $t \in \mathbb{R}^n$ let $e_t(x) = e^{2\pi(x,t)}$. It follows that $\hat{V}_0$ has a orthonormal basis of the form $\{e_t\chi_K\}_{t \in \mathcal{T}}$. From this we get two things. The first is that $(K, \mathcal{T})$ is a spectral pair and thus K is a $\mathcal{T}$-tile. The second is that $\hat{V}_0 = L^2(K)$ and by the SMRA structure, we get that $\hat{V}_{-1} \subset \hat{V}_0$ which implies that $B^{-1}K \subset K$.

Assume now that $B^{-1}K \subset K$ and that K is a $\mathcal{T} - tile$.

Set $\hat{V}_0 = L^2(K)$ and $\hat{V}_j = L^2(B^j K)$. Since $B^{-1}K \subset K$, it follows that $\hat{V}_j \subset \hat{V}_{j+1}$. The other conditions are easy to verify. Thus $K \subset \mathbb{R}^n$ is a scaling set. $\square$

The next theorem gives in a constructive way, the existence of SMRA wavelets.

THEOREM 3.5. If $K \subset \mathbb{R}^n$ is a scaling set, then

i) $\hat{V}_0 = L^2(K)$ and $\hat{V}_j = L^2(B^j K)$

ii) $\{\psi^i = \check{\chi}_{\Omega_i}\}_{i=1}^{q-1}$ is a SMRA multiwavelet, where $\Omega_i = BK_i$.

PROOF. i) It follows from the theorem above.

ii) By lemma 1, $\Omega_i \sim K$. This implies that $(\Omega_i, \mathcal{T})$ is a spectral pair. Then

$$\{\hat{\psi}^i_{o,t}\}_{t \in \mathcal{T}}$$

is an orthonormal basis for $L^2(\Omega_i)$.

Set $\widehat{W}_{0,i} = L^2(\Omega_i)$ for $i = 1, ..., q - 1$. By construction,

$$BK = K \cup \bigcup_{i=1}^{q-1} \Omega_i.$$

Therefore

$$\widehat{V}_1 = L^2(BK) = L^2(K) \oplus \bigoplus_{i=1}^{q-1} L^2(\Omega_i) = \widehat{V}_0 \oplus \widehat{W}_{0,1} \oplus ... \oplus \widehat{W}_{0,q-1}.$$

So

$$V_1 = V_0 \oplus \bigoplus_{i=1}^{q-1} W_{0,i}$$

and for any $j \in \mathcal{Z}$, we have

$$V_{j+1} = V_j \oplus \bigoplus_{i=1}^{q-1} W_{j,i} \,.$$

Thus, $\{\psi^i = \mathcal{F}^{-1}\chi_{\Omega_i}\}_{i=1}^{q-1}$ is a SMRA multi-wavelet. $\qquad\square$

4. Multiresolution and Coxeter Groups

We now go back to the notation in Section 2. In particular Δ is a root system in $\mathbb{R}^n$, Π a system of simple roots, $W = < r_{\alpha_i} | \alpha_i \in \Pi >$ the corresponding a finite Coxeter group and $C = C(\Pi)$ be the corresponding positive cone which is a fundamental domain for the action of W on $\mathbb{R}^n$. Let A be such that with $B = A^T$ we have $B(\alpha_j^*) = a_j \alpha_j^*$, $a_j > 1$, $j = 1, \ldots, n$. The more general case $B(\alpha_j) = a_j \alpha_{\pi(j)}^*$ where, $\pi : \{1, \ldots, n\} \to \{1, \ldots, n\}$ is a permutation, is handled in similar way. Let

$$P = \{ \sum_{i=1}^{n} t_i \alpha_i^* | 0 < t_i \le s_i \},$$

where s_i are such that $|P| = 1$. Note that P is a n dimensional parallelepiped and a $B\mathbb{Z}^n$ tile.

Indeed, if $z \in \mathbb{Z}^n$, then

$$
\begin{aligned}
P + Rz &= \{ \sum_{i=1}^{n} t_i \alpha_i^* + \sum_{i=1}^{n} n_i \alpha_i^* | 0 < t_i \le s_i, \, n_i \in \mathbb{Z} \} \\
&= \{ \sum_{i=1}^{n} (t_i + n_i)\alpha_i^* | 0 < t_i \le s_i, \, n_i \in \mathbb{Z} \}
\end{aligned}
$$

so

$$|P \cap (P + Rz)| = 0.$$

We have

$$B^{-1}P = \{ \sum_{i=1}^{n} a_i^{-1} t_i \alpha_i^* | o < t_i \le s_i \} \subset P \,,$$

since $0 < a_i^{-1} t_i \le s_i$. Moreover,

$$(\sum_{i=1}^{n} t_i \alpha_i^*, \alpha_m) = \sum_{i=1}^{n} t_i (\alpha_i^*, \alpha_m) = t_m > 0 \,.$$

Thus, $B^{-1}P \subset P \subset C(\Pi)$.

THEOREM 4.1. *Let P, B be as above, $P_i = B^{-1}P + B^{-1}v_i$ and $\Omega_i = BP_i$. Then $\{\psi^i = \check{\chi}_{\Omega_i}\}_{i=1}^{q-1}$ is a SMRA multiwavelet associated to the multiresolution $V_j = L^2_{B^j P}(\mathbb{R}^n)$.*

PROOF. As shown above, P is a $R\mathbb{Z}^n$-tile and $B^{-1}P \subset P$ and so P is a scaling set. Thus, by theorem 4.8, $\{\psi^i = \check{\chi}_{\Omega_i}\}_{i=1}^{q-1}$ is a SMRA multi-wavelet. $\qquad\square$

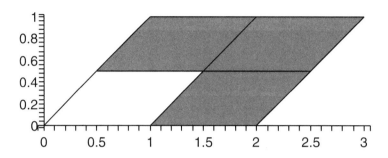

FIGURE 3. SMRA wavelet sets in $\mathbb{R}^2$

EXAMPLE 4.2. Let the group $\mathcal{D} = \{R_{\frac{2\pi}{m}}^k\}_{k=0}^{m-1}$ act on $\mathbb{R}^2$ and let $D = \{t_1(1,0) + t_2(\cot 2\pi/m, 1), 0 < t_{1,2}\}$ be the fundamental domain of this action. Let $B = 2\mathrm{id}_2$ and $P = \{t_1(1,0) + t_2(\cot 2\pi/m, 1), t_{1,2} \in [0,1]\}$. Let

$$\Omega_1 = P + (1,0),$$
$$\Omega_2 = P + (1 + \cot 2\pi/m, 1),$$
$$\Omega_3 = P + (\cot 2\pi/m, 1).$$

Then $\{\psi^i = \check{\chi}_{\Omega_i}\}_{i=1}^3$ is a SMRA multiwavelet, see Figure 3.

EXAMPLE 4.3. Let $\Omega = < r_a, r_b, r_c >$ be a Coxeter group, where a, b, c are as described in Example 2.8. Then the fundamental domain for the action of Ω on $\mathbb{R}^3$ is

$$D = \{t_a a^* + t_b b^* + t_c c^* | 0 < t_a, t_b, t_c\}.$$

Let $P = \{t_a a^* + t_b b^* + t_c c^* | 0 < t_a < s_a, 0 < t_b < s_b, 0 < t_c < s_c\}$, such that $|P| = 1$.

Let $B = 2\mathrm{id}_3$. Then $\det(B) = 2^3 = 8$ and so there are 7 MRA wavelet sets, see Figure 4.

REMARK 4.4. We would like to remark that the constructions in this section can easily be reduced to the case $\alpha_i = e_i$, where e_i is the standard basis for $\mathbb{R}^n$. This is done by using the linear map $\alpha_j \mapsto e_j$.

5. Wavelet Sets and Coxeter Groups

We now explain how the construction of wavelet sets mentioned in Section 1 can be done.

THEOREM 5.1. Let P be as above and let $F = BP \setminus P$. Define

$$W_{1,1} = (P \setminus B^{-1}P) + \alpha_i^*$$
$$W_{2,1} = B^{-2}[F \setminus (P + \alpha_i^*)]$$
$$\cdots$$
$$W_{1,n} = [(B^{-n+1}P \setminus B^{-n}P) \setminus W_{2,n-1}] + \alpha_i^*$$

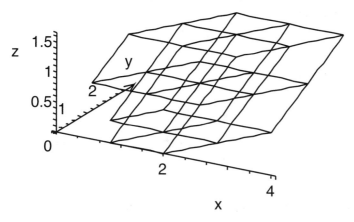

FIGURE 4. SMRA wavelet sets in $\mathbb{R}^3$

$$W_{2,n} = B^{-n-1}\{[(B^{-n+1}P \setminus B^{-n}P) + \alpha_i^*] \setminus W_{1,n}\}.$$

Then

$$P = \bigcup_n^\infty W_{2,n} \bigcup \bigcup_n^\infty (W_{1,n} - \alpha_i^*)$$

$$F = \bigcup_n^\infty W_{1,n} \bigcup \bigcup_n^\infty B^{n+1}W_{1,n}.$$

Moreover, if we let

$$W = \bigcup_{j=1,2} \bigcup_n^\infty W_{j,n},$$

then W is a wavelet set.

PROOF. We have shown above that P is a $R\mathbb{Z}^n$-tile. On the other hand,

$$BF \cap F = \emptyset,$$

and

$$\bigcup_{n \in \mathbb{Z}} B^n F = D,$$

so F is a multiplicative tiling. By definition,

$$W \sim_{R\mathbb{Z}^n} P,$$

and

$$W \sim_B F.$$

Thus, W is a wavelet set. $\square$

6. Coxeter Groups and Sampling

In this final section we discuss how the results in Section 4 are related to sampling theory. First we note that

$$L^2(\mathbb{R}^n) = \bigoplus_{w \in W} L^2_{C(w\Pi)}(\mathbb{R}^n) \,.$$

Thus each $f \in L^2(\mathbb{R}^n)$ can be decomposed as

(6.1) $$f = \sum_{w \in W} f_w,$$

where $f_w = \mathcal{F}^{-1}(\chi_{C(w\Pi)}\mathcal{F}(f))$ is the orthogonal projection of f onto $L^2_{C(w\Pi)}(\mathbb{R}^n)$. The function f_w contains the exact frequency information of f in the direction of the cone $C(w\Pi)$.

We note that $C(w\Pi)^o$ is an open convex cone such that $C(w\Pi)^o \cap -C(w\Pi)^o = \emptyset$, thus the dual cone $C_w \neq \emptyset$ is an open convex cone. In fact

$$
\begin{aligned}
C_w := [C(w\Pi)^o]^* &= \{y \in \mathbb{R}^n \mid (\forall \lambda \in C(w\Pi)^o)\,(y,\lambda) > 0\} \\
&= w\{\sum_{i=1}^n t_j \alpha_j \mid t_j > 0,\, j = 1, \ldots, n\} \\
&= w[C(\Pi)^o]^* \neq \{0\}
\end{aligned}
$$

is an open convex cone, and the function f_w extends to a holomorphic function on the tube domain

$$T(C_w) = \mathbb{R}^n + iC_w \,.$$

This holomorphic extension is given by

$$
\begin{aligned}
F_w(x + iy) &= \int_{C(w\Pi)} \mathcal{F}(f)(\lambda) e^{2\pi i(x+iy,\lambda)}\, d\lambda \\
&= \int_{C(w\Pi)} \mathcal{F}(f)(\lambda) e^{-2\pi(y,\lambda)} e^{2\pi i(x,\lambda)}\, d\lambda
\end{aligned}
$$

and

(6.2) $$f_w(x) = \lim_{y \to 0} F_w(x + iy)$$

where the limit is taken in $L^2(\mathbb{R}^n)$, see [**23**] for details. Thus f_w is in the Hardy space $H^2(T(C_w))$ and the equations (6.1) and (6.2) give us a decomposition of f as L^2-limit of $\#W$-holomorphic functions.

We can then sample an approximate version of each f_w (and then f) using the following simple version of the Whittaker-Shannon-Kotel'nikov sampling theorem [**22, 24**]. We include a short proof using the idea of spectral sets which we have not seen elsewhere in the literature, even if it is well known. Note that if the Fourier transform of f is supported in a set of finite measure, then f is continuous so $f(x)$ is well defined for all $x \in \mathbb{R}^n$.

THEOREM 6.1. *(WSK-sampling theorem for spectral sets) Let $P \subset \mathbb{R}^n$ be measurable, $0 < |P| < \infty$ and such that there exists a discrete set $\Gamma \subset \mathbb{R}^n$, such that* *the functions* $\{e_\gamma\chi_P\}_{\gamma\in\Gamma}$ *form a orthogonal basis for* $L^2(P)$, *i.e.,* P *is similar to a spectral set. Let*

$$\varphi = |P|^{-1/2}\mathcal{F}^{-1}\chi_P\,.$$

Then we have that for all $f \in L_P^2(\mathbb{R}^n)$,

$$(\mathcal{F}(f), e_\gamma) = f(-\gamma)$$

and

$$f(x) = |P|^{-1}\sum_{\gamma\in\Gamma}f(-\gamma)\varphi(x+\gamma) = |P|^{-1}\sum_{\gamma\in-\Gamma}f(\gamma)\varphi(x-\gamma)\,.$$

Furthermore, if $B \in \mathrm{GL}(n,\mathbb{R})$ *and* $L = (B^{-1})^T$. *Then*

$$\begin{aligned}
f(x) &= \frac{1}{|P|}\sum_{\gamma\in\Gamma}f(-L\gamma)\varphi(B^Tx+\gamma)\\
&= \frac{1}{|P|}\sum_{\gamma\in\Gamma}f(-L\gamma)\varphi(B^T(x+L\gamma))\\
&= |P|^{-1}\sum_{\gamma\in L\Gamma}f(-\gamma)\varphi\circ B^T(x+\gamma)
\end{aligned}$$

for all $f \in L_{BP}^2(\mathbb{R}^n)$.

PROOF. We have

$$(\mathcal{F}(f), e_\gamma\chi_P) = \int_{\mathbb{R}^n}\mathcal{F}(f)(\lambda)e^{-2\pi(\lambda,\gamma)}\,d\lambda = f(-\gamma)\,.$$

Furthermore, as $\{|P|^{-1/2}e_\gamma\chi_P\}_{\gamma\in\Gamma}$ is an orthonormal basis for $L^2(P)$,

$$\begin{aligned}
\mathcal{F}(f)(\lambda) &= |P|^{-1}\sum_{\gamma\in\Gamma}(\mathcal{F}(f), e_\gamma\chi_P)(e_\gamma\chi_P)(\lambda)\\
&= |P|^{-1}\sum_{\gamma\in\Gamma}f(-\gamma)(e_\gamma\chi_P)(\lambda)\,.
\end{aligned}$$

Hence

$$\begin{aligned}
f(x) &= \sum_{\gamma\in\Gamma}f(-\gamma)\mathcal{F}(e_\gamma\chi_P)(x)\\
&= \sum_{\gamma\in\Gamma}f(-\gamma)\varphi(x+\gamma)\,.
\end{aligned}$$

Let now $f \in L_{BP}^2(\mathbb{R}^n)$. As $\mathcal{F}(f\circ L)(\lambda) = |\det L|^{-1}\mathcal{F}(f)(B\lambda)$ it follows that $f\circ L \in L_P^2(\mathbb{R}^n)$. Hence

$$\begin{aligned}
f(x) &= (f\circ L)(B^Tx)\\
&= |P|^{-1}\sum_{\gamma\in\Gamma}(f\circ L)(-\gamma)\varphi(B^Tx+\gamma)\\
&= |P|^{-1}\sum_{\gamma\in\Gamma}f(-L\gamma)|\det B|\varphi(B^T(x+L\gamma))
\end{aligned}$$

and the claim follows. $\qquad\qquad\square$

We now go back to the situation in Section 4. Let $\Pi = \{\alpha_1, \ldots, \alpha_n\}$ be a system of simple roots and $\Pi^* = \{\alpha_1^*, \ldots, \alpha_n^*\}$ its dual basis. If we let $P = \{\sum_{j=1}^n t_j \alpha_j^* \mid 0 < t_j \le s_j\}$, where the numbers s_j are chosen so that $|P| = 1$, then P is a spectral set and $\{e_\gamma \chi_P\}_\gamma$ is a orthonormal basis of $L^2(P)$ where $\Gamma = \mathbb{Z}\Pi$. Let $B(\alpha_j^*) = a_j \alpha_j^*$ be as before and $V_j = L^2_{B^j P}(\mathbb{R}^n)$. By Theorem 6.1 we therefore get:

THEOREM 6.2. *Let* $\varphi = \mathcal{F}^{-1} \chi_P$ *and* $L = (B^{-1})^T$. *Then, if* $f \in V_j = L^2_{B^j P}(\mathbb{R}^n)$ *we have*

$$
\begin{aligned}
f(x) &= \sum_{\gamma \in \Gamma} f(-L^j \gamma) \varphi((B^j)^T x + \gamma) \\
&= \sum_{\gamma \in \Gamma} f(-L^j \gamma) \varphi((B^j)^T (x + L^j \gamma)) \\
&= \sum_{\gamma \in L^j \Gamma} f(-\gamma) \varphi \circ B^T (x + \gamma).
\end{aligned}
$$

References

[1] A. Aldroubi, Carlos Cabrelli, and Ursula M. Molter. Wavelets on irregular grids with arbitrary dilation matrices and frame atoms for $L^2(\mathbb{R}^d)$. *Appl. Comput. Harmon. Anal.*, 17(2):119–140, 2004.

[2] P. Aniello, G. Cassinelli, E. De Vito, A. Levrero. Wavelet transforms and discrete frames associated to semidirect products. *J. Math. Phys.*, 39:3965–3973, 1998.

[3] L. Baggett, A. Carey, W. Moran, P. Ohring. General Existence Theorem for Orthonormal Wavelet. An Abstract Approach. *Publ. Res. Inst. Mth. Sci. (1)*, 31:95–111, 1995.

[4] D. Bernier, K. F. Taylor. Wavelets from square-integrable representations. *SIAM J. Math. Anal.*, 27:594–608, 1996.

[5] N. Bourbaki. *Lie groups and Lie algebras. Chapters 4–6.* Elements of Mathematics (Berlin). Springer-Verlag, Berlin, 2002. Translated from the 1968 French original by Andrew Pressley.

[6] M. Dobrescu, and G. Ólafsson. Wavelet sets with and without groups. *Cont. Math.*, 2005.

[7] R. Fabec and G. Ólafsson. The continuous wavelet transform and symmetric spaces. *Acta Appl. Math.*, 77(1):41–69, 2003.

[8] H. Führ. Wavelet frames and admissibility in higher dimensions. *J. Math. Phys.*, 37:6353–6366, 1996.

[9] H. Führ. Continuous wavelet transforms with Abelian dilation groups. *J. Math. Phys.*, 39:3974–3986, 1998.

[10] P. E. Gunnells. Cells in coxeter groups. *Notices of the AMS*, 53(5):528–535, 2006.

[11] C.E. Heil, and D.F. Walnut. Continuous and discrete wavelet transform. *J. Math. Phys.*, 39:3974–3986, 1998.

[12] S. Helgason. *Differential geometry, Lie groups, and symmetric spaces*, volume 80 of *Pure and Applied Mathematics*. Academic Press Inc. [Harcourt Brace Jovanovich Publishers], New York, 1978.

[13] J.E. Humphreys . *Reflection Groups and Coxeter Groups*. Cambridge Studies in Advanced Mathematics, 1992.

[14] R.S. Laugesen, N. Weaver, G.L. Weiss, E.N. Wilson. A characterization of the higher dimensional groups associated with continuous wavelets. *J. Geom. Anal.*, 12:89–102, 2002.

[15] P. R. Massopust. *Fractal functions, fractal surfaces, and wavelets*. Academic Press Inc., San Diego, CA, 1994.

[16] G. Ólafsson. Continuous action of Lie groups on $\mathbb{R}^n$ and frames. *Int. J. Wavelets Multiresolut. Inf. Process.*, 3(2):211–232, 2005.

[17] G. Ólafsson and A. Pasquale. Support properties and holmgren's uniqueness theorem for differential operators with hyperplane singularities. *To apppear in J. Funct. Anal.*

[18] Gestur Ólafsson and Angela Pasquale. A Paley-Wiener theorem for the Θ-hypergeometric transform: the even multiplicity case. *J. Math. Pures Appl. (9)*, 83(7):869–927, 2004.

[19] G. Ólafsson, D. Speegle. Wavelets, wavelet sets, and linear actions on $\mathbb{R}^n$. In C. Heil, P.E.T. Jorgensen, D.R. Larson, editor, *Wavelets, Frames and Operator Theory (College Park, MD,*

2003), volume 345 of *Contemporary Mathematics*, pages 253–282. AMS, Providence, RI, 2003.

[20] E. M. Opdam. Root systems and hypergeometric functions. IV. *Compositio Math.*, 67(2):191–209, 1988.

[21] E. M. Opdam. An analogue of the Gauss summation formula for hypergeometric functions related to root systems. *Math. Z.*, 212(3):313–336, 1993.

[22] C. E. Shannon. A mathematical theory of communication. *Bell System Tech. J.*, 27:379–423, 623–656, 1948.

[23] Elias M. Stein and Guido Weiss. *Introduction to Fourier analysis on Euclidean spaces.* Princeton University Press, Princeton, N.J., 1971. Princeton Mathematical Series, No. 32.

[24] Ahmed I. Zayed. *Advances in Shannon's sampling theory.* CRC Press, Boca Raton, FL, 1993.

E-mail address: `mihaela.dobrescu@furman.edu`

DEPARTMENT OF MATHEMATICS, CHRISTOPHER NEWPORT UNIVERSITY, NEWPORT NEWS, VA 23606, USA

E-mail address: `olafsson@math.lsu.edu`

DEPARTMENT OF MATHEMATICS, LOUISIANA STATE UNIVERSITY, BATON ROUGE, LA 70803, USA

Contemporary Mathematics
Volume **451**, 2008

Operator Theory and Modulation Spaces

Christopher Heil and David Larson

ABSTRACT. This is a "problems" paper. We isolate some connections between operator theory and the theory of modulation spaces that were stimulated by a question of Feichtinger's regarding integral and pseudodifferential operators. We discuss several problems inspired by this question, and give a reformulation of the original question in operator-theoretic terms. A detailed discussion of the background and context for these problems is included, along with a solution of the problem for the case of finite-rank operators.

1. Introduction

The purpose of this article is to present some connections between two subareas of modern analysis: operator theory and the theory of modulation spaces. The Oberwolfach mini-workshop on *Wavelets, Frames, and Operator Theory*, which took place in February 2004, had as one of its central aims the forging of direct and indirect connections between these two areas. After Hans Feichtinger gave out a particular problem on modulation spaces in a workshop problem session, the two authors of this article set out to give a reformulation of the problem in operator-theoretic terms, in order to promote connections. The results of our discussions are given here. It is hoped that this article will promote the development of new inroads into both of these subjects. We think that we have isolated some interesting research problems, whose solution could conceivably impact mathematics beyond operator theory and beyond modulation space theory.

In Section 2 we give the statements of several problems in operator theory inspired by Feichtinger's original question, as well as an operator-theoretic reformulation of his question. Then in Section 3 we present some background and explain the precise relationship of the original question to the problems discussed in Section 2.

The form of the problems that we state in Section 2 appear to be natural ones accessible to specialists in operator theory, but of a type that perhaps would not have been considered by specialists without an external motivation. We think of

1991 *Mathematics Subject Classification.* Primary 45P05, 47B10; Secondary 42C15.

Key words and phrases. Frames, Gabor systems, modulation spaces, orthonormal bases, positive operators, spectral representation, time-frequency analysis, trace-class operators, Wilson bases.

The first author was partially supported by NSF Grant DMS-0139261.

The second author was partially supported by NSF Grant DMS-0139386.

them as *problems in operator theory that are motivated by a problem in modulation space theory.*

1.1. Notation. Throughout, H will denote an infinite-dimensional separable Hilbert space, with norm $\| \cdot \|$ and inner product $\langle \cdot, \cdot \rangle$. If x and y are vectors in H, then $x \otimes y$ will denote the rank one operator defined by $(x \otimes y)z = \langle z, y \rangle \, x$. The operator norm of $x \otimes y$ is then just the product of $\|x\|$ and $\|y\|$.

If $A \in \mathcal{B}(H)$ is a compact operator, then A has a countable set of nonnegative singular values $\{s_n(A)\}_{n=1}^{\infty}$, which we arrange in nonincreasing order. These can be defined by spectral theory to be the square-roots of the eigenvalues of the positive, self-adjoint operator A^*A, i.e., $s_n(A) = \lambda_n(A^*A)^{1/2}$. If A is self-adjoint then $s_n(A) = |\lambda_n(A)|$, and if A is positive (the case we are concerned with in this article), then $s_n(A) = \lambda_n(A)$. We say that A is *trace-class* if

$$\|A\|_{\mathcal{I}_1} = \sum_{n=1}^{\infty} s_n(A) < \infty.$$

The space $\mathcal{I}_1$ of trace-class operators is a Banach space under the norm $\| \cdot \|_{\mathcal{I}_1}$, see [**Sim79**], [**DS88**].

2. Some Problems in Operator Theory

Below we will introduce some terminology and provide a characterization of postive finite-rank operators that have a certain property, and then present several new problems.

2.1. Definitions and Observations. Fix an orthonormal basis $\mathcal{E} = \{e_k\}_{k \in \mathbf{N}}$ for H, and define

$$(2.1) \qquad H^1 = \Big\{ f \in H : \|f\|_1 = \sum_{k=1}^{\infty} |\langle f, e_k \rangle| < \infty \Big\}.$$

We have the following facts.

LEMMA 2.1.

(a) H^1 *is a dense subspace of* H.

(b) H^1 *is a Banach space with respect to the norm* $\| \cdot \|_1$.

(c) H^1 *is isomorphic to* ℓ^1.

(d) $\|e_k\|_1 = \|e_k\| = 1$ *for every* k.

(e) *If* $f \in H^1$, *then* $f = \sum_{k=1}^{\infty} \langle f, e_k \rangle \, e_k$, *with convergence of this series in both norms* $\| \cdot \|$ *and* $\| \cdot \|_1$.

Let T be any positive trace-class operator in $\mathcal{B}(H)$. Let $\{\lambda_k\}$ denote the positive eigenvalues of T (either finitely many or converging to zero if infinitely many). Then the spectral representation of T is

$$(2.2) \qquad T = \sum_k \lambda_k \, (g_k \otimes g_k) = \sum_k h_k \otimes h_k,$$

where $\{g_k\}$ are orthonormal eigenvectors of T and $h_k = \lambda_k^{1/2} g_k$. Since the eigenvalues of T coincide with its singular values, the trace of T coincides with its trace-class norm. In particular,

$$(2.3) \qquad \sum_k \|h_k\|^2 \;=\; \sum_k \lambda_k \;=\; \mathrm{trace}(A) \;=\; \|A\|_{\mathcal{I}_1} \;<\; \infty.$$

Let us say that T is of *Type A* with respect to the orthonormal basis $\mathcal{E}$ if, for the eigenvectors $\{h_k\}$ as above, we have that

$$(2.4) \qquad \sum_k \|h_k\|_1^2 \;<\; \infty.$$

Note that this is just the (somewhat unusual) formula displayed above for the trace of T with the H^1-norm used in place of the usual Hilbert space norm of the vectors $\{h_k\}$.

Let us also say that T is of *Type B* with respect to the orthonormal basis $\mathcal{E}$ if there is *some* sequence of vectors $\{v_k\}$ in H such that

$$\sum_k \|v_k\|_1^2 \;<\; \infty \qquad \text{and} \qquad T = \sum_k v_k \otimes v_k,$$

where the convergence of this series is in the strong operator topology. Note that the definition of Type B does depend on the choice of orthonormal basis, for reasons which will become apparent later.

EXAMPLE 2.2. Not every positive trace-class operator is of Type A. For example, choose any vector $f \in H \setminus H^1$. Then $T = f \otimes f$ is positive and trace-class (in fact, has rank one), but is clearly not of Type A with respect to $\mathcal{E}$. In particular, positive finite-rank operators need not be of Type A.

In fact, we can characterize all finite-rank operators that are of Type A or Type B. To do this, we need the following lemma.

LEMMA 2.3. *Let T be a positive operator on H such that $T = \sum_k v_k \otimes v_k$, where the series has either finitely or countably many terms and converges in the strong operator topology. Then $\overline{\mathrm{range}(T)} = \overline{\mathrm{span}}\{v_k\}$.*

PROOF. Let P be the orthogonal projection of H onto $\overline{\mathrm{range}(T)}$, and let $P^\perp = I - P$. Then

$$0 \;=\; P^\perp T P^\perp \;=\; \sum_k P^\perp v_k \otimes P^\perp v_k.$$

But each projection $P^\perp v_k \otimes P^\perp v_k$ is a positive operator, so this implies that $P^\perp v_k = 0$ for all k. Hence $v_k \in P(H) = \overline{\mathrm{range}(T)}$, so $\overline{\mathrm{span}}\{v_k\} \subseteq \mathrm{range}(T)$.

For the converse inclusion, suppose that $\overline{\mathrm{span}}\{v_k\}$ was a proper subset of $\overline{\mathrm{range}(T)}$. Then we could find a unit vector $z \in \overline{\mathrm{range}(T)}$ that is perpendicular to each v_k. Let $z_j = T w_j \in \mathrm{range}(T)$ be such that $z_j \to z$, and let $Q = z \otimes z$ be the orthogonal projection onto the span of z. Then $Q z_j \to Q z = z$, but for each j we also have

$$Q z_j \;=\; Q T w_j \;=\; \sum_k \langle w_j, v_k \rangle Q v_k \;=\; 0,$$

so this implies $z = 0$, which is a contradiction. $\square$

Now we can characterize the finite-rank operators that are of Type A or Type B.

PROPOSITION 2.4. *Let T be a positive finite-rank operator. Then the following statements are equivalent.*

(a) *T is of Type A.*

(b) *T is of Type B.*

(c) *Each eigenvector of T corresponding to a nonzero eigenvalue belongs to H^1.*

(d) *$\text{range}(T) \subseteq H^1$.*

PROOF. (a) $\Leftrightarrow$ (c) and (a) $\Rightarrow$ (b) are clear.

(b) $\Rightarrow$ (d). Assume that $T = \sum_k v_k \otimes v_k$ where $\sum_k \|v_k\|_1^2 < \infty$. Since T has finite rank, it has closed range. Lemma 2.3 therefore implies that $\text{range}(T) = \overline{\text{span}}\{v_k\} \subseteq H^1$.

(d) $\Rightarrow$ (a). Suppose that $\text{range}(T) \subseteq H^1$, and let $T = \sum_k h_k \otimes h_k$ be the spectral representation of T as in (2.2). Since T has finite rank, this is a finite sum. Further, Lemma 2.3 implies that $h_k \in \overline{\text{range}(T)} \subset H^1$ for each k. Since there are only finitely many k, it follows that $\sum_k \|h_k\|_1^2 < \infty$, so T is of Type A. $\square$

In particular, the operator of Example 2.2 is neither of Type A nor of Type B.

2.2. Problems. Now we will give a set of problems related to the definition of Type A and Type B operators.

As shown in Proposition 2.4, Type A and Type B are equivalent for positive finite-rank operators. Our first problem asks if this is true in general.

PROBLEM 2.5. If T is of Type B with respect to an orthonormal basis $\mathcal{E}$, must it be of Type A with respect to $\mathcal{E}$? $\diamond$

We expect that the answer to this problem is negative, but this leads immediately to the following problem.

PROBLEM 2.6. Let $\mathcal{E}$ be an orthonormal basis for H. Find a characterization of all positive trace-class operators T that are of Type B with respect to $\mathcal{E}$. $\diamond$

Our next problem comes closer to an operator-theoretic formulation of the question of Feichtinger. This problem asks if a particular class of operators are of Type A. Note that for this class of operators, each eigenvector of T corresponding to a nonzero eigenvalue belongs to H^1 (compare Proposition 2.4).

PROBLEM 2.7. Let $\mathcal{E} = \{e_k\}_{k \in \mathbf{N}}$ be an orthonormal basis for H. Fix scalars $\{c_{mn}\}_{m,n \in \mathbf{N}} \in \ell^1$ such that $c_{mn} = \overline{c_{nm}}$ for all m, n. Each operator $e_m \otimes e_n$ is trace-class, with trace-class norm

$$\|e_m \otimes e_n\|_{\mathcal{I}_1} = \|e_m\| \, \|e_n\| = 1.$$

Therefore, we can define $T \colon H \to H$ by

$$T = \sum_{m=1}^{\infty} \sum_{n=1}^{\infty} c_{mn} \, (e_m \otimes e_n).$$

This series converges in the strong operator topology and also absolutely in trace-class norm, because

$$(2.5) \qquad \|T\|_{\mathcal{I}_1} \leq \sum_{m=1}^{\infty} \sum_{n=1}^{\infty} |c_{mn}| \, \|e_m \otimes e_n\|_{\mathcal{I}_1} = \sum_{m=1}^{\infty} \sum_{n=1}^{\infty} |c_{mn}| < \infty.$$

Thus $T \in \mathcal{I}_1$. Further, the condition $c_{mn} = \overline{c_{nm}}$ implies that T is self-adjoint, and for simplicity we will also assume that T is positive.

If we write the spectral representation of T as in (2.2), then for each eigenvalue λ_k we have

$$\lambda_k h_k = T h_k = \sum_{m=1}^{\infty} \sum_{n=1}^{\infty} c_{mn} \langle h_k, e_m \rangle e_n.$$

Since $\|h_k\| = \lambda_k^{1/2} > 0$, since $\|e_m\| = 1$, and since $\|e_n\|_1 = 1$, we therefore have

$$
\begin{aligned}
|\lambda_k| \, \|h_k\|_1 = \|T h_k\|_1 &\leq \sum_{m=1}^{\infty} \sum_{n=1}^{\infty} |c_{mn}| \, |\langle h_k, e_m \rangle| \, \|e_n\|_1 \\
&\leq \sum_{m=1}^{\infty} \sum_{n=1}^{\infty} |c_{mn}| \, \|h_k\| \, \|e_m\| \, \|e_n\|_1 \\
&= \lambda_k^{1/2} \sum_{m=1}^{\infty} \sum_{n=1}^{\infty} |c_{mn}| < \infty.
\end{aligned}
$$

Thus $\|h_k\|_1 < \infty$ for each k, and in particular we have $h_k \in H^1$ for every k.

Problem: Is this operator of Type A? That is, must we have $\sum_k \|h_k\|_1^2 < \infty$? Unfortunately, the calculation above only tells us that

$$\sum_k \|h_k\|_1^2 \leq C \sum_k \frac{1}{\lambda_k},$$

and the right-hand side above is infinite if there are infinitely many eigenvalues. $\diamond$

A positive solution to Problem 2.7 would imply a positive solution to the original problem of Feichtinger. In particular, the following is an equivalent reformulation of his question.

PROBLEM 2.8. Let $H = L^2(\mathbf{R})$ and let $\mathcal{E}$ be a *Wilson orthonormal basis* for $L^2(\mathbf{R})$ (defined precisely in Section 3.5 below). If T is one of the operators defined in Problem 2.7 with respect to a Wilson basis, must T be of Type A? $\diamond$

Thus, the operator-theoretic problem of Problem 2.7 extracts the essence of Feichtinger's question without reference to the specific structure of a Wilson basis. A counterexample to Problem 2.7 would not necessarily settle the original question, for it could be the case that the particular structure of the Wilson bases plays a role. That is, if this problem really just depends on having an orthonormal basis, then it is a purely operator-theoretic question, while if it depends more explicitly on the particular functional properties of Wilson bases then is becomes more specifically a question about the modulation spaces. In any case, Problem 2.7 establishes a potentially interesting connection between operator theory and modulation space theory.

3. Background and Setting

In the remainder of this article we will attempt to present some background on the modulation spaces and then give the original formulation of Feichtinger's question.

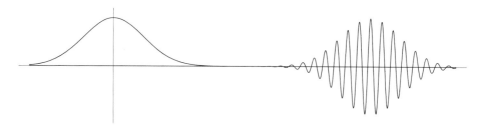

FIGURE 1. Window g and time-frequency shift $g_{x,\omega}$ (real part).

3.1. The STFT. Given a *window function* $g \in L^2(\mathbf{R})$ and given x, $\omega \in \mathbf{R}$, let $g_{x,\omega}$ be the *time-frequency shift* of g defined by

$$g_{x,\omega}(t) = e^{2\pi i \omega t} g(t - x)$$

(see Figure 1). The *short-time Fourier transform* (STFT) or *continuous Gabor transform* of a function $f \in L^2(\mathbf{R})$ with respect to the window g is

$$V_g f(x, \omega) = \langle f, g_{x,\omega} \rangle = \int f(t) \overline{g(t - x)} \, e^{-2\pi i \omega t} \, dt.$$

A standard fact is that V_g is a multiple of an isometry from $L^2(\mathbf{R})$ into $L^2(\mathbf{R}^2)$. Specifically, if $\|g\|_2 = 1$, then we have the norm equality
(3.1)

$$\|f\|_2 = \|V_g f\|_2 = \left(\iint |V_g f(x, \omega)|^2 \, dx \, d\omega \right)^{1/2} = \left(\iint |\langle f, g_{x,\omega} \rangle|^2 \, dx \, d\omega \right)^{1/2},$$

and we also have the formal inversion formula

$$f = \iint V_g f(x, \omega) \, g_{x,\omega} \, dx \, d\omega.$$

The inversion formula represents a function f as a superposition of "notes" $g_{x,\omega}$, with the value of $V_g f(x, \omega)$ determining the "amount" of the note $g_{x,\omega}$ that is present in f. At least qualitatively, $V_g f(x, \omega)$ represents the amount of frequency ω present in f at time x, and hence we say that $V_g f$ is a *time-frequency representation* of f. See [**Grö01**] for precise interpretations of these remarks and the inversion formula.

3.2. Modulation Spaces. The modulation spaces were invented and extensively investigated by Feichtinger, with some of the main references being [**Fei81**], [**Fei89**], [**FG89a**], [**FG89b**], [**FG97**], [**Fei03**]. For a detailed development of the theory of modulation spaces we refer to the original literature mentioned above and to Gröchenig's text [**Grö01**, Ch. 11–13]. For a personal historical account of the development of the modulation spaces, including the Feichtinger algebra in particular, see Feichtinger's recent article [**Fei06**].

The modulation spaces are defined by imposing a different norm on the STFT in place of the usual L^2 norm.

DEFINITION 3.1. Fix $1 \leq p \leq 2$, and let ϕ be any nonzero Schwartz-class function (for example, the Gaussian function e^{-x^2}). Then the *modulation space*

$M^p(\mathbf{R})$ consists of all functions $f \in L^2(\mathbf{R})$ such that

$$\|f\|_{M^p} = \|V_\phi f\|_p = \left(\iint |V_\phi f(x, \omega)|^p \, dx \, d\omega \right)^{1/p}$$

$$= \left(\iint |\langle f, \phi_{x,\omega} \rangle|^p \, dx \, d\omega \right)^{1/p} < \infty.$$

The definition of M^p is independent of the choice of ϕ in the sense that each different choice of ϕ defines an equivalent norm for the same set M^p. Each modulation space is a Banach space. Even M^∞, which is "nearly" all of the space of tempered distributions, is a Banach space, whereas the space of tempered distributions $\mathcal{S}'(\mathbf{R})$ is only a Fréchet space.

By equation (3.1), we have that $M^2 = L^2$. For other p, the space M^p is not L^p. Instead, the M^p norm measures the L^p norm *of the STFT of f*, not the L^p norm of f itself. Thus the M^p norm is quantifying the quality of joint time-frequency concentration that f possesses.

REMARK 3.2. a. Because ϕ is taken to be a Schwartz-class function, "inner products" $\langle f, \phi_{x,\omega} \rangle$ are defined not only when $f \in L^2(\mathbf{R})$ but whenever f is a tempered distribution. Thus, we can define modulation spaces not only when $p \leq 2$ but also for $p > 2$ by defining

$$M^p(\mathbf{R}) = \left\{ f \in \mathcal{S}'(\mathbf{R}) : \|f\|_{M^p} = \left(\iint |\langle f, \phi_{x,\omega} \rangle|^p \, dx \, d\omega \right)^{1/p} < \infty \right\}.$$

When $p > 2$, the space M^p is a superset of L^2. In fact, we have the inclusions

$$\mathcal{S}(\mathbf{R}) \subsetneq M^1(\mathbf{R}) \subsetneq M^2(\mathbf{R}) = L^2(\mathbf{R}) \subsetneq M^\infty(\mathbf{R}) \subsetneq \mathcal{S}'(\mathbf{R}).$$

b. The M^p spaces defined above are only the simplest examples of the modulation spaces; we can define other modulation spaces by imposing other norms on the Gabor coefficients. For example, if $1 \leq p, q \leq \infty$ and $v \colon \mathbf{R} \to (0, \infty)$ is a weight function, then the modulation space $M_v^{p,q}(\mathbf{R})$ consists of all tempered distributions f for which the norm

$$\|f\|_{M_v^{p,q}} = \|V_g f\|_{L_v^{p,q}} = \left(\int \left(\int |V_g f(x, \omega)|^p \, v(x, \omega)^p \, dx \right)^{q/p} d\omega \right)^{1/q}$$

is finite, with the usual adjustments if p or q is infinite. These more complicated spaces have many important applications. For example, the space $M_v^{\infty,1}$ plays an important role in modeling transmission channels for wireless communications [**SB03**], [**SH03**], [**Str06**], and in the theory of pseudodifferential operators [**GH99**], [**BO04**].

c. Among the modulation spaces, the space M^1 plays an especially important role. This space is often called the *Feichtinger algebra*, and it corresponds to the space H^1 in the statement of the problems in Section 2. The Feichtinger algebra is sometimes denoted by S_0 instead of M^1, to denote that it is a particular *Segal algebra*. The Feichtinger algebra has many interesting properties. For example, it is a Banach algebra under two operations: pointwise products and convolution. Also, M^1 is invariant under the Fourier transform, as is L^2, the Schwartz space $\mathcal{S}$, the tempered distributions $\mathcal{S}'$, and indeed each of the spaces M^p for $1 \leq p \leq \infty$. There are many equivalent characterizations of M^1; for example, it is the minimal non-trivial Banach space contained in L^1 that is isometrically invariant under both translations and modulations.

d. If we substitute translations and dilations for time-frequency shifts (i.e., we use the continuous wavelet transform instead of the STFT), then the analogue of the modulation spaces (obtained by imposing norms on the continuous wavelet transform) are the Besov or Triebel–Lizorkin spaces. The norms of the Besov and Triebel–Lizorkin spaces quantify the smoothness properties of functions, while the norms of the modulation spaces quantify the time-frequency concentration of functions. The spaces L^p for $p \neq 2$ are Triebel–Lizorkin spaces, but they are not modulation spaces. Both of the Besov/Triebel–Lizorkin and the modulation space classes are special cases of the general *coorbit theory* developed in [**FG89a**], [**FG89b**].

3.3. Gabor Frames. Gabor frames provide natural basis-like expansions of elements of the Hilbert space $L^2(\mathbf{R})$. We refer to the texts [**Dau92**], [**Grö01**], [**You01**], or [**Chr03**] for more information on frames and Gabor frames in particular.

Fix any *window function* $g \in L^2(\mathbf{R})$ and any α, $\beta > 0$. The *Gabor system* generated by g, α, and β is

$$\mathcal{G}(g, \alpha, \beta) = \{e^{2\pi i \beta n t} g(t - \alpha k)\}_{k,n \in \mathbf{Z}} = \{g_{\alpha k, \beta n}\}_{k,n \in \mathbf{Z}}.$$

If there exist constants A, $B > 0$ such that

$$\forall f \in L^2(\mathbf{R}), \quad A \|f\|_2^2 \leq \sum_{k,n \in \mathbf{Z}} |\langle f, g_{\alpha k, \beta n} \rangle|^2 \leq B \|f\|_2^2,$$

then $\mathcal{G}(g, \alpha, \beta)$ is called a *Gabor frame*, and A, B are *frame bounds*. In this case there exists a *canonical dual window* $\tilde{g} \in L^2(\mathbf{R})$ such that $\mathcal{G}(\tilde{g}, \alpha, \beta)$ is a frame and we have the basis-like *frame expansions*

$$(3.2) \quad \forall f \in L^2(\mathbf{R}), \quad f = \sum_{k,n \in \mathbf{Z}} \langle f, \tilde{g}_{\alpha k, \beta n} \rangle \, g_{\alpha k, \beta n} = \sum_{k,n \in \mathbf{Z}} \langle f, g_{\alpha k, \beta n} \rangle \, \tilde{g}_{\alpha k, \beta n},$$

where these series converge unconditionally in L^2-norm.

If we can take $A = B = 1$ then we call $\mathcal{G}(g, \alpha, \beta)$ a *Parseval Gabor frame*. This case is especially simple since the dual frame coincides with the original frame, and we have the orthonormal basis-like expansions

$$\forall f \in L^2(\mathbf{R}), \quad f = \sum_{k,n \in \mathbf{Z}} \langle f, g_{\alpha k, \beta n} \rangle \, g_{\alpha k, \beta n}.$$

It is easy to construct Gabor frames, see for example the "painless nonorthogonal expansions" of [**DGM86**]. We have the following facts.

(a) If $\alpha \beta > 1$ then $\mathcal{G}(g, \alpha, \beta)$ cannot be a frame, and in fact is incomplete in $L^2(\mathbf{R})$.

(b) If $\alpha \beta = 1$ then $\mathcal{G}(g, \alpha, \beta)$ is a Gabor frame if and only if it is a Riesz basis for $L^2(\mathbf{R})$, i.e., the image of an orthonormal basis under a continuous invertible map. Further, if $\alpha \beta = 1$ then $\mathcal{G}(g, \alpha, \beta)$ is a Parseval Gabor frame if and only if it is an orthonormal basis for $L^2(\mathbf{R})$.

(c) If $\alpha\beta < 1$, then any Gabor frame $\mathcal{G}(g, \alpha, \beta)$ is *overcomplete*, i.e., it contains a complete proper subset. In particular, such a frame is neither orthonormal nor a basis, and the coefficients in the frame expansions in (3.2) are not unique.

These facts are part of the *Density Theorem* for Gabor frames; see [**Hei06**] for a detailed survey of and references for this theorem.

Unfortunately, the *Balian–Low Theorem* implies that if $\mathcal{G}(g, \alpha, \beta)$ is a Riesz basis for $L^2(\mathbf{R})$, then the window g cannot be jointly well-localized in the time-frequency plane—either g or $\hat{g}$ must decay slowly (see [**BHW95**] for a survey of the Balian–Low Theorem). As a consequence, in practice we must usually deal with overcomplete Gabor frames. As long as $\alpha\beta < 1$, we can construct Gabor frames or Parseval Gabor frames with g extremely nice, e.g., Schwartz-class or even infinitely differentiable and compactly supported.

3.4. Gabor Frames and the Modulation Spaces. If $\mathcal{G}(\phi, \alpha, \beta)$ is a frame for $L^2(\mathbf{R})$ and ϕ is a "nice" function (e.g., ϕ Schwartz-class or indeed any $\phi \in M^1$), then the Gabor frame expansions given in (3.2) converge not only in L^2 but *in all the modulation spaces*, as follows (see [**Grö01**, Ch. 11–13] for proofs).

THEOREM 3.3. *Assume that*

(a) $\phi \in M^1(\mathbf{R})$, *and*
(b) $\mathcal{G}(\phi, \alpha, \beta)$ *is a frame for $L^2(\mathbf{R})$.*

Then the following statements hold.

(i) *The dual window $\tilde{\phi}$ belongs to $M^1(\mathbf{R})$.*

(ii) *For every $1 \le p \le \infty$ we have that*

$$\forall f \in M^p(\mathbf{R}), \quad f = \sum_{k,n\in\mathbf{Z}} \langle f, \tilde{\phi}_{\alpha k, \beta n}\rangle\, \phi_{\alpha k, \beta n} = \sum_{k,n\in\mathbf{Z}} \langle f, \phi_{\alpha k, \beta n}\rangle\, \tilde{\phi}_{\alpha k, \beta n},$$

where these series converge unconditionally in the norm of M^p (weak* *convergence if $p = \infty$).*

(iii) *For every $1 \le p \le \infty$ the Gabor frame coefficients provide an* equivalent norm *for the modulation space M^p, i.e.,*

$$(3.3) \qquad \|f\|_{M^p} = \left(\sum_{k,n\in\mathbf{Z}} |\langle f, \phi_{\alpha k, \beta n}\rangle|^p \right)^{1/p}$$

is an equivalent norm for M^p.

Thus, $\mathcal{G}(\phi, \alpha, \beta)$ is a *Banach frame* for M^p in the sense of [**Grö91**], [**CHL99**]. However, we emphasize that rather than constructing a Banach frame for a single particular space, Theorem 3.3 says that any Gabor frame for the Hilbert space $L^2(\mathbf{R})$ whose window lies in M^1 is *simultaneously* a Banach frame for *every modulation space.*

The fact that if $\phi \in M^1$ then $\tilde{\phi} \in M^1$ as well was proved by Gröchenig and Leinert [**GL04**], using deep results about symmetric Banach algebras (for the case that $\alpha\beta$ is rational, this result was obtained earlier by Feichtinger and Gröchenig in [**FG97**]). A different proof based on the concept of *localized frames* was given in [**BCHL06**]. That proof also extends to *irregular Gabor frames*, whose index set is not a lattice, and whose dual frame will not itself be a Gabor frame.

3.5. Wilson Bases. The equivalent norm for M^p given in equation (3.4) suggests that we should have $M^p \cong \ell^p$. While this is true, it does not follow from the facts we have presented so far. The issue is that Gabor frames need not be bases, and hence we cannot use them to define an isomorphism from M^p to ℓ^p. For example, even for the case $p = 2$ we know that for an overcomplete Gabor frame $\mathcal{G}(\phi, \alpha, \beta)$, the range of the *analysis mapping* $f \mapsto \{\langle f, \phi_{\alpha k, \beta n}\rangle\}_{k,n \in \mathbf{Z}}$ is only a proper subspace of ℓ^2. Moreover, because of the Balian–Low Theorem, we cannot get around this by trying to construct a nice ϕ so that the Gabor frame is a Riesz basis—no such nice ϕ exists. And ϕ must be nice (specifically $\phi \in M^1$) in order to apply Theorem 3.3 to conclude that the Gabor frame coefficients will provide an equivalent norm for M^p and that the Gabor frame expansions will converge in M^p. A frame with $\phi \in L^2$ that is not in M^1 will provide frame expansions for L^2, but those frame expansions will not converge in M^p.

Fortunately, there does exist a remarkable construction of an orthonormal basis for L^2, called a *Wilson basis*, which is simultaneously an unconditional basis for every modulation space. Wilson bases were first suggested by Wilson in [**Wil87**]. The fact that they provide orthonormal bases for $L^2(\mathbf{R})$ was rigorously proved by Daubechies, Jaffard, and Journé [**DJJ91**]. Feichtinger, Gröchenig, and Walnut proved that Wilson bases are unconditional bases for all of the modulation spaces [**FGW92**]. We also mention that the *local sine and cosine bases* of Coifman and Meyer [**CM91**] include many examples of Wilson and wavelet bases, and that the lapped transforms of Malvar [**Mal90**] are closely related. We refer to the original literature and to Sections 8.5 and 12.3 of [**Grö01**] for more details and for proofs of the results below.

The construction of a Wilson basis starts with a "twice redundant" Parseval Gabor frame $\mathcal{G}(g, \frac{1}{2}\mathbf{Z} \times \mathbf{Z})$ whose generator satisfies a symmetry condition, then forms linear combinations of elements, namely,

$$M_n T_{\frac{k}{2}} g \pm M_{-n} T_{\frac{k}{2}} g,$$

and finally "magically" extracts from the set of these linear combinations a subset which forms an orthonormal basis for $L^2(\mathbf{R})$. Moreover, if the original window g has sufficient joint concentration in the time-frequency plane, then a Wilson basis will be an unconditional basis not only for $L^2(\mathbf{R})$, but for all the modulation spaces. This is summarized in the following result, see [**Grö01**, Thm. 8.5.1] and [**Grö01**, Thm. 8.5.1] for proof.

THEOREM 3.4. *Assume that $\mathcal{G}(g, \frac{1}{2}\mathbf{Z} \times \mathbf{Z})$ is a Parseval Gabor frame for $L^2(\mathbf{R})$, and that $g(x) = \overline{g(-x)}$. Define*

$$\psi_{k,0} = T_k g, \qquad k \in \mathbf{Z},$$

and

$$\psi_{k,n}(x) = \begin{cases} \sqrt{2}\cos(2\pi n x)\, g(x - \frac{k}{2}), & \text{if } k + n \text{ is even}, \\ \sqrt{2}\sin(2\pi n x)\, g(x - \frac{k}{2}), & \text{if } k + n \text{ is odd}, \end{cases}$$

and set

$$\mathcal{W}(g) = \left\{\psi_{k,n}\right\}_{k \in \mathbf{Z},\, n \geq 0}.$$

Then $\mathcal{W}(g)$ is an orthonormal basis for $L^2(\mathbf{R})$.

If in addition we have $g \in M^1(\mathbf{R})$, then the following further statements hold.

(i) *For every* $1 \leq p \leq \infty$ *we have that*

$$\forall f \in M^p(\mathbf{R}), \quad f = \sum_{k \in \mathbf{Z}} \sum_{n \geq 0} \langle f, \psi_{kn} \rangle \, \psi_{kn},$$

where the series converges unconditionally in the norm of M^p *(weak* convergence if* $p = \infty$*).*

(ii) *For every* $1 \leq p \leq \infty$ *the Wilson basis coefficients provide an* equivalent norm *for the modulation space* M^p, *i.e.,*

(3.4)
$$\|f\|_{M^p} = \left(\sum_{k \in \mathbf{Z}} \sum_{n \geq 0} |\langle f, \phi_{\alpha k, \beta n} \rangle|^p \right)^{1/p}$$

is an equivalent norm for M^p.

Consequently, $f \mapsto \{\langle f, \psi_{kn} \rangle\}_{k \in \mathbf{Z}, n \geq 0}$ defines an isomorphism of M^p onto ℓ^p. In particular, if $H = L^2(\mathbf{R})$ and $\mathcal{E} = \mathcal{W}(g)$ is a Wilson basis for $L^2(\mathbf{R})$, then the space H^1 defined by equation (2.1) is precisely the modulation space $M^1(\mathbf{R})$. When we consider Wilson bases in this paper, we assume that they are constructed from M^1 windows.

By forming tensor products, the Wilson basis construction can be extended to create unconditional bases for the modulation spaces in higher dimensions.

The definition of the Wilson bases is rather technical, and the procedure behind it is in some sense "magical" and is not well-understood. For example, it is not known if it is possible to start with a "three times" redundant Gabor Parseval frame $\mathcal{G}(g, \frac{1}{3}\mathbf{Z} \times \mathbf{Z})$ and somehow create an orthonormal basis in the spirit of the Wilson bases.

3.6. Integral Operators. Now we return to the setting of Feichtinger's original problem. Given a *kernel function* $k \in L^2(\mathbf{R}^2)$, the corresponding *integral operator* is

(3.5)
$$Tf(x) = \int k(x, y) \, f(y) \, dy.$$

In terms of the kernel, T is self-adjoint if

$$k(x, y) = \overline{k(y, x)}.$$

Because $k \in L^2(\mathbf{R}^2)$, we know that T is a compact mapping of $L^2(\mathbf{R})$ onto itself. In fact, we have the equivalence that

$$k \in L^2(\mathbf{R}^2) \quad \Longleftrightarrow \quad T \text{ is a Hilbert–Schmidt operator.}$$

While such a characterization is not known for the trace-class operators, we will prove below a simple sufficient condition, namely that if k lies in the two-dimensional version of the Feichtinger algebra, i.e., $k \in M^1(\mathbf{R}^2)$, then T is a trace-class operator. This result was proved in [**Grö96**] and [**GH99**], and in fact is only a special case of more general theorems proved in [**GH99**]. For a survey of the role that modulation spaces play in the theory of integral and pseudodifferential operators, see [**Hei03**].

THEOREM 3.5. *If* $k \in M^1(\mathbf{R}^2)$, *then the corresponding integral operator* T *defined by* (3.5) *is trace-class, i.e.,* $T \in \mathcal{I}_1$.

PROOF. Let $\mathcal{W}(g)$ be a Wilson orthonormal basis for $L^2(\mathbf{R})$ such that $g \in M^1(\mathbf{R})$. By Theorem 3.4, $\mathcal{W}(g)$ is also an unconditional basis for $M^1(\mathbf{R})$. For simplicity of notation, let us index this basis as $\mathcal{W}(g) = \{w_n\}_{n \in \mathbf{N}}$.

Now construct an orthonormal basis for $L^2(\mathbf{R}^2)$ by forming tensor products, i.e., set

$$(3.6) \qquad W_{mn}(x,y) = w_m(x)\,\overline{w_n(y)}.$$

Then $\mathcal{U} = \{W_{mn}\}_{m,n \in \mathbf{N}}$ is both an orthonormal basis for $L^2(\mathbf{R}^2)$ and an unconditional basis for $M^1(\mathbf{R}^2)$. Therefore, since $k \in M^1(\mathbf{R}^2)$, we have that

$$(3.7) \qquad k = \sum_{m,n \in \mathbf{Z}} \langle k, W_{mn} \rangle\, W_{mn},$$

with convergence of the series in M^1-norm, and furthermore

$$(3.8) \qquad \|k\|_{M^1} = \sum_{m,n \in \mathbf{Z}} |\langle k, W_{mn} \rangle| < \infty.$$

Substituting the expansion (3.7) into the definition of the integral operator in (3.5) yields

$$
\begin{aligned}
(3.9) \qquad Tf(x) &= \int_{\mathbf{R}} \sum_{m,n \in \mathbf{Z}} \langle k, W_{mn} \rangle\, W_{mn}(x,y)\, f(y)\, dy \\
&= \sum_{m,n \in \mathbf{Z}} \langle k, W_{mn} \rangle \int_{\mathbf{R}} w_m(x)\,\overline{w_n(y)}\, f(y)\, dy \\
&= \sum_{m,n \in \mathbf{Z}} \langle k, W_{mn} \rangle \langle f, w_n \rangle\, w_m(x) \\
&= \sum_{m,n \in \mathbf{Z}} \langle k, W_{mn} \rangle\, (w_m \otimes w_n)(f)(x).
\end{aligned}
$$

The interchanges in order can all be justified because of the absolute convergence of the series implied by (3.8). Therefore we conclude that

$$(3.10) \qquad T = \sum_{m,n \in \mathbf{Z}} \langle k, W_{mn} \rangle\, (w_m \otimes w_n).$$

Since each operator $w_m \otimes w_n$ belongs to $\mathcal{I}_1$ and the scalars $\langle k, W_{mn} \rangle$ are summable, the series (3.10) converges absolutely in $\mathcal{I}_1$, and therefore $T \in \mathcal{I}_1$. $\square$

Theorem 3.5 can also be formulated in terms of pseudodifferential operators. In particular, the hypothesis that the kernel k of T written as an integral operator belongs to M^1 is equivalent to the hypothesis that the symbol σ of T written as a pseudodifferential operator belongs to M^1. We refer to [**Hei03**] for discussion along these lines.

3.7. The Problem. At last we come to the actual original question of Feichtinger, which is the following question about integral operators whose kernel belongs to $M^1(\mathbf{R}^2)$.

PROBLEM 3.6. Let T be a positive integral operator whose kernel k lies in $M^1(\mathbf{R}^2)$. Let equation (2.2) be the spectral representation of T. Must it be true that

$$\sum_{n=1}^{\infty} \|h_n\|_{M^1}^2 < \infty? \qquad \Diamond$$

We close by showing why Problem 2.8 is an equivalent reformulation of Problem 3.6.

PROPOSITION 3.7. *Problem 2.8 and Problem 3.6 are equivalent.*

PROOF. We simply have to show that every operator of the type considered in Problem 2.8 is an operator of the type considered in Problem 3.6, and vice versa.

Suppose that T is an operator of the type considered in Problem 3.6, i.e., T is be a positive integral operator whose kernel k lies in $M^1(\mathbf{R}^2)$. Then by using a Wilson basis we can, as in (3.10), write

$$(3.11) \qquad T = \sum_{m,n\in\mathbf{Z}} c_{mn}\,(w_m \otimes w_n),$$

where the scalars $c_{mn} = \langle k, W_{mn} \rangle$ are summable. Hence T is exactly the type of operator considered in Problem 2.8.

Conversely, suppose that T is an operator of the type considered in Problem 2.8. That is, T is a positive operator of the form in (3.11) where the scalars c_{mn} are summable and $\{w_n\}_{n\in\mathbf{N}}$ is a Wilson orthonormal basis generated by an M^1 window. Let W_{mn} be the tensor product functions defined in (3.6). Then since $\{W_{mn}\}_{m,n\in\mathbf{N}}$ is an orthonormal basis for $L^2(\mathbf{R}^2)$, we have that

$$(3.12) \qquad k = \sum_{m,n\in\mathbf{Z}} c_{mn}W_{mn} \in L^2(\mathbf{R}^2).$$

As in the calculations in (3.9), it follows that k is the kernel of the operator T (and hence T is a Hilbert–Schmidt operator).

Moreover, each W_{mn} belongs to $M^1(\mathbf{R})$, so since the c_{mn} are summable, we have that the series in (3.12) converges absolutely in M^1-norm. Hence the kernel k belongs to $M^1(\mathbf{R})$, and therefore T is a trace-class operator of exactly the type considered in Problem 3.6. $\qquad\square$

References

[BHW95] J. J. Benedetto, C. Heil, and D. F. Walnut, *Differentiation and the Balian–Low Theorem*, J. Fourier Anal. Appl., **1** (1995), 355–402.

[BO04] A. Bényi and K. Okoudjou, *Bilinear pseudodifferential operators on modulation spaces*, J. Fourier Anal. Appl., **10** (2004), 301–313.

[BCHL06] R. Balan, P. G. Casazza, C. Heil, and Z. Landau, *Density, overcompleteness, and localization of frames, II. Gabor frames*, J. Fourier Anal. Appl., to appear (2006).

[CHL99] P. G. Casazza, D. Han, and D. R. Larson, *Frames for Banach spaces*, in: *The Functional and Harmonic Analysis of Wavelets and Frames* (San Antonio, TX, 1999), Contemp. Math., Vol. 247, L. W. Baggett and D. R. Larson, eds., Amer. Math. Soc., Providence, RI (1999), pp. 149–182.

[Chr03] O. Christensen, *An Introduction to Frames and Riesz Bases*, Birkhäuser, Boston, 2003.

[CM91] R. R. Coifman and Y. Meyer, *Remarques sur l'analyse de Fourier à fenêtre*, C. R. Acad. Sci. Paris Sér. I Math., **312** (1991), 259–261.

[Dau92] I. Daubechies, *Ten Lectures on Wavelets*, SIAM, Philadelphia, 1992.

[DGM86] I. Daubechies, A. Grossmann, and Y. Meyer, *Painless nonorthogonal expansions*, J. Math. Phys., **27** (1986), 1271–1283.

[DJJ91] I. Daubechies, S. Jaffard, and J.-L. Journé, *A simple Wilson orthonormal basis with exponential decay*, SIAM J. Math. Anal., **22** (1991), 554–573.

[DS88] N. Dunford and J. T. Schwartz, *Linear Operators, Part II*, Wiley, New York, 1988.

[Fei81] H. G. Feichtinger, *On a new Segal algebra*, Monatsh. Math., **92** (1981), 269–289.

[Fei89] H. G. Feichtinger, *Atomic characterizations of modulation spaces through Gabor-type representations*, Rocky Mountain J. Math., **19** (1989), 113–125.

[Fei03] H. G. Feichtinger, *Modulation spaces on locally compact Abelian groups*, in: *Wavelets and their Applications* (Chennai, January 2002), M. Krishna, R. Radha and S. Thangavelu, eds., Allied Publishers, New Delhi (2003), pp. 1–56.

[Fei06] H. G. Feichtinger, *Modulation spaces: Looking back and ahead*, Sampl. Theory Signal Image Process., **5** (2006), 109–140.

[FG89a] H. G. Feichtinger and K. Gröchenig, *Banach spaces related to integrable group representations and their atomic decompositions, I*, J. Funct. Anal., **86** (1989), 307–340.

[FG89b] H. G. Feichtinger and K. Gröchenig, *Banach spaces related to integrable group representations and their atomic decompositions, II*, Monatsh. Math., **108** (1989), 129–148.

[FG97] H. G. Feichtinger and K. Gröchenig, *Gabor frames and time-frequency analysis of distributions*, J. Funct. Anal., **146** (1997), 464–495.

[FGW92] H. G. Feichtinger, K. Gröchenig, and D. Walnut, *Wilson bases and modulation spaces*, Math. Nachr., **155** (1992), 7–17.

[Grö91] K. Gröchenig, *Describing functions: Atomic decompositions versus frames*, Monatsh. Math., **112** (1991), 1–42.

[Grö96] K. Gröchenig, *An uncertainty principle related to the Poisson summation formula*, Studia Math., **121** (1996), 87–104.

[Grö01] K. Gröchenig, *Foundations of Time-Frequency Analysis*, Birkhäuser, Boston, 2001.

[GH99] K. Gröchenig and C. Heil, *Modulation spaces and pseudodifferential operators*, Integral Equations Operator Theory, **34** (1999), 439–457.

[GL04] K. Gröchenig and M. Leinert, *Wiener's lemma for twisted convolution and Gabor frames*, J. Amer. Math. Soc., **17** (2004), 1–18.

[Hei03] C. Heil, *Integral operators, pseudodifferential operators, and Gabor frames*, in: *Advances in Gabor Analysis*, H. G. Feichtinger and T. Strohmer, eds., Birkhäuser, Boston (2003), pp. 153–169.

[Hei06] C. Heil, *History and evolution of the Density Theorem for Gabor frames*, preprint (2006).

[Mal90] H. S. Malvar, *Lapped transforms for efficient transform/subband coding*, IEEE Trans. Acoustics, Speech, Signal Proc., **38** (1990), 969–978.

[Sim79] B. Simon, *Trace ideals and their applications*, Cambridge University Press, Cambridge, 1979.

[Str06] T. Strohmer, *Pseudodifferential operators and Banach algebras in mobile communications*, Appl. Comput. Harmon. Anal., **20** (2006), 237–249.

[SB03] T. Strohmer and S. Beaver, *Optimal OFDM system design for time-frequency dispersive channels*, IEEE Trans. Communications, **51** (2003), 1111–1122.

[SH03] T. Strohmer and R. W. Heath, Jr., *Grassmannian frames with applications to coding and communication*, Appl. Comput. Harmon. Anal., **14** (2003), 257–275.

[Wil87] K. G. Wilson, *Generalized Wannier functions*, preprint (1987).

[You01] R. Young, *An Introduction to Nonharmonic Fourier Series*, Revised First Edition, Academic Press, San Diego, 2001.

(C. Heil) School of Mathematics, Georgia Institute of Technology, Atlanta, Georgia 30332 USA

E-mail address: heil@math.gatech.edu

(D. Larson) Department of Mathematics, Texas A&M University College Station, Texas 77843 USA

E-mail address: larson@math.tamu.edu

Contemporary Mathematics
Volume **451**, 2008

FRAME ANALYSIS AND APPROXIMATION IN REPRODUCING KERNEL HILBERT SPACES

PALLE E. T. JORGENSEN

ABSTRACT. We consider frames F in a given Hilbert space, and we show that every F may be obtained in a constructive way from a reproducing kernel and an orthonormal basis in an ambient Hilbert space. The construction is operator-theoretic, building on a geometric formula for the analysis operator defined from F. Our focus is on the infinite-dimensional case where *a priori* estimates play a central role, and we extend a number of results which are known so far only in finite dimensions. We further show how this approach may be used both in constructing useful frames in analysis and in applications, and in understanding their geometry and their symmetries.

1. INTRODUCTION

Frames are redundant bases which turn out in certain applications to be more flexible than the better known orthonormal bases (ONBs) in Hilbert space. The frames allow for more symmetries than ONBs do, especially in the context of signal analysis, and of wavelet constructions; see, e.g., [CoDa93, BDP05, Dut06]. Since frame bases (although containing redundancies) still allow for efficient algorithms, they have found many applications, even in finite dimensions; see, for example, [BeFi03, CaCh03, Chr03, Eld02, FJKO05, VaWa05].

As is well known, when a vector f in a Hilbert space $\mathcal{H}$ is expanded in an orthonormal basis B, there is then automatically an associated Parseval identity. In physical terms, this identity typically reflects a *stability* feature of a decomposition based on the chosen ONB B. Specifically, Parseval's identity reflects a conserved quantity for a problem at hand, for example, energy conservation in quantum mechanics.

The theory of frames (see Definitions 6.1) begins with the observation that there are useful vector systems which are in fact not ONBs but for which a Parseval formula still holds. In fact, in applications it is important to go beyond ONBs. While this viewpoint originated in signal processing (in connection with frequency bands, aliasing, and filters), the subject of frames appears now to be of independent interest in mathematics.

On occasion, we may have a system of vectors S in $\mathcal{H}$ for which Parseval's identity is still satisfied, but such that a generalized Parseval's identity might only hold up to a fixed constant c of scale. (For example, in sampling theory, a scale might be introduced as a result of "oversampling".) In this case, we say that the constant

2000 *Mathematics Subject Classification*. 33C50, 42C15, 46E22, 47B32.

Key words and phrases. tight frames, Parseval frames, Bessel sequences, Hilbert space, reproducing kernel, Shannon, polar decomposition.

This material is based upon work supported by the U.S. National Science Foundation under grants DMS-0139473 (FRG) and DMS-0457581.

c scales the expansion. Suppose a system of vectors S in a given Hilbert space $\mathcal{H}$ allows for an expansion, or decomposition of every f in $\mathcal{H}$, but the analogue of Parseval's identity holds only up to a fixed constant c of scale. In that case, we say that S is a *tight frame* with frame constant c. So the special case $c = 1$ is the case of a Parseval frame. For precise definitions of these terms, we refer to Section 2 below, or to the book literature, e.g., [Chr03].

Aside from applications, at least three of the other motivations for frame theory come from: (1) wavelets, e.g., [CoDa93], [BJMP05], and [BJMP06]; (2) from non-harmonic Fourier expansions [DuSc52]; and (3) from computations with polynomials in several variables, and their generalized orthogonality relations [DuXu01].

While frames already have impressive uses in signal processing (see, e.g., [ALTW04, Chr99]), they have recently [CCLV05, CKL04] been shown to be central in our understanding of a fundamental question in operator algebras, the Kadison–Singer conjecture. We refer the reader to [CFTW06] for up-to-date research, and to [Chr99, KaRi97, Nel57, Nel59] for background.

In all these cases, the authors work with recursive algorithms, and the issue of *stability* plays a crucial role. Stability, however, may obtain in situations that are much more general than the context of traditional ONBs, or even tight frames. In fact, stability may apply even when we have only *a priori* estimates, as opposed to identities: for example, when the scaled version of Parseval's identity is replaced with a pair of estimates, a fixed lower bound and an upper bound; see (6.1) below. If such bounds exist, they are called lower and upper frame bounds.

If a system S of vectors in a Hilbert space $\mathcal{H}$ satisfies such a pair of *a priori* estimates, we say that S is simply a *frame*. And if such an estimate holds only with an *a priori* upper bound, we say that S is a *Bessel sequence*. It is known (see, e.g., [AkGl93]) that for a fixed Hilbert space $\mathcal{H}$, the various classes of frames S in $\mathcal{H}$ may be obtained from some ambient Hilbert space $\mathcal{K}$ and an orthonormal basis B in $\mathcal{K}$, i.e., when the pair $(S, \mathcal{H})$ is given, there are choices of $\mathcal{K}$ such that the frame S may be obtained from applying a certain bounded operator T to a suitable ONB B in $\mathcal{K}$. Passing from the given structure in $\mathcal{H}$ to the ambient Hilbert space is called *dilation* in operator theory. The properties of the operator T which does the job depend on the particular frame in question. For example, if S is a Parseval frame, then T will be a projection of the ambient Hilbert space $\mathcal{K}$ onto $\mathcal{H}$. But this operator-theoretic approach to frame theory has been hampered by the fact that the ambient Hilbert space is often an elusive abstraction. Starting with a frame S in a fixed Hilbert space $\mathcal{H}$, then by dilation, or extension, we pass to an ambient Hilbert space $\mathcal{K}$. In this paper we make concrete the selection of the "magic" operator $T : \mathcal{K} \to \mathcal{H}$ which maps an ONB in $\mathcal{K}$ onto S. While existence is already known, the building of a dilation system $(\mathcal{K}, T, \text{ONB})$ is often rather non-constructive, and the various methods for getting $\mathcal{K}$ are fraught with choices that are not unique.

Nonetheless, it was shown recently [Dut04b, Dut06] that when the dilation approach is applied to Parseval frames of wavelets in $\mathcal{H} = L^2(\mathbb{R})$, i.e., to wavelet bases which are not ONBs, then the ambient Hilbert space $\mathcal{K}$ can be made completely explicit, and the constructions are algorithmic. Moreover, the "inflated" ONB in $\mathcal{K}$ then takes the form of a traditional ONB-wavelet basis, a so-called "super-wavelet". For details, see [BDP05, Dut04b], and also the papers [KoLa04, BJMP05, BJMP06].

It is the purpose of the present paper to show that the techniques which work well in this restricted context, "super-wavelets" and redundant wavelet frames, apply to a more general and geometric context, one which is motivated in turn by extension principles in probability theory; see, e.g., [PaSc72], [Dut04a], and [Jor06].

A key idea in our present approach is the use of reproducing Hilbert spaces, and their reproducing kernels in the sense of [Aro50]. See also [Nel57, Nel59] for an attractive formulation. We show that for every Hilbert space $\mathcal{H}$, and every frame S in $\mathcal{H}$ (even if S is merely a Bessel sequence), there is a way of constructing the ambient Hilbert space $\mathcal{K}$ in such a way that the operator T has a concrete reproducing kernel.

Finally, we mention that a recent paper [VaWa05] serves as a second motivation for our work; in fact [VaWa05] contains finite-dimensional cases of two of our present theorems. These results in [VaWa05] are Theorems 2.9 and 6.4 in that paper: The results in [VaWa05] are concerned with symmetries of tight frames, and with associated families of unitary representations of the symmetry groups. It turns out that this approach to symmetry is natural in the context of operator theory; see, e.g., [PaSc72], [Dut04a], [Dut06].

2. PRELIMINARY NOTIONS AND DEFINITIONS

Let S be a countable set, finite or infinite, and let $\mathcal{H}$ be a complex or real Hilbert space. We shall be interested in a class of spanning families of vectors $(\mathbf{v}(s))$ in $\mathcal{H}$ indexed by points $s \in S$. Their properties will be defined precisely below, and the families are termed *frames*. The simplest instance of this is when $\mathcal{H} = \ell^2(S) =$ the Hilbert space of all square-summable sequences, i.e., all $f \colon S \to \mathbb{C}$ such that $\sum_{s \in S} |f(s)|^2 < \infty$. In that case, set

$$(2.1) \qquad \langle\, f_1 \mid f_2 \,\rangle := \sum_{s \in S} \overline{f_1(s)}\, f_2(s)$$

for all $f_1, f_2 \in \ell^2(S)$.

It is then immediate that the delta functions $\{\, \delta_s \mid s \in S \,\}$ given by

$$(2.2) \qquad \delta_s(t) = \begin{cases} 1, & t = s, \\ 0, & t \in S \setminus \{s\}, \end{cases}$$

form an *orthonormal basis* (ONB) for $\mathcal{H}$, i.e., that

$$(2.3) \qquad \langle\, \delta_{s_1} \mid \delta_{s_2} \,\rangle = \begin{cases} 1 & \text{if } s_1 = s_2 \text{ in } S, \\ 0 & \text{if } s_1 \neq s_2, \end{cases}$$

and that this is a maximal orthonormal family in $\mathcal{H}$. Moreover,

$$(2.4) \qquad f = \sum_{s \in S} f(s)\, \delta_s \qquad \text{for all } f \in \ell^2(S).$$

It also is immediate from (2.1) that Parseval's formula

$$(2.5) \qquad \|f\|^2 = \sum_{s \in S} |\langle\, \delta_s \mid f \,\rangle|^2$$

holds for all $f \in \ell^2(S)$.

We shall consider pairs $(S, \mathcal{H})$ and indexed families

$$(2.6) \qquad \{\, \mathbf{v}(s) \mid s \in S \,\} \subset \mathcal{H}$$

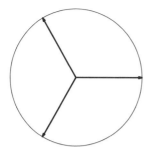
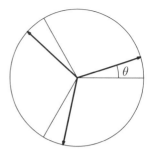

FIGURE 1. Two illustrations for $n = 3$

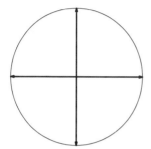
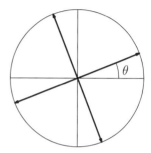

FIGURE 2. Two illustrations for $n = 4$

such that for some $c \in \mathbb{R}_+$, the identity

$$(2.7) \qquad \|f\|^2 = c \sum_{s \in S} |\langle \mathbf{v}(s) \mid f \rangle|^2$$

holds for all $f \in \mathcal{H}$.

When a Hilbert space $\mathcal{H}$ is given, our main result states that solutions to (2.7) exist if and only if $\mathcal{H}$ is isometrically embedded in $\ell^2(S)$. But we further characterize these embeddings, and we use this in understanding the geometry of tight frames (details below).

Definition 2.1. Let $\left(S, \mathcal{H}, c, (\mathbf{v}(s))_{s \in S}\right)$ be as above. We shall say that this system constitutes a *tight frame* with frame constant c if (2.7) holds.

(Note that if $(\mathbf{v}(s))_{s \in S}$ satisfies (2.7), then the scaled system $(\sqrt{c}\,\mathbf{v}(s))_{s \in S}$ has the property with frame constant one.)

Example 2.2. (S finite.) Let $\mathcal{H}$ be the two-dimensional real Hilbert space, and let $n \geq 3$. Set $S := \{1, 2, \ldots, n\} =: S_n$, and

$$(2.8) \qquad \mathbf{v}(s) := \begin{pmatrix} \cos\left(\dfrac{2\pi s}{n}\right) \\[2mm] \sin\left(\dfrac{2\pi s}{n}\right) \end{pmatrix}, \qquad s \in S.$$

Then it is easy to see that this constitutes a tight frame with frame constant $c = \frac{2}{n}$. Examples are presented in Figures 1 and 2.

Definition 2.3. Let $\mathcal{H}$ and $\mathcal{K}$ be Hilbert spaces over $\mathbb{C}$ or $\mathbb{R}$, and let $V : \mathcal{H} \to \mathcal{K}$ be a linear mapping. We say that V is an *isometry*, and that $\mathcal{H}$ is *isometrically embedded* in $\mathcal{K}$ (via V) if

$$(2.9) \qquad\qquad \|Vf\|_{\mathcal{K}} = \|f\|_{\mathcal{H}}, \qquad f \in \mathcal{H}.$$

Given a linear operator $V : \mathcal{H} \to \mathcal{K}$, we then denote the adjoint operator $V^* : \mathcal{K} \to \mathcal{H}$. It is easy to see that V is *isometric* iff $V^*V = I_{\mathcal{H}}$, where $I_{\mathcal{H}}$ denotes the identity operator in $\mathcal{H}$. Moreover, if V is isometric then $P = P_V = VV^* : \mathcal{K} \to \mathcal{K}$ is a projection, i.e.,

$$(2.10) \qquad\qquad P = P^* = P^2$$

holds, and the subspace

$$(2.11) \qquad\qquad P\mathcal{K} \subset \mathcal{K}$$

may be identified with $\mathcal{H}$ via the isometric embedding.

We state our next result only in the case of frame constant $c = 1$, but as noted it easily generalizes.

Theorem 2.4. *Let S be a countable set, and let $\mathcal{H}$ be a Hilbert space over $\mathbb{C}$ (or $\mathbb{R}$).*

Then the following two conditions are equivalent:

(i) *There is a tight frame $\{\mathbf{v}(s) \mid s \in S\} \subset \mathcal{H}$ with frame constant $c = 1$;*
(ii) *$\mathcal{H}$ is isometrically embedded (as a closed subspace) in $\ell^2(S)$.*

Proof. The details of proof will follow in the next section.   □

Remark 2.5. Tight frames with frame constant equal to one are called *Parseval frames*.

3. Proof of Theorem 2.4

Let the data be as specified in the theorem: A pair $(S, \mathcal{H})$ is given where S is a set and $\mathcal{H}$ is a Hilbert space. The conclusion is that (i) the S-tight-frame property for $\mathcal{H}$ is equivalent to (ii) the (isometric) embedding of $\mathcal{H}$ into $\ell^2(S)$.

Assume (i), and let $\{\mathbf{v}(s) \mid s \in S\} \subset \mathcal{H}$ be the frame system which is asserted. Now define

$$(3.1) \qquad\qquad V : \mathcal{H} \ni f \longmapsto (\langle \mathbf{v}(s) \mid f \rangle_{\mathcal{H}})_{s \in S}.$$

This operator is called the *analysis operator* for the frame.

Equivalently, the analysis operator V may be written in terms of the canonical ONB $(\delta_s)_{s \in S}$ for $\ell^2(S)$ as follows:

$$(3.2) \qquad\qquad Vf = \sum_{s \in S} \langle \mathbf{v}(s) \mid f \rangle \delta_s \qquad (\in \ell^2(S)).$$

Our first two observations are that the range of V is contained in $\ell^2(S)$, and that V is isometric. But note that both these conclusions are immediate consequences of identity (2.7) for the special case $c = 1$.

We formalize this in the following lemma. (The proof of converse implication in Theorem 2.4 will be resumed after the lemma.)

Lemma 3.1. *Let $(\mathbf{v}(s))_{s \in S}$ be a system in the Hilbert space $\mathcal{H}$, and let V be defined by (3.1). Then (2.7) holds if and only if $\sqrt{c}\,V$ is isometric.*

An easy calculation yields the formula for the adjoint operator:

$$(3.3) \qquad V^* \left((c_s)_{s \in S} \right) = \sum_{s \in S} c_s \mathbf{v}(s), \qquad (c_s) \in \ell^2(S).$$

Hence

$$(3.4) \qquad f = V^* V f = \sum_{s \in S} \langle \mathbf{v}(s) \mid f \rangle \mathbf{v}(s)$$

holds for all $f \in \mathcal{H}$.

Moreover, the projection operator $P := V V^*$ is given by the formula

$$(3.5) \qquad (P(c_s))_t = \sum_{s \in S} \langle \mathbf{v}(t) \mid \mathbf{v}(s) \rangle_{\mathcal{H}} c_s;$$

in other words, P has a concrete matrix representation in the Hilbert space $\ell^2(S)$. Specifically, P is represented as multiplication on column vectors $(c_s)_{s \in S}$, and the matrix of P is

$$(3.6) \qquad P(t,s) = \langle \mathbf{v}(t) \mid \mathbf{v}(s) \rangle_{\mathcal{H}}.$$

To prove the converse, assume (ii). Hence, there is an isometry $V : \mathcal{H} \to \ell^2(S)$. As we noted, this means that $P := V V^*$ is a projection in $\ell^2(S)$, and that

$$(3.7) \qquad P \ell^2(S) = \{ V f \mid f \in \mathcal{H} \}.$$

Now observe that $\mathbf{v}(s) := P(\delta_s)$, where δ_s for each $s \in S$ is the delta function of (2.2). We claim that (2.7) holds for $c = 1$. To see this, let $f \in \mathcal{H}$ be given. Then

$$\begin{aligned} \|f\|^2 = \|V f\|^2 &\underset{\text{by } (2.5)}{=} \sum_{s \in S} |\langle \delta_s \mid V f \rangle|^2 \\ &= \sum_{s \in S} |\langle \delta_s \mid P V f \rangle|^2 \\ &= \sum_{s \in S} |\langle P \delta_s \mid V f \rangle|^2 \\ &= \sum_{s \in S} |\langle \mathbf{v}(s) \mid f \rangle|^2, \end{aligned}$$

which is the desired conclusion. $\qquad \square$

Remark 3.2. Even when no restricting assumptions are placed on a given system of vectors $(\mathbf{v}(s))_{s \in S}$, the function P from (3.6) is always positive semidefinite; see Definition 5.1. Moreover, the following converse implication holds (Theorem 5.2 below), known as the *reconstruction principle*: Every positive semidefinite function has the form (3.6) for some Hilbert space and an associated system of vectors. As we show in Section 6, when this is specialized, we find that there is a graduated system of frame properties which may or may not hold for a given system of vectors $(\mathbf{v}(s))_{s \in S}$. Moreover we show that each of these properties reflects a corresponding axiom for the associated function (3.6), Gramian, or correlation matrix. Our results in Sections 5 and 6 below serve three purposes: (1) one is to use reproducing kernels [Aro50, Nel57, PaSc72] and their spectral theory to study classes of frames and their symmetries; (2) another (e.g., Theorems 5.5 and 6.2) is to use an idea from operator theory to establish results about frame deformations and their stability; and finally (3) these results serve to tie the operator theory to more current applications.

Corollary 3.3. *For $i = 1, 2$, consider two systems $(S_i, \mathcal{H}_i)$ of sets and Hilbert spaces, and assume that $(\mathbf{v}_i\,(s_i))_{s_i \in S_i}$ is a pair of tight frames with respective frame constants c_i. Then*

$$(3.8) \qquad \mathbf{w}\,(s_1, s_2) := \mathbf{v}_1\,(s_1) \otimes \mathbf{v}_2\,(s_2)$$

defines a tight frame for the tensor-product Hilbert space $\mathcal{H}_1 \otimes \mathcal{H}_2$ with frame constant $c = c_1 c_2$.

Proof. The result follows immediately from the theorem and Lemma 3.1. To see this, note that if V_i, $i = 1, 2$, are the operators defined in (3.1) above, then both

$$(3.9) \qquad \sqrt{c_i}\, V_i \colon \mathcal{H}_i \longrightarrow \ell^2\,(S_i)$$

are isometric embeddings. It then follows that

$$(3.10) \qquad \sqrt{c_1}\, V_1 \otimes \sqrt{c_2}\, V_2 = \sqrt{c_1 c_2}\, V_1 \otimes V_2$$

is an isometric embedding of $\mathcal{H}_1 \otimes \mathcal{H}_2$ into $\ell^2\,(S_1 \times S_2)$. From the fact that each V_i is defined from the system $(\mathbf{v}_i\,(s_i))_{s_i \in S_i}$, it follows readily that the tensor operator $V_1 \otimes V_2$ is defined from the system in (3.8).

Specifically,

$$(3.11) \quad \mathbf{w}\,(s_1, s_2) = (V_1 \otimes V_2)^*\,(\delta_{s_1} \otimes \delta_{s_2}) = V_1^* \delta_{s_1} \otimes V_2^* \delta_{s_2} = \mathbf{v}_1\,(s_1) \otimes \mathbf{v}_2\,(s_2),$$

which is the desired result. $\qquad\qquad\square$

Definition 3.4. Let $(S, \mathcal{H})$ be a pair as above: S is a set and $\mathcal{H}$ is a Hilbert space. Then the matrix

$$(3.12) \qquad k\,(t, s) := \langle\, \mathbf{v}\,(t) \mid \mathbf{v}\,(s)\,\rangle_{\mathcal{H}}, \qquad s, t \in S,$$

is called the *Gramian*, or the *Gram matrix*.

Corollary 3.5. *Let the pair $(S, \mathcal{H})$ be as in the statement of Theorem 2.4. Let $\{\,\mathbf{v}\,(s) \mid s \in S\,\}$ be a system of vectors in $\mathcal{H}$ and let $k\,(t, s) = \langle\,\mathbf{v}\,(t) \mid \mathbf{v}\,(s)\,\rangle$ be the corresponding Gram matrix. Then $(\mathbf{v}\,(s))_{s \in S}$ is a tight frame with frame constant c if and only if the following two conditions hold:*

(a) $k\,(t, s) = \overline{k\,(s, t)}$, *and*
(b) $\displaystyle\sum_{t \in S} k\,(s_1, t)\, k\,(t, s_2) = c^{-1} k\,(s_1, s_2)$ *for all $s_1, s_2 \in S$.*

Proof. It is immediate that the stated conditions are equivalent to the matrix P in (3.6) defining a projection when $P = cK$ and $K\,(t, s) := \langle\,\mathbf{v}\,(t) \mid \mathbf{v}\,(s)\,\rangle$. As a consequence, we see that the two conditions (a) and (b) are a restatement of (2.10), i.e., the definition of a projection. (We shall consider only orthogonal projections P, i.e., operators P where both of the conditions in (2.10) are assumed.) $\qquad\square$

Remark 3.6. As an application, we are now able to verify the assertion in Example 2.2. First note that the Gram matrix of (2.8) is

$$(3.13) \qquad k\,(s_1, s_2) = \cos\left(\frac{2\pi\,(s_1 - s_2)}{n}\right), \qquad s_1, s_2 \in S_n.$$

Hence property (a) in the corollary is immediate. To verify (b), note that

$$
\sum_{t \in S_n} k(s_1, t)\, k(t, s_2) = \sum_{t \in S_n} \cos\left(\frac{2\pi(s_1 - t)}{n}\right) \cos\left(\frac{2\pi(t - s_2)}{n}\right)
$$

$$
= \frac{1}{2} \sum_{t \in S_n} \left(\cos\left(\frac{2\pi(s_1 - s_2)}{n}\right) + \cos\left(\frac{2\pi(s_1 + s_2 - 2t)}{n}\right)\right)
$$

$$
= \frac{n}{2} \sum_{t \in S_n} \cos\left(\frac{2\pi(s_1 - s_2)}{n}\right) = c^{-1} k(s_1, s_2)
$$

with $c = \frac{2}{n}$.

In the last step, we used the trigonometric formula

$$
\sum_{l=1}^{n} \cos\left(\frac{4\pi l}{n} - \theta\right) = 0.
$$

(We refer to Figures 1 to 2 for simple illustrations.) $\qquad \square$

Corollary 3.7. *Let the pair $(S, \mathcal{H})$ be as in Corollary 3.5 and Theorem 2.4. Suppose $\{\mathbf{v}(s) \mid s \in S\}$ is a tight frame in $\mathcal{H}$ with frame constant c. Then every f in $\mathcal{H}$ has the representation*

$$
(3.14) \qquad\qquad f = c \sum_{s \in S} \langle\, \mathbf{v}(s) \mid f\,\rangle\, \mathbf{v}(s),
$$

where the sum converges in the norm of the Hilbert space $\mathcal{H}$.

Proof. In view of the argument in the proof of Corollary 3.5, we may reduce to the case where the frame constant c is one, i.e., $c = 1$. (In general, the Gram matrix $K(t, s) = \langle\, \mathbf{v}(t) \mid \mathbf{v}(s)\,\rangle_{\mathcal{H}}$ satisfies $K = c^{-1}P$ where P is a projection in the Hilbert space $\ell^2(S)$.) The reduction to the case $c = 1$ means that

$$
(3.15) \qquad\qquad V : \mathcal{H} \ni f \longmapsto (\langle\, \mathbf{v}(s) \mid f\,\rangle)_{s \in S} \in \ell^2(S)
$$

is isometric. The projection $P := VV^*$ is multiplication by the matrix

$$
(\langle\, \mathbf{v}(t) \mid \mathbf{v}(s)\,\rangle)_{s, t \in S}.
$$

Since P is the projection onto the range of V in (3.15) we conclude that

$$
(3.16) \qquad\qquad \langle\, \mathbf{v}(t) \mid f\,\rangle = \sum_{s \in S} \langle\, \mathbf{v}(t) \mid \mathbf{v}(s)\,\rangle \langle\, \mathbf{v}(s) \mid f\,\rangle,
$$

where the convergence is in $\ell^2(S)$. But the vectors $\{\mathbf{v}(t) \mid t \in S\}$ span a dense subspace in $\mathcal{H}$, so we get the desired formula

$$
(3.17) \qquad\qquad f = \sum_{s \in S} \langle\, \mathbf{v}(s) \mid f\,\rangle\, \mathbf{v}(s),
$$

now referring to the norm and the inner product in $\mathcal{H}$. Recall that on the range of V, the respective inner products of $\mathcal{H}$ and of $\ell^2(S)$ coincide. $\qquad \square$

Corollary 3.8. *Let the pair $(S, \mathcal{H})$ be as in Corollary 3.7, and suppose that*

$$
\{\mathbf{v}(s) \mid s \in S\}
$$

*is a tight frame for $\mathcal{H}$ with frame constant c. Further suppose that some $\xi: S \to \mathbb{C}$
satisfies*

(3.18)
$$f = c \sum_{s \in S} \xi_s \mathbf{v}(s),$$

where the sum is convergent in $\mathcal{H}$. Then

(3.19)
$$\sum_{s \in S} |\xi_s|^2 \geq \sum_{s \in S} |\langle \mathbf{v}(s) \mid f \rangle|^2.$$

Proof. With the assumptions in the corollary, apply the mapping V from (3.15) to
both sides in (3.18). We get the formula

(3.20)
$$(P\xi)_t = (\langle \mathbf{v}(t) \mid f \rangle_{\mathcal{H}}) \in \ell^2(S).$$

Hence

(3.21)
$$\xi = P\xi + (I - P)\xi$$

is an orthogonal decomposition, i.e.,

(3.22)
$$\|\xi\|_{\ell^2}^2 = \|P\xi\|_{\ell^2}^2 + \|(I - P)\xi\|_{\ell^2}^2,$$

and the conclusion (3.19) is immediate.   □

4. Shannon's example

In the Hilbert space $L^2(\mathbb{R})$ we will consider the usual Fourier transform

(4.1)
$$\hat{f}(t) := \int_{\mathbb{R}} e^{-i2\pi tx} f(x) \, dx,$$

where convergence is understood in the sense of L^2. The familiar interpolation
formula of Shannon [Ash90] applies to band-limited functions, i.e., to functions f
on $\mathbb{R}$ such that the Fourier transform $\hat{f}$ is of compact support. For the present
purpose, pick the following normalization

(4.2)
$$\operatorname{supp}\left(\hat{f}\right) \subset \left[-\frac{1}{2}, \frac{1}{2}\right],$$

and let $\mathcal{H}$ denote the subspace in $L^2(\mathbb{R})$ defined by this support condition. In
particular, $\mathcal{H}$ is the range of the projection operator P in $L^2(\mathbb{R})$ defined by

(4.3)
$$(Pf)(x) := \int_{-\frac{1}{2}}^{\frac{1}{2}} e^{i2\pi xt} \hat{f}(t) \, dt.$$

Shannon's interpolation formula applies to $f \in \mathcal{H}$, and it reads:

(4.4)
$$f(x) = \sum_{n \in \mathbb{Z}} f(n) \frac{\sin \pi(x - n)}{\pi(x - n)}.$$

Definitions 4.1. Let $S \subset \mathbb{R}$, and set

(4.5)
$$\mathbf{v}(s)(x) := \mathbf{v}(s, x) = \frac{\sin \pi(x - s)}{\pi(x - s)}, \qquad s \in S.$$

Hence if we take as index set $S := \mathbb{Z}$, then we may observe that the functions on
the right-hand side in Shannon's formula (4.4) are $\mathbf{v}(n)$ frame vectors, $n \in \mathbb{Z}$. We
shall be interested in other index sets S, so-called sets of *sampling points*. We shall

view the functions $\mathbf{v}(s)$ as vectors in $\mathcal{H}$. The inner product in $\mathcal{H}$ will be that which is induced from $L^2(\mathbb{R})$, i.e.,

$$(4.6) \qquad \langle f_1 \mid f_2 \rangle := \int_{\mathbb{R}} \overline{f_1(x)}\, f_2(x)\, dx.$$

The following is well known but is included as an application of Corollary 3.5. It is also an example of a pair $(S, \mathcal{H})$ where S is *infinite*.

Proposition 4.2. *Let $S \subset \mathbb{R}$ be a fixed discrete subgroup, and assume that $\mathbb{Z} \subset S$. Then $\{\mathbf{v}(s) \mid s \in S\}$ is a tight frame in $\mathcal{H}$ if and only if the group index $(S:\mathbb{Z})$ is finite, and in that case the frame constant c is $c = (S:\mathbb{Z})^{-1}$. For the Gram matrix, we have:*

$$(4.7) \qquad K(s_1, s_2) = \begin{cases} \dfrac{\sin \pi(s_1 - s_2)}{\pi(s_1 - s_2)} & \text{for } s_1, s_2 \in S,\ s_1 \neq s_2, \\ 1 & \text{if } s_1 = s_2. \end{cases}$$

Proof. Formula (4.7) for the Gram matrix follows from Fourier transform and the following computation of the inner products:

$$(4.8) \qquad \langle \mathbf{v}(s_1) \mid \mathbf{v}(s_2) \rangle_{\mathcal{H}} = \int_{\mathbb{R}} \frac{\sin \pi(x - s_1)}{\pi(x - s_1)} \frac{\sin \pi(x - s_2)}{\pi(x - s_2)}\, dx.$$

A second computation shows that $K(s_1, s_2) = \langle \mathbf{v}(s_1) \mid \mathbf{v}(s_2) \rangle$ satisfies the two conditions (a) and (b) of Corollary 3.5, i.e.,

$$(4.9) \qquad K(s_1, s_2) = K(s_2, s_1),$$

and

$$(4.10) \qquad \sum_{t \in S} K(s_1, t)\, K(t, s_2) = (S:\mathbb{Z})\, K(s_1, s_2) \qquad \text{for all } s_1, s_2 \in S.$$

The argument behind this formula uses the known fact that the functions

$$\{\mathbf{v}(n) \mid n \in \mathbb{Z}\}$$

in (4.4) form an ONB in $\mathcal{H}$; in particular, that

$$(4.11) \qquad \langle \mathbf{v}(n_1) \mid \mathbf{v}(n_2) \rangle_{\mathcal{H}} = \delta_{n_1, n_2} \qquad \text{for } n_1, n_2 \in \mathbb{Z}.$$

Since $\mathbb{Z} \subset S$, we get the following summation formula:

$$\sum_{t \in S} K(s_1, t)\, K(t, s_2) = \sum_{t \in S} \langle \mathbf{v}(s_1) \mid \mathbf{v}(t) \rangle_{\mathcal{H}} \langle \mathbf{v}(t) \mid \mathbf{v}(s_2) \rangle_{\mathcal{H}}$$

$$= \sum_{k \in S/\mathbb{Z}} \sum_{n \in \mathbb{Z}} \langle \mathbf{v}(s_1) \mid \mathbf{v}(k+n) \rangle_{\mathcal{H}} \langle \mathbf{v}(k+n) \mid \mathbf{v}(s_2) \rangle_{\mathcal{H}}$$

$$= \sum_{k \in S/\mathbb{Z}} \sum_{n \in \mathbb{Z}} \frac{\sin \pi(s_1 - k - n)}{\pi(s_1 - k - n)} \frac{\sin \pi(n - (s_2 - k))}{\pi(n - (s_2 - k))}$$

$$= \sum_{k \in S/\mathbb{Z}} \frac{\sin \pi(s_1 - s_2)}{\pi(s_1 - s_2)}$$

$$= (S:\mathbb{Z}) \langle \mathbf{v}(s_1) \mid \mathbf{v}(s_2) \rangle_{\mathcal{H}} = (S:\mathbb{Z})\, K(s_1, s_2),$$

which is the desired formula (4.10). The remaining conclusions in the proposition now follow immediately from Corollary 3.5. $\qquad \square$

Remark 4.3. The significance of using a larger subgroup S, i.e., $\mathbb{Z} \subset S$, in a modified version of Shannon's interpolation formula (4.4) is that a larger (discrete) group represents *"oversampling"*. However, note that the oversampling changes the frame constant.

As a contrast showing stability, we now recast a result on oversampling from [BJMP05] in the present context. It is for tight frames of *wavelet bases* in $L^2(\mathbb{R})$, and it represents an instance of stability: a case when oversampling leaves invariant the frame constant.

Proposition 4.4. *Let $\psi \in L^2(\mathbb{R})$, and suppose that the family*

$$(4.12) \qquad \psi_{j,k}(x) := 2^{j/2}\psi\left(2^j x - k\right), \qquad j, k \in \mathbb{Z},$$

is a Parseval frame in $L^2(\mathbb{R})$. Let $p \in \mathbb{N}$ be odd, $p > 1$, and set

$$(4.13) \qquad \tilde{\psi}_p(x) := \frac{1}{p}\psi\left(\frac{x}{p}\right),$$

and

$$(4.14) \qquad \tilde{\psi}_{p,j,k}(x) := 2^{j/2}\tilde{\psi}_p\left(2^j x - k\right), \qquad j, k \in \mathbb{Z}.$$

Then the "oversampled" family (4.14) is again a Parseval frame in the Hilbert space $L^2(\mathbb{R})$.

Proof. We refer the reader to the argument in Section 2 of [BJMP05], and to Example 6.9 below. $\square$

5. Symmetries

A basic fact of Hilbert space is that permutations induce unitary operators. By this we mean that a permutation of the vectors in an orthonormal basis for a Hilbert space $\mathcal{H}$ induces a unitary transformation U in $\mathcal{H}$. In this section we explore generalizations of this to frames.

Definition 5.1. Let S be a set and let $k \colon S \times S \to \mathbb{C}$ be a function. We say that k is *positive definite*, or more precisely positive semidefinite, if the following holds for all finite sums:

$$(5.1) \qquad \sum_{s \in S}\sum_{t \in S}\bar{\xi}_s \xi_t k(s,t) \geq 0.$$

While the following result is known, it is not readily available in the literature, at least not precisely in the form in which we need it. Thus we include here its statement to save readers from having to track it down.

Theorem 5.2. (Kolmogorov, Parthasarathy–Schmidt [PaSc72])

(a) *Let $k \colon S \times S \to \mathbb{C}$ be positive definite. Then there are a Hilbert space $\mathcal{H}$ and vectors $\{\mathbf{v}(s) \mid s \in S\} \subset \mathcal{H}$ such that*

$$(5.2) \qquad k(s,t) = \langle \mathbf{v}(s) \mid \mathbf{v}(t)\rangle_{\mathcal{H}}, \qquad s, t \in S,$$

and $\mathcal{H}$ is the closed linear span of $\{\mathbf{v}(s) \mid s \in S\}$.

(b) *Let $\pi \colon S \to S$ be a bijection. Then*

$$(5.3) \qquad U_\pi \colon \mathbf{v}(s) \longmapsto \mathbf{v}(\pi(s))$$

extends to a unitary operator in $\mathcal{H}$ $(=: \mathcal{H}(k))$ if and only if

$$(5.4) \qquad k(\pi(s), \pi(t)) = k(s,t) \qquad \text{for all } s, t \in S.$$

(c) *If* $\lambda \colon S \to \mathbb{T} = \{\, z \in \mathbb{C} \mid |z| = 1 \,\}$ *and if* π *is as in* (b), *then*

(5.5) $$U_{\pi,\lambda} \colon \mathbf{v}\,(s) \longmapsto \lambda\,(s)\,\mathbf{v}\,(\pi\,(s))$$

extends to a unitary operator in $\mathcal{H}$ *if and only if*

(5.6) $$\overline{\lambda\,(s)}\,\lambda\,(t)\,k\,(\pi\,(s)\,,\pi\,(t)) = k\,(s,t) \qquad \textit{for all } s,t \in S.$$

Proof. The reader is referred to [PaSc72] for details. □

In fact, there are important applications of Theorem 5.2 to frames with continuous index set. The best known (also included in [PaSc72]) is the example when $\mathbf{v}\,(t)$ is a Wiener process, i.e., a mathematical realization of Einstein's Brownian motion.

Remark 5.3. As an application of Theorem 5.2(b), note that in each of the Figures 1 and 2, the frames are unitarily equivalent as the angle θ varies. But they are inequivalent as n varies from 3 to 4. More generally, turning to Example 2.2, if n is fixed and a new system $(\mathbf{v}_\theta\,(s))$ is defined by translating the argument in (2.8) by θ, then the assignment $\mathbf{v}\,(s) \to \mathbf{v}_\theta\,(s)$ extends to a unitary operator in the two-dimensional Hilbert space $\mathcal{H}$. But as n varies, we get families that are not unitarily equivalent. In fact, it follows from Remark 3.6 that the examples from Figure 1 are not equivalent to those in Figure 2, even when equivalence is defined in the less restrictive sense of (5.5) in Theorem 5.2(c), i.e., allowing a phase factor in the transformation of the respective systems of frame vectors.

For simplicity, return here to the case when the set S is assumed countable and discrete.

Definition 5.4. (Following [Nel57, Nel59].) A closed subspace $\mathcal{H}$ in $\ell^2\,(S)$ is said to be a *reproducing kernel subspace* if for every $s \in S$, the mapping

(5.7) $$\mathcal{H} \ni f \longmapsto f\,(s) \in \mathbb{C}$$

is continuous. Note that by Riesz's lemma this means that there is for each s a unique element $\mathbf{v}\,(s) \in \mathcal{H}$ such that

(5.8) $$f\,(s) = \langle\, \mathbf{v}\,(s) \mid f \,\rangle.$$

And by Schwarz's inequality we get

(5.9) $$|f\,(s)| \leq \|\mathbf{v}\,(s)\|_{\mathcal{H}}\,\|f\|_{\mathcal{H}}.$$

Theorem 5.5. *Let the pair* $(S, \mathcal{H})$ *be as in the statement of Theorem 2.4. Specifically, we assume that there is a tight frame* $(\mathbf{v}\,(s))_{s \in S}$ *for* $\mathcal{H}$ *with frame constant* c.

Then it follows that $\mathcal{H}$ *is a reproducing kernel Hilbert space, and that*

(5.10) $$|f\,(s)| \leq k\,(s,s)^{1/2}\,\|f\|_{\mathcal{H}} \qquad \textit{for all } s \in S \textit{ and } f \in \mathcal{H},$$

where $k\,(s,t) = \langle\, \mathbf{v}\,(s) \mid \mathbf{v}\,(t) \,\rangle$ *is the Gram matrix of* $\mathcal{H}$.

Proof. Let $(S, \mathcal{H})$ be as stated. Then by Corollary 3.7, we have the representation

(5.11) $$f = c \sum_{s \in S} \langle\, \mathbf{v}\,(s) \mid f \,\rangle\,\mathbf{v}\,(s),$$

referring to the isometric embedding $\mathcal{H} \xhookrightarrow{\;\simeq\;} \ell^2\,(S)$. The Gram matrix of $(\mathbf{v}\,(s))_{s \in S}$ induces an operator K, and $P = cK$ is the projection of $\ell^2\,(S)$ onto the subspace

$\mathcal{H}$. Moreover, $\mathbf{v}(s) = P(\delta_s)$ for all $s \in S$. Now apply both sides of (5.11) to some point $t \in S$. Via the isometric embedding, we know that

$$(5.12) \qquad \mathcal{H} = \left\{ f \in \ell^2(S) \mid Pf = f \right\}.$$

Hence, if $f \in \mathcal{H}$, we get $f(t) = \langle \delta_t \mid f \rangle = \langle \delta_t \mid Pf \rangle = \langle P\delta_t \mid f \rangle = \langle \mathbf{v}(t) \mid f \rangle$; and by (5.11),

$$(5.13) \qquad f(t) = c \sum_{s \in S} \langle \mathbf{v}(s) \mid f \rangle \underbrace{\langle \mathbf{v}(t) \mid \mathbf{v}(s) \rangle}_{k(t,s)}.$$

An application of Corollary 3.5, and of Schwarz for $\ell^2(S)$, now yields

$$|f(t)| \leq c \left(\sum_{s \in S} |\langle \mathbf{v}(s) \mid f \rangle|^2 \right)^{1/2} \left(\sum_{s \in S} |k(t,s)|^2 \right)^{1/2}$$
$$= c \cdot c^{-1/2} \|f\|_{\mathcal{H}} \, c^{-1/2} \, k(t,t)^{1/2}$$
$$= \|f\|_{\mathcal{H}} \, k(t,t)^{1/2}. \qquad \qquad \square$$

Remark 5.6. It is clear that the converse to the theorem also holds: If $\mathcal{H} \overset{\simeq}{\hookrightarrow} \ell^2(S)$ is a reproducing kernel Hilbert space, then for each $s \in S$, and $f \in \mathcal{H}$, $f(s) = \langle \mathbf{v}(s) \mid f \rangle_{\mathcal{H}}$ holds for a unique family $(\mathbf{v}(s))_{s \in S}$ in $\mathcal{H}$. This family will be a tight frame with frame constant one. If $c \in (0,1)$, then $\mathbf{w}_c(s) := c^{-1/2}\mathbf{v}(s)$ will define a tight frame with frame constant c.

6. More general frames

Since the condition (2.7) which defines tight frames is rather rigid, it is of interest to consider how it can be relaxed in a way which still makes it useful.

As before, we will consider a pair $(S, \mathcal{H})$, where S is a fixed countable set, and where $\mathcal{H}$ is a Hilbert space

Definitions 6.1. A system of vectors $(\mathbf{v}(s))_{s \in S}$ in $\mathcal{H}$ is called a *frame* for $\mathcal{H}$ if there are constants $0 < A_1 \leq A_2 < \infty$ such that

$$(6.1) \qquad A_1 \|f\|^2 \leq \sum_{s \in S} |\langle \mathbf{v}(s) \mid f \rangle|^2 \leq A_2 \|f\|^2 \qquad \text{for all } f \in \mathcal{H}.$$

It is called a *Bessel sequence* if only the estimate on the right-hand side in (6.1) is assumed, i.e., if for some finite constant A,

$$(6.2) \qquad \sum_{s \in S} |\langle \mathbf{v}(s) \mid f \rangle|^2 \leq A \|f\|^2 \qquad \text{for all } f \in \mathcal{H}.$$

Recall that when (6.2) is assumed, then the analysis operator $V = V_{(\mathbf{v}(s))}$ given by

$$(6.3) \qquad \mathcal{H} \ni f \overset{V}{\longmapsto} (\langle \mathbf{v}(s) \mid f \rangle)_{s \in S} \in \ell^2(S)$$

is well defined and bounded. Hence, the adjoint operator $V^*: \ell^2(S) \to \mathcal{H}$ is bounded as well, and

$$(6.4) \qquad V^*(\xi_s) = \sum_{s \in S} \xi_s \mathbf{v}(s) \qquad \text{for all } (\xi_s) \in \ell^2(S),$$

where the sum on the right-hand side in (6.4) is convergent in $\mathcal{H}$ for all $(\xi_s) \in \ell^2(S)$.

The tight frames for which $A_1 = A_2 = 1$ are called *Parseval frames*.

Theorem 6.2. *Let $(S, \mathcal{H})$ be as above, and let $(\mathbf{v}(s))_{s \in S}$ be a Bessel sequence with Bessel constant A.*

(a) *Then the closed span $\mathcal{H}_{\text{in}}$ of $(\mathbf{v}(s))_{s \in S}$ contains a derived Parseval frame.*

(b) *The derived Parseval frame is a Parseval frame for $\mathcal{H}$ if and only if $(\mathbf{v}(s))_{s \in S}$ is a frame for $\mathcal{H}$, i.e., iff $\mathcal{H}_{\text{in}} = \mathcal{H}$.*

(c) *In the general case when $(\mathbf{v}(s))_{s \in S}$ is a Bessel sequence, the operator $W := V(V^*V)^{-1/2}$ is well defined and isometric on $\mathcal{H}_{\text{in}}$, and*

$$(6.5) \qquad \mathbf{w}(s) := (V^*V)^{-1/2} \mathbf{v}(s), \qquad s \in S,$$

is a Parseval frame in $\mathcal{H}_{\text{in}}$.

Proof. It is easy to see that both V and V^* are bounded; V is everywhere defined on $\mathcal{H}$, and V^* everywhere defined on $\ell^2(S)$. For the operator norms, we have $\|V\| = \|V^*\| = \|V^*V\|^{1/2} \leq \sqrt{A}$. The *initial space* of V, $\mathcal{H}(V)$, is defined as

$$(6.6) \qquad \mathcal{H} \ominus \{ f \in \mathcal{H} \mid Vf = 0 \} = \overline{R(V^*)},$$

where the over-bar stands for norm-closure in $\mathcal{H}$, and where $R(V^*)$ denotes the range of V^*.

We let $\mathcal{H}_{\text{in}}$ denote the closed span of the given Bessel sequence $(\mathbf{v}(s))_{s \in S}$. Our claim is that

$$(6.7) \qquad \mathcal{H}_{\text{in}} = \mathcal{H}(V).$$

To see this, note that (6.4) implies the inclusion $(\supseteq)$ in (6.7). But all finite linear combinations $\sum_s \xi_s \mathbf{v}(s)$ are contained in $R(V^*)$, and by closure, we get $\mathcal{H}_{\text{in}} \subseteq \overline{R(V^*)}$, which is the second inclusion in (6.7). Hence (6.7) holds.

We now apply the polar decomposition (from operator theory), see [KaRi97], to the operator V. The conclusion is that there is a *partial isometric* W such that

$$(6.8) \qquad V = W(V^*V)^{1/2} = (VV^*)^{1/2} W$$

and such that the *initial space* of W is $\mathcal{H}_{\text{in}}$ and the *final space* of W is $\overline{R(V)}$. Implied in this are the following assertions:

(i) Both of the operators W^*W and WW^* are projections;

(ii) The range of W^*W is $\mathcal{H}_{\text{in}}$;

(iii) The range of WW^* is $\overline{R(V)}$;

(iv) The operator V^*V is selfadjoint, and $(V^*V)^{1/2}$ is defined by the spectral theorem applied to V^*V.

Now apply Theorem 5.5 to the restriction operator

$$(6.9) \qquad W : \mathcal{H}_{\text{in}} \longrightarrow \ell^2(S).$$

In view of (ii) above, this is an isometry. Recall $W^*W\mathcal{H} = \mathcal{H}_{\text{in}}$ by (6.7) and the polar decomposition (6.8). As a result, we get that the vectors $\mathbf{w}(s) = W^*\delta_s$, for $s \in S$, form a Parseval frame for $\mathcal{H}_{\text{in}}$. But we also have the formula $V^*\delta_s = \mathbf{v}(s)$ as an application of (6.4). Using now the two facts (6.9) and (6.8), we get $\mathbf{w}(s) = W^*\delta_s = (V^*V)^{-1/2} V^*\delta_s = (V^*V)^{-1/2} \mathbf{v}(s)$, which is the desired formula (6.5) from the conclusion in the theorem. The remaining conclusions stated in the theorem are already implied by the reasoning above. $\qquad \square$

Remark 6.3. The operators V^*V and VV^*. Except for the point $\lambda = 0$, the two operators V^*V and VV^* have the same spectrum. But in the general case, this spectrum could be discrete or continuous; or it could even be singular.

The fact that the spectrum minus $\{0\}$ is the same follows from (6.8). In the *frame case*, it follows from (6.1) that the spectrum is contained in the interval $[A_1, A_2]$. But if $(\mathbf{v}(s))_{s \in S}$ is only a Bessel sequence, (6.2), the best that can be said is that the spectrum is contained in $[0, A]$. In fact, the lower bound in $\mathrm{spectrum}(V^*V)$ is also the best lower frame bound, referring to (6.1).

Computationally, however, the two operators V^*V and VV^* are quite different. The first one operates in $\mathcal{H}$, while the second one maps $\ell^2(S)$ into itself.

The formula

$$(6.10) \qquad V^*Vf = \sum_{s \in S} \langle \mathbf{v}(s) \mid f \rangle \, \mathbf{v}(s) \qquad \text{for } f \in \mathcal{H}$$

shows that V^*V serves to decompose vectors in $\mathcal{H}$ relative to the given frame $(\mathbf{v}(s))_{s \in S}$. In contrast, VV^* has an explicit matrix representation:

Proposition 6.4. *Relative to the ONB $(\delta_s)_{s \in S}$, VV^* is simply multiplication with the Gram matrix*

$$(6.11) \qquad \left(\langle \mathbf{v}(s) \mid \mathbf{v}(t) \rangle_{\mathcal{H}} \right)_{s, t \in S}.$$

Proof. To compute the (s, t)-matrix entry for the matrix which represents the operator VV^*, we use the inner product in $\ell^2(S)$ as follows: The (s, t)-matrix entry is

$$(6.12) \qquad \langle \delta_s \mid VV^* \delta_t \rangle_{\ell^2} = \langle V^* \delta_s \mid V^* \delta_t \rangle_{\mathcal{H}} = \langle \mathbf{v}(s) \mid \mathbf{v}(t) \rangle_{\mathcal{H}},$$

which is the Gram matrix. $\qquad \square$

Example 6.5. Example 2.2 revisited for $n = 3$.

In the case of Example 2.2 (see also Fig. 1) for $n = 3$, VV^* is the 3×3 matrix

$$\begin{pmatrix} 1 & -\frac{1}{2} & -\frac{1}{2} \\ -\frac{1}{2} & 1 & -\frac{1}{2} \\ -\frac{1}{2} & -\frac{1}{2} & 1 \end{pmatrix},$$

while V^*V may be represented as

$$\begin{pmatrix} \frac{3}{2} & 0 \\ 0 & \frac{3}{2} \end{pmatrix},$$

i.e., a 2×2 matrix. Moreover, $\mathrm{spec}(VV^*) = \{0, \frac{3}{2}\}$, and $\mathrm{spec}(V^*V) = \{\frac{3}{2}\}$.

Corollary 6.6. *Consider a system $\left(\mathcal{H}, S, (\mathbf{v}(s))_{s \in S} \right)$ as above, and let*

$$V = V_{(\mathbf{v}(s))} \colon \mathcal{H} \longrightarrow \ell^2(S)$$

be the associated analysis operator. Then the respective lower and upper frame estimates (6.1) are equivalent to

$$(6.13) \qquad \mathrm{spec}\left(\left(\langle \mathbf{v}(s) \mid \mathbf{v}(t) \rangle \right) \right) \setminus \{0\} \subset [A_1, A_2],$$

where $\left(\langle \mathbf{v}(s) \mid \mathbf{v}(t) \rangle \right)_{s, t \in S}$ is the Gram matrix in (6.11).

Proof. Since we noted in Remark 6.3 that

$$(6.14) \qquad \qquad \operatorname{spec}(V^*V) \setminus \{0\} = \operatorname{spec}(VV^*) \setminus \{0\},$$

the assertion in (6.13) follows immediately from Proposition 6.4. Specifically, the two estimates in (6.1) are equivalent to the following system of operator inequalities:

$$(6.15) \qquad \qquad A_1 I_{\mathcal{H}} \leqq V^*V \leqq A_2 I_{\mathcal{H}},$$

where the ordering "$\leqq$" in (6.15) refers to the familiar ordering on the set of self-adjoint operators [KaRi97]. Now an application of the spectral theorem [KaRi97] to V^*V shows that (6.15) is also equivalent to the set-containment $\operatorname{spec}(V^*V) \subseteq [A_1, A_2]$. In fact, the lower estimate in (6.1) holds iff $\operatorname{spec}(V^*V) \subseteq [A_1, \infty)$, while the upper estimate is equivalent to the containment $\operatorname{spec}(V^*V) \subseteq [0, A_2]$. $\qquad \square$

While Corollary 6.6 is rather abstract, it has a variety of specific uses which serve to make it clear how certain initial frame systems are transformed into more detailed frames having additional structure. A case in point is the transformation of frames used in the analysis of signals into frequency bands, for example mutually non-interfering low-pass and high-pass bands. A convenient tool for accomplishing this is the discrete wavelet transform. As noted in, for example, [Chr03] and [Jor06], the discrete wavelet transform serves to create nested families of resolution subspaces (details below) in the Hilbert spaces of sequences which are used in our representation of time-series, or of speech signals. When more sub-bands are introduced, this same idea works for the analysis of images.

By a discrete wavelet transform with perfect reconstruction we shall mean (following [Jor06]) a Hilbert space $\mathcal{H}$ and a system of operators F_0, F_1 in $\mathcal{H}$ such that

$$(6.16) \qquad F_i F_j^* = \delta_{i,j} I, \quad \sum_{i=0}^{1} F_i^* F_i = I, \quad \text{and} \quad F_0^n f \underset{n \to \infty}{\longrightarrow} 0 \qquad \text{for all } f \in \mathcal{H}.$$

As further noted in [Jor06], each system of quadrature mirror filters from standard signal processing defines an operator system (F_i) as in (6.16).

Remarks 6.7. (a) The subscript convention for the two operators F_0 and F_1 in (6.16) comes from engineering. The index value $i = 0$ corresponds to a chosen low-pass filter followed by downsampling, while F_1 is the operation of a high-pass filtering followed by downsampling. Hence "low" is indexed by zero. This index convention is not related to that of the two frame bounds A_i in (6.1). These two numbers are usually called A_1 and A_2.

(b) [Referring to the sum-formula in (6.16).] In stating our quadrature conditions (6.16) we have for reasons of clarity restricted attention to the simplest case: dyadic and orthogonal filters. (The dyadic case refers to the dyadic wavelets from Proposition 4.4 above.) But our present discussion (both in Sections 4 and 6) easily generalizes to systems with more than two bands, and even to less restrictive quadrature conditions. In case of more, say N bands, the indexing of the operators F_i is $i = 0, 1, \ldots, N-1$. And instead of using the operator F_i^* in the sum in (6.16) at the i'th place, we may for some applications rather use a second operator G_i (not equal to F_i^*); see for example [JoKr03] for details on these more general systems.

Corollary 6.8. *Let $(\mathbf{v}(s))_{s \in S}$ be a frame with frame bounds A_i, $i = 1, 2$, in a Hilbert space $\mathcal{H}$, and let (F_i) be a discrete wavelet transform system (as in (6.16)).*

Set

(6.17) $\mathbf{v}(k, s) := F_0^{*\,k} F_1^* \mathbf{v}(s), \qquad s \in S,\ k = 0, 1, 2, \ldots.$

Then the subdivided vector system $(\mathbf{v}(k, s))$ is a frame in $\mathcal{H}$ with the same frame bounds A_i as $(\mathbf{v}(s))$. Moreover the Gramian of $(\mathbf{v}(k, s))$ has the form $I \otimes G$ where I is the (infinite) identity matrix, and G is the Gramian for $(\mathbf{v}(s))$.

Moreover, if two frames $(\mathbf{v}(s))$ and $(\mathbf{v}'(s))$ are unitarily equivalent in the sense of Theorem 5.2, then two refined systems $(\mathbf{v}(k, s))$ and $(\mathbf{v}'(k, s))$ resulting from (6.17) are also unitarily equivalent.

Proof. Since in general, by Corollary 6.6, the two frame bounds (upper and lower) coincide with the spectral bounds for the corresponding Gramian, we only need to study the Gramian of the new system $(\mathbf{v}(k, s))$ in (6.17). Recall that the Gramian for the new system is the matrix of inner products, with the only modification to the formula in Theorem 5.5 being that the row and column indices are now double indices.

The significance of the third condition from (6.16) is related to the kind of subspace structure in $\mathcal{H}$ which models *resolutions* of signals. Since $F_0 F_0^* = I$, we get a nested system of projections

(6.18) $P_k := F_0^{*\,k} F_0^k, \qquad k = 0, 1, \ldots,$

and

(6.19) $\cdots P_{k+1} \le P_k \le \cdots \le P_1 \le P_0 = I.$

Recall that the ordering of projections coincides with the associated ordering of the range subspaces $\mathcal{H}_k := P_k \mathcal{H}$ in $\mathcal{H}$, i.e.,

(6.20) $\cdots \mathcal{H}_{k+1} \subseteq \mathcal{H}_k \subseteq \cdots \subseteq \mathcal{H}_1 \subseteq \mathcal{H}.$

The third condition in (6.16) is equivalent to the assertion that

(6.21) $\displaystyle\bigcap_{k \in \{0, 1, \ldots\}} \mathcal{H}_k = \{\mathbf{0}\}.$

Specifically,

(6.22) $\langle \mathbf{v}(j, s) \mid \mathbf{v}(k, t) \rangle = \left\langle F_0^{*\,j} F_1^* \mathbf{v}(s) \mid F_0^{*\,k} F_1^* \mathbf{v}(t) \right\rangle = \delta_{j,k} \langle \mathbf{v}(s) \mid \mathbf{v}(t) \rangle.$

Since the last expression is the tensor product of I with G, the result follows.

The last part of the conclusion in the corollary about preservation of unitary equivalence in passing to the refinement $(\mathbf{v}(s)) \to (\mathbf{v}(k, s))$ follows from this and Theorem 5.2. $\square$

Example 6.9. (Following the terminology from (4.12) in Proposition 4.4.) Suppose some dyadic wavelet $(\psi_{j,k})$ comes from a scaling function φ. Set $S := \mathbb{Z}$, and

(6.23) $\mathbf{v}(k) := \varphi(x - k), \qquad k \in \mathbb{Z}.$

Then there are known conditions on such a scaling function φ for the frame estimates (6.1) to hold; see, e.g., [CoDa93]. To apply the operator system (6.16) from above, choose the Hilbert space $\mathcal{H}$ be the closed subspace in $L^2(\mathbb{R})$ spanned by the integral translates $\{\varphi(x - k) \mid k \in \mathbb{Z}\}$. Following [Jor06] and (6.17) above, we may then construct operators F_0, F_1 and introduce the associated resolution frame $(\mathbf{v}(j, k))$, double-indexed as in Corollary 6.8 as follows:

Set

$$(6.24) \qquad \mathbf{v}\,(j,k) := F_0^{*\,j}\,F_1^*\,\mathbf{v}\,(k) = \psi_{-j-1,k} \qquad \text{for } j = 0,1,2,\ldots, \text{ and } k \in \mathbb{Z}.$$

Using the corollary, we then conclude that for each j, the functions $\{\,\mathbf{v}\,(j,k) \mid k \in \mathbb{Z}\,\}$ generate the relative complement subspace $\mathcal{H}_j \ominus \mathcal{H}_{j+1}$ from the nested resolution system we introduced there. By this we mean that for each j, $\mathcal{H}_j \ominus \mathcal{H}_{j+1}$ is the closed linear span of $\{\,\mathbf{v}\,(j,k) \mid k \in \mathbb{Z}\,\}$.

References

[AkGl93]   N.I. Akhiezer and I.M. Glazman, *Theory of Linear Operators in Hilbert Space*, Dover, New York, 1993, reprint of the 1961–63 publication (Frederick Ungar Publishing Co., New York) of a translation by Merlynd Nestell; a corrected and augmented Russian edition ("Vishcha Shkola", Kharkov, 1977–78) has also been translated by E.R. Dawson and published in the series Monographs and Studies in Mathematics, vol. 9, Pitman (Advanced Publishing Program), Boston–London, 1981.

[ALTW04]   A. Aldroubi, D. Larson, W.-S. Tang, and E. Weber, *Geometric aspects of frame representations of abelian groups*, Trans. Amer. Math. Soc. **356** (2004), 4767–4786.

[Aro50]   N. Aronszajn, *Theory of reproducing kernels*, Trans. Amer. Math. Soc. **68** (1950), 337–404.

[Ash90]   R.B. Ash, *Information Theory*, Dover, New York, 1990, corrected reprint of the original 1965 Interscience/Wiley edition.

[BJMP05]   L.W. Baggett, P.E.T. Jorgensen, K.D. Merrill, and J.A. Packer, *Construction of Parseval wavelets from redundant filter systems*, J. Math. Phys. **46** (2005), no. 8, 083502, 28 pp., doi:10.1063/1.1982768.

[BJMP06]   _____, *A non-MRA C^r frame wavelet with rapid decay*, Acta Appl. Math. **Online First** (2006), 20 pp., doi:10.1007/s10440-005-9011-4.

[BeFi03]   J.J. Benedetto and M. Fickus, *Finite normalized tight frames*, Adv. Comput. Math. **18** (2003), 357–385.

[BDP05]   S. Bildea, D.E. Dutkay, and G. Picioroaga, *MRA super-wavelets*, New York J. Math. **11** (2005), 1–19.

[CaCh03]   P.G. Casazza and O. Christensen, *Gabor frames over irregular lattices*, Adv. Comput. Math. **18** (2003), 329–344.

[CCLV05]   P.G. Casazza, O. Christensen, A.M. Lindner, and R. Vershynin, *Frames and the Feichtinger conjecture*, Proc. Amer. Math. Soc. **133** (2005), 1025–1033.

[CFTW06]   P.G. Casazza, M. Fickus, J.C. Tremain, and E. Weber, *The Kadison–Singer problem in mathematics and engineering: a detailed account*, Operator Theory, Operator Algebras, and Applications (GPOTS 2005) (D. Han, P.E.T. Jorgensen, and D. Larson, eds.), Contemp. Math., American Mathematical Society, Providence, to appear.

[CKL04]   P.G. Casazza, G. Kutyniok, and M.C. Lammers, *Duality principles in frame theory*, J. Fourier Anal. Appl. **10** (2004), 383–408.

[Chr99]   O. Christensen, *Operators with closed range, pseudo-inverses, and perturbation of frames for a subspace*, Canad. Math. Bull. **42** (1999), 37–45.

[Chr03]   _____, *An Introduction to Frames and Riesz Bases*, Applied and Numerical Harmonic Analysis, Birkhäuser, Boston, 2003.

[CoDa93]   A. Cohen and I. Daubechies, *Nonseparable bidimensional wavelet bases*, Rev. Mat. Iberoamericana **9** (1993), 51–137.

[DuSc52]   R.J. Duffin and A.C. Schaeffer, *A class of nonharmonic Fourier series*, Trans. Amer. Math. Soc. **72** (1952), 341–366.

[DuXu01]   C.F. Dunkl and Y. Xu, *Orthogonal Polynomials of Several Variables*, Encyclopedia of Mathematics and its Applications, vol. 81, Cambridge University Press, Cambridge, 2001.

[Dut04a]   D.E. Dutkay, *Positive definite maps, representations and frames*, Rev. Math. Phys. **16** (2004), no. 4, 451–477.

[Dut04b]   _____, *The local trace function for super-wavelets*, Wavelets, Frames, and Operator Theory (Focused Research Group Workshop, College Park, Maryland, January 15–21,

2003) (C. Heil, P.E.T. Jorgensen, and D. Larson, eds.), Contemp. Math., vol. 345, American Mathematical Society, Providence, 2004, pp. 115–136.

[Dut06]   _____, *Low-pass filters and representations of the Baumslag–Solitar group*, Trans. Amer. Math. Soc., to appear, http://www.arxiv.org/abs/math.CA/0407344.

[Eld02]   Y.C. Eldar, *Least-squares inner product shaping*, Linear Algebra Appl. **348** (2002), 153–174.

[FJKO05]   M. Fickus, B.D. Johnson, K. Kornelson, and K.A. Okoudjou, *Convolutional frames and the frame potential*, Appl. Comput. Harmon. Anal. **19** (2005), 77–91.

[Jor06]   P.E.T. Jorgensen, *Analysis and Probability: Wavelets, Signals, Fractals*, Grad. Texts in Math., vol. 234, Springer-Verlag, New York, to appear 2006.

[JoKr03]   P.E.T. Jorgensen and D.W. Kribs, *Wavelet representations and Fock space on positive matrices*, J. Funct. Anal. **197** (2003), 526–559.

[KaRi97]   R.V. Kadison and J.R. Ringrose, *Fundamentals of the Theory of Operator Algebras, Vol. I: Elementary Theory*, Graduate Studies in Mathematics, vol. 15, American Mathematical Society, Providence, 1997, reprint of the 1983 original Academic Press edition.

[KoLa04]   K.A. Kornelson and D.R. Larson, *Rank-one decomposition of operators and construction of frames*, Wavelets, Frames and Operator Theory (College Park, MD, 2003) (C. Heil, P.E.T. Jorgensen, and D.R. Larson, eds.), Contemp. Math., vol. 345, American Mathematical Society, Providence, 2004, pp. 203–214.

[Nel57]   E. Nelson, *Kernel functions and eigenfunction expansions*, Duke Math. J. **25** (1957), 15–27.

[Nel59]   E. Nelson, *Correction to "Kernel functions and eigenfunction expansions"*, Duke Math. J. **26** (1959), 697–698.

[PaSc72]   K. R. Parthasarathy and K. Schmidt, *Positive Definite Kernels, Continuous Tensor Products, and Central Limit Theorems of Probability Theory*, Lecture Notes in Mathematics, vol. 272, Springer-Verlag, Berlin-New York, 1972.

[VaWa05]   R. Vale and S. Waldron, *Tight frames and their symmetries*, Constr. Approx. **21** (2005), 83–112.

DEPARTMENT OF MATHEMATICS, THE UNIVERSITY OF IOWA, 14 MACLEAN HALL, IOWA CITY, IA 52242-1419, U.S.A.

E-mail address: jorgen@math.uiowa.edu

URL: http://www.math.uiowa.edu/~jorgen

Contemporary Mathematics
Volume **451**, 2008

Operator-valued Frames on C*-Modules

Victor Kaftal, David Larson, Shuang Zhang

ABSTRACT. Frames on Hilbert C*-modules have been defined for unital C*-algebras by Frank and Larson [**5**] and operator-valued frames on a Hilbert space have been studied in [**8**]. The goal of this paper is to introduce operator-valued frames on a Hilbert C*-module for a σ-unital C*-algebra. Theorem 1.4 reformulates the definition given in [**5**] in terms of a series of rank-one operators converging in the strict topology. Theorem 2.2. shows that the frame transform and the frame projection of an operator-valued frame are limits in the strict topology of a series in the multiplier algebra and hence belong to it. Theorem 3.3 shows that two operator-valued frames are right similar if and only if they share the same frame projection. Theorem 3.4 establishes an one-to-one correspondence between Murray-von Neumann equivalence classes of projections in the multiplier algebra and right similarity equivalence classes of operator-valued frames and provides a parametrization of all Parseval operator-valued frames on a given Hilbert C*-module. Left similarity is then defined and Proposition 3.9 establishes when two left unitarily equivalent frames are also right unitarily equivalent.

Introduction

Frames on a Hilbert space are collections of vectors satisfying the condition

$$a\|\xi\|^2 \leq \sum_{j \in \mathbb{J}} |<\xi, \xi_j>|^2 \leq b\|\xi\|^2$$

for some positive constants a and b and all vectors ξ. This notion has been naturally extended by Frank and Larson [**5**] to countable collections of vectors in a Hilbert C*-module for a unital C*-algebra satisfying an analogous defining property (see below **1.1** for the definitions). Most properties of frames on a Hilbert space hold also for Hilbert C*-modules, often have quite different proofs, but new phenomena do arise.

2000 *Mathematics Subject Classification*. Primary 42C15, 46C05, 46L05.

Key words and phrases. Hilbert C*-modules, modular frames, Parseval frames, similarity.

The research of the first and third named author was supported in part by grants of the Charles Phelps Taft Research Center. The research of the second author was supported in part by a grant from the NSF. Part of this work was accomplished while the first named author participated in the NSF supported Workshop in Linear Analysis and Probability at Texas A&M University.

A different generalization where frames are no longer vectors in a Hilbert space but operators on a Hilbert space is given in [**8**] with the purpose of providing a natural framework for multiframes, especially for those obtained from a unitary system, e.g, a discrete group representation. Operator-valued frames both generalize vector frames and can be decomposed into vector frames.

The goal of this article is to introduce the notion of operator-valued frame on a Hilbert C*-module. Since the frame transform of Frank and Larson permits to identify a vector frame on an arbitrary Hilbert C*-module with a vector frame on the standard Hilbert C*-module $\ell^2(\mathcal{A})$ of the associated C*-algebra $\mathcal{A}$, for simplicity's sake we confine our definition directly to frames on $\ell^2(\mathcal{A})$. When the associated C*-algebra is σ-unital, it well known (see [**9**]) that the algebra of bounded adjointable operators on $\ell^2(\mathcal{A})$ can be identified with the multiplier algebra $M(\mathcal{A}\otimes\mathcal{K})$ of $\mathcal{A}\otimes\mathcal{K}$, about which a good deal is known. Since reference are mainly formulated in terms of right-modules, we treat $\ell^2(\mathcal{A})$ as a right module (i.e., as 'row vectors').

A frame on $\ell^2(\mathcal{A})$ is thus defined as a collection of operators $\{A_j\}_{j\in\mathbb{J}}$ with $A_j \in E_0 M(\mathcal{A}\otimes\mathcal{K})$ for a fixed projection $E_0 \in M(\mathcal{A}\otimes\mathcal{K})$ for which

$$aI \leq \sum_{j\in\mathbb{J}} A_j^* A_j \leq bI$$

for some positive constants a and b, where I is the identity of $M(\mathcal{A}\otimes\mathcal{K})$ and the series converges in the strict topology of $M(\mathcal{A}\otimes\mathcal{K})$.

We will show in Theorem 1.4 how to associate (albeit not uniquely) to a vector frame in the sense of [**5**] an operator-valued frame. When $\mathcal{A}$ is unital, we will decompose in Section 3.10) every operator-valued frame (albeit not uniquely into vector frames (i.e., a multiframe) Some properties of operator-valued frames on a Hilbert C*-module track fairly well the properties of operator-valued and vector-valued frames on a Hilbert space. Often, the key difference in the proofs is the need to express objects like the *frame transform* or the *frame projection* as series of elements of $M(\mathcal{A}\otimes\mathcal{K})$ that converge in the strict topology, and hence, belong to $M(\mathcal{A}\otimes\mathcal{K})$.

We illustrate some commonalities and differences with the Hilbert space case by considering in particular three topics. That the dilation approach of Han and Larson in [**6**], which was extended to operator-valued frames on Hilbert spaces in [**8**], has a natural analog for operator-valued frames on Hilbert C*-modules if the frame transform is defined to have values inside the same Hilbert C*-module instead of into an ampliation of it.

Similarity of frames can also be defined and characterized as in the Hilbert space case, but now there is also a similarity from the left and we compare the two notions.

Finally, there is a natural composition of operator-valued frames - a new operation that has no vector frame analog and that illustrates the 'multiplicity' of operator-valued frames.

In this paper we have explored the analogs of some of the properties of Hilbert space frames and much work remains to be done.

1. Operator-valued frames

1.1. Frames and operator-valued frames on a Hilbert space.

A frame on a Hilbert space $\mathcal{H}$ is a collection of vectors $\{\xi_j\}_{j\in\mathbb{J}}$ indexed by a countable set $\mathbb{J}$ for which there exist two positive constants a and b such that for all $\xi \in \mathcal{H}$,

$$a\|\xi\|^2 \le \sum_{j\in\mathbb{J}} |<\xi,\xi_j>|^2 \le b\|\xi\|^2.$$

Equivalently,

$$aI \le \sum_{j\in\mathbb{J}} \xi_j \otimes \xi_j \le bI,$$

where I is the identity of $\mathcal{B}(\mathcal{H})$, $\eta \otimes \xi$ is the rank-one operator defined by $(\eta \otimes \xi)\zeta := <\zeta,\xi> \eta$ and the series converges in the *strong operator topology* (pointwise convergence). The above condition can be rewritten as

$$aI \le \sum_{j\in\mathbb{J}} A_j^* A_j \le bI$$

where $A_j := \eta \otimes \xi_j$ for some arbitrary fixed unit vector $\eta \in \mathcal{H}$. It is thus equivalent to the series $\sum_{j\in\mathbb{J}} A_j^* A_j$ converging in the strong operator topology to a bounded invertible operator. Notice that the convergence of the numerical and the operatorial series are unconditional.

This reformulation naturally leads to the more general notion of operator-valued frames $\{A_j\}_{j\in\mathbb{J}}$ on a Hilbert space $\mathcal{H}$ in [8], namely a collection of operators $A_j \in B(H, H_0)$, with ranges in a fixed Hilbert space H_0 (not necessarily of dimension one) for which the series $\sum_{j\in\mathbb{J}} A_j^* A_j$ converges in the strong operator topology to a bounded invertible operator. Operator-valued frames on a Hilbert space can be decomposed into, and hence identified with, multiframes.

Frames with values in a Hilbert C*-module have been introduced in [5] and then studied in [7], [13], and others. So much of the Hilbert space frame theory carries over, that one could argue that frame theory finds a natural general setting in Hilbert C*-modules. We will show in Theorem 1.4 that frames on a Hilbert C*-module can be equivalently defined in terms of rank-one operators on the module. This leads naturally to the definition of general operator-valued frames on a Hilbert C*-module. Before giving the formal definitions, we recall for the readers' convenience some relevant background about Hilbert C*-modules.

1.2. Hilbert C*-modules ([2,Ch.13], [9]).

Let $\mathcal{A}$ be a C*-algebra. Then a Hilbert (right) C*-$\mathcal{A}$-module is a pair $(\mathcal{H}, < .,. >)$, with $\mathcal{H}$ a (right) module over $\mathcal{A}$ and $< .,. >$ a binary operation from $\mathcal{H}$ into $\mathcal{A}$, that satisfies the following six axioms, similar to those of Hilbert spaces, except that for right modules the linearity occurs for the second and not the first component of the inner product. For $\xi, \eta, \eta_1, \eta_2 \in \mathcal{H}$ and $a \in \mathcal{A}$

(i) $< \xi, \eta_1 + \eta_2 > = < \xi, \eta_1 > + < \xi, \eta_2 >$;

(ii) $< \xi, \eta a > = < \xi, \eta > a$;

(iii) $< \xi, \eta >^* = < \eta, \xi >$;

(iv) $< \xi, \xi > \geq 0$;

(v) $< \xi, \xi > = 0 \iff \xi = 0$;

(vi) $(\mathcal{H}, \| \cdot \|)$ is complete, where $\|\xi\| := \| < \xi, \xi > \|^{1/2}$

The classic example of Hilbert (right) $\mathcal{A}$-module and the only one we will consider in this paper is the *standard module* $\mathcal{H}_\mathcal{A} := \ell^2(\mathcal{A})$, the space of all sequences $\{a_i\} \subset \mathcal{A}$ such that $\sum_{i=1}^\infty a_i^* a_i$ converges in norm to a positive element of $\mathcal{A}$. $\ell^2(\mathcal{A})$ is endowed with the natural linear (A-module) structure and right $\mathcal{A}$-multiplication, and with the $\mathcal{A}$-valued inner product defined by

$$< \{a_i\}, \{b_i\} > = \sum_{i=1}^\infty a_i^* b_i,$$

where the sum converges in norm by the Schwartz Inequality ([**2**] or [**9**]).

A map T from $\mathcal{H}_\mathcal{A}$ to $\mathcal{H}_\mathcal{A}$ is called a (linear) bounded operator on $\mathcal{H}_\mathcal{A}$, if $T(\lambda\xi) + T(\mu\eta) = T(\lambda\xi + \mu\eta)$, $T(\xi a) = T(\xi)a$ for all $\xi, \eta \in \mathcal{H}_\mathcal{A}$, $\lambda, \mu \in \mathbb{C}$, and $a \in \mathcal{A}$, and if

$$\|T\| := \sup\{\|T\xi\| \mid \xi \in \mathcal{H}_\mathcal{A}, \|\xi\| \leq 1\} < \infty.$$

Not every bounded operator T has a bounded adjoint T^*, namely

$$< T^*\xi, \eta > = < \xi, T\eta > \quad \text{for all} \quad \xi, \ \eta \ \in \mathcal{H}_\mathcal{A},$$

as there is no Riesz Representation Theorem for general Hilbert C*-modules. Nevertheless, there are abundant operators on $\mathcal{H}_\mathcal{A}$ whose adjoints exist and the collection of bounded adjointable operators is denoted by $\mathcal{B}(\mathcal{H}_\mathcal{A})$. Then, $\mathcal{B}(\mathcal{H}_\mathcal{A})$ is a C*-algebra (see [**9**] or [**2**, Ch. 13]). Notice that if $\mathcal{A} = \mathbb{C}$, then $\mathcal{H}_\mathcal{A} = \ell^2$ and $\mathcal{B}(\mathcal{H}_\mathcal{A}) = \mathcal{B}(\ell^2)$. Some of the properties of $\mathcal{B}(\ell^2)$ extend naturally to $\mathcal{B}(\mathcal{H}_\mathcal{A})$. For each pair of elements ξ and η in $\mathcal{H}_\mathcal{A}$, a bounded 'rank-one' operator is defined by

$$\theta_{\xi,\eta}(\zeta) = \xi < \eta, \zeta > \quad \text{for all } \zeta \in \mathcal{H}_\mathcal{A}.$$

The closed linear span of all rank-one operators is denoted by $\mathcal{K}(\mathcal{H}_\mathcal{A})$. When $\mathcal{A} = \mathbb{C}$, $\mathcal{K}(\mathcal{H}_\mathcal{A})$ coincides with the ideal $\mathcal{K}$ of all compact operators on ℓ^2. $\mathcal{K}(\mathcal{H}_\mathcal{A})$ is always a closed ideal of $\mathcal{B}(\mathcal{H}_\mathcal{A})$, but contrary to the separable infinite dimensional Hilbert space case, in general it is not unique (e.g., see [**2**] or [**9**]).

The analog of the strong*-topology on $\mathcal{B}(\ell^2)$ is the *strict topology* on $\mathcal{B}(\mathcal{H}_\mathcal{A})$ defined by

$$\mathcal{B}(\mathcal{H}_\mathcal{A}) \ni T_\lambda \longrightarrow T \text{ strictly if } \|(T_\lambda - T)S\| \to 0 \text{ and } \|S(T_\lambda - T)\| \to 0 \ \forall S \in \mathcal{K}(\mathcal{H}_\mathcal{A}).$$

We will use the following elementary properties: $T_\lambda \longrightarrow T$ strictly iff $T_\lambda^* \longrightarrow T^*$ strictly, and either of these convergences implies $BT_\lambda \longrightarrow BT$ and $T_\lambda B \longrightarrow TB$ strictly for all $B \in \mathcal{B}(\mathcal{H}_\mathcal{A})$. Also, if $T_\lambda \longrightarrow T$ strictly and $S_\lambda \longrightarrow S$ strictly, then $T_\lambda S_\lambda \longrightarrow TS$ strictly.

There is an alternative view of the objects $\mathcal{B}(\mathcal{H}_\mathcal{A})$ and $\mathcal{K}(\mathcal{H}_\mathcal{A})$. Embed the tensor product $\mathcal{A} \otimes \mathcal{K}$ into its Banach space double dual $(\mathcal{A} \otimes \mathcal{K})^{**}$, which, as is well known, is a W*-algebra ([**12**]). The multiplier algebra of $\mathcal{A} \otimes \mathcal{K}$, denoted by $M(\mathcal{A} \otimes \mathcal{K})$, is defined as the collection

$$\{T \in (\mathcal{A} \otimes \mathcal{K})^{**} : \quad TS, ST \in \mathcal{A} \otimes \mathcal{K} \quad \forall \, S \, \in \, \mathcal{A} \otimes \mathcal{K}\}.$$

Equipped with the norm of $(\mathcal{A} \otimes \mathcal{K})^{**}$, $M(\mathcal{A} \otimes \mathcal{K})$ is a C*-algebra. Assuming that $\mathcal{A}$ is σ-unital, we will frequently apply the following two *-isomorphisms without further reference:

$$\mathcal{B}(\mathcal{H}_{\mathcal{A}}) \cong M(\mathcal{A} \otimes \mathcal{K}) \quad \text{and} \quad \mathcal{K}(\mathcal{H}_{\mathcal{A}}) \cong \mathcal{A} \otimes \mathcal{K} \quad [\text{Ka1}].$$

The algebra $\mathcal{B}(\mathcal{H}_{\mathcal{A}})$ is technically hard to work with, while $M(\mathcal{A} \otimes \mathcal{K})$ is more accessible due to many established results. More information on the subject can be found in the sample references [**9**] and [**2**], among many others. Although most properties hold with appropriate modifications also for left modules, since the original theory was developed by Kasparov for right Hilbert C*-modules ([**9**]), the results found in the literature are often formulated for right modules. This is the reason why our definition of frames is given for right modules instead of left modules as in [**5**].

To avoid unnecessary complications, from now on, we assume that $\mathcal{A}$ is a σ-unital C*-algebra.

1.3. Vector Frames on Hilbert C*-modules

According to [**5**], a (vector) frame on the Hilbert C*-module $\mathcal{H}_{\mathcal{A}}$ of a σ-unital C*-algebra $\mathcal{A}$ is a collection of elements $\{\xi_j\}_{j \in \mathbb{J}}$ in $\mathcal{H}_{\mathcal{A}}$ for which there are two positive scalars a and b such that for all $\xi \in \mathcal{H}_{\mathcal{A}}$,

$$a < \xi, \xi > \leq \sum_{j \in \mathbb{J}} < \xi, \xi_j, > < \xi_j, \xi > \leq b < \xi, \xi >,$$

where the convergence is in the norm of the C*-algebra $\mathcal{A}$. The following theorem permits us to reformulate this definition in terms of rank-one operators. Notice that $< \xi, \xi_j, > < \xi_j, \xi > = < \theta_{\xi_j, \xi_j} \xi, \xi >$.

1.4. Theorem Let $\mathcal{A}$ be a σ-unital C*-algebra. Then the collection $\{\xi_j\}_{j \in \mathbb{J}}$ in the Hilbert C*-module $\mathcal{H}_{\mathcal{A}}$ is a frame if and only if the series $\sum_{j \in \mathbb{J}} \theta_{\xi_j, \xi_j}$ converges in the strict topology to a bounded invertible operator in $B(\mathcal{H}_{\mathcal{A}})$.

First we need the following elementary facts. For the readers' convenience we present their proofs.

1.5 Lemma Assume that $\eta, \eta', \xi, \xi' \in \mathcal{H}_{\mathcal{A}}$. Then the following hold:

(i) $\theta_{\xi, \eta} \theta_{\eta', \xi'} = \theta_{\xi < \eta, \eta' >, \xi'}$.

(ii) $\theta_{\xi, \eta}^* = \theta_{\eta, \xi}$.

(iii) $\theta_{\xi, \eta}^* \theta_{\xi, \eta} = \theta_{\eta < \xi, \xi >, \eta} = \theta_{\eta < \xi, \xi >^{\frac{1}{2}}, \eta < \xi, \xi >^{1/2}}$.

(iv) If $T \in B(\mathcal{H}_{\mathcal{A}})$, then $T\theta_{\xi, \eta} = \theta_{T\xi, \eta}$

(v) $\|\theta_{\xi, \eta}\| = \|\xi < \eta, \eta >^{1/2}\| = \| < \xi, \xi >^{1/2} < \eta, \eta >^{1/2} \|$. In particular, if $\mathcal{A}$ is unital and $< \eta, \eta > = I$, then $\|\theta_{\xi, \eta}\| = \|\xi\|$.

(vi) $\|\theta_{\xi, \eta}\| \leq \|\eta\| \|\xi\|$

(vii) The rank-one operator $\theta_{\eta, \eta}$ is a projection if and only if $< \eta, \eta >$ is a projection, if and only if $\eta = \eta < \eta, \eta >$. [**5**, Lemma 2.3]

PROOF. (i) For any $\gamma \in \mathcal{H}_\mathcal{A}$, one has

$$\theta_{\xi,\eta}\theta_{\eta',\xi'}\gamma = \xi < \eta, \theta_{\eta',\xi'}\gamma >$$
$$= \xi < \eta, \eta' < \xi', \gamma >>$$
$$= \xi < \eta, \eta' >< \xi', \gamma >$$
$$= \theta_{\xi<\eta,\eta'>,\xi'}\gamma.$$

(ii)

$$< \theta_{\xi,\eta}^* \xi', \eta' > = < \xi', \theta_{\xi,\eta}\eta' >$$
$$= < \xi', \xi < \eta, \eta' >>$$
$$= < \xi', \xi >< \eta, \eta' >$$
$$= < \eta < \xi, \xi' >, \eta' >$$
$$= < \theta_{\eta,\xi}\xi', \eta' > .$$

(iii) The first identity follows from (i) and (ii). Moreover

$$\theta_{\eta<\xi,\xi>,\eta}\xi' = \eta < \xi, \xi >< \eta, \xi' >$$
$$= \eta < \xi, \xi >^{\frac{1}{2}} (< \eta < \xi, \xi >^{\frac{1}{2}}>, \xi' >$$
$$= \theta_{\eta<\xi,\xi>^{\frac{1}{2}},\eta<\xi,\xi>^{1/2}}\xi'.$$

(iv) follows directly from the definition.

(v)

$$\|\theta_{\xi,\eta}\| = \sup\{\|\theta_{\xi,\eta}\gamma\| \mid \|\gamma\| = 1\}$$
$$= \sup\{\| < \xi < \eta, \gamma >, \xi < \eta, \gamma >> \|^{1/2} \mid \|\gamma\| = 1\}$$
$$= \sup\{\| < \eta, \gamma >^* < \xi, \xi >< \eta, \gamma > \|^{1/2} \mid \|\gamma\| = 1\}$$
$$= \sup\{\| < \xi, \xi >^{1/2} < \eta, \gamma > \| \mid \|\gamma\| = 1\}$$
$$= \sup\{\| < \eta < \xi, \xi >^{1/2}, \gamma > \| \mid \|\gamma\| = 1\}$$
$$= \|\eta < \xi, \xi >^{1/2} \|$$
$$= \| < \xi, \xi >^{1/2} < \eta, \eta >< \xi, \xi >^{1/2} \|^{1/2}$$
$$= \| < \xi, \xi >^{1/2} < \eta, \eta >^{1/2} \|$$

(vi) is obvious

(vii) For completeness we add a short proof. By (ii), $\theta_{\eta,\eta}$ is a projection if and only if

$$0 = \theta_{\eta,\eta} - \theta_{\eta,\eta}\theta_{\eta,\eta} = \theta_{\eta,\eta} - \theta_{\eta<\eta,\eta>,\eta} = \theta_{\eta-\eta<\eta,\eta>,\eta}$$

and by (v), this condition is equivalent to $(\eta - \eta < \eta, \eta >) < \eta, \eta >^{1/2} = 0$. If $< \eta, \eta >$ is a projection, then

$$< \eta - \eta < \eta, \eta >, \eta - \eta < \eta, \eta >> = < \eta, \eta > -2 < \eta, \eta >^2 + < \eta, \eta >^3 = 0,$$

hence $< \eta - \eta < \eta, \eta >= 0$, and thus $\theta_{\eta,\eta}$ is a projection. Conversely, if $(\eta - \eta < \eta, \eta >) < \eta, \eta >^{1/2} = 0$ then

$$< \eta, (\eta - \eta < \eta, \eta >) < \eta, \eta >^{1/2} > = < \eta, \eta >^{3/2} - < \eta, \eta >^{5/2} = 0,$$

whence $< \eta, \eta >$ is a projection.

$\square$

Notice that equality in (vi) may fail. For instance, if $\xi = \{p, 0, 0, ..., \}$ and $\eta = \{q, 0, 0, ..., \}$ where p, $q \in \mathcal{A}$ are orthogonal non-zero projections, then $< \xi, \xi > = p = < \xi, \xi >^{1/2}$ and $< \eta, \eta > = q = < \eta, \eta >^{1/2}$ hence $< \xi, \xi >^{1/2} < \eta, \eta >^{1/2} = 0$.

PROOF OF THEOREM 1.4. Assume first that the series $\sum_{j \in \mathbb{J}} \theta_{\xi_j, \xi_j}$ converges in the strict topology to some operator $D_A \in B(\mathcal{H}_A)$. Set $T_{\mathbb{F}} = \sum_{j \in \mathbb{F}} \theta_{\xi_j, \xi_j} - D_A$ for any finite subset $\mathbb{F}$ of $\mathbb{J}$. Then the net $\{T_{\mathbb{F}}\}$ converges to 0 in the strict topology. Using the equality $\|T_{\mathbb{F}} \theta_{\xi, \eta}\| = \|T_{\mathbb{F}} < \eta, \eta >^{1/2} \|$ for all ξ, η from Lemma 1.5 (v), it follows that $\|T_{\mathbb{F}} \xi a\| \to 0$ for all positive $a \in \mathcal{A} \otimes \mathcal{K}$. But then,

$$\|T_{\mathbb{F}} \xi\| \le \|T_{\mathbb{F}} \xi a\| + \|T_{\mathbb{F}}(\xi - a\xi)\|$$
$$\le \|T_{\mathbb{F}} \xi a\| + \sup\{\|T_{\mathbb{F}}\| \|(\xi - a\xi)\|$$
$$= \|T_{\mathbb{F}} \xi a\| + \sup\{\|T_{\mathbb{F}}\| \|\{< \xi, \xi > -a < \xi, \xi > - < \xi, \xi > a + a < \xi, \xi > a)\}\|^{1/2}$$
$$\le \|T_{\mathbb{F}} \xi a\| + \sup\{\|T_{\mathbb{F}}\| \{\|(< \xi, \xi > -a < \xi, \xi >)\| + \|a\| \| < \xi, \xi > -a < \xi, \xi > \|\}^{1/2}.$$

Since every C*-algebra $\mathcal{A}$ has a positive approximate identity, one can choose $a > 0$ such that

$$\sup\{\|T_{\mathbb{F}}\| \{\|(< \xi, \xi > -a < \xi, \xi >)\| + \|a\| \| < \xi, \xi > -a < \xi, \xi > \|\}^{1/2} < \epsilon.$$

For that a, $\|T_{\mathbb{F}} \xi a\| < \epsilon$ for all $\mathbb{F} \supset G$ for some finite subset G of $\mathbb{J}$. This shows that $\|T_{\mathbb{F}} \xi\| \to 0$. Consequently, the series $\sum_{j \in \mathbb{J}} \theta_{\xi_j, \xi_j} \xi$ converges in the norm of $\mathcal{H}_A$ to $D_A \xi$, and hence,

$$\sum_{j \in \mathbb{J}} < \theta_{\xi_j, \xi_j} \xi, \xi > = \sum_{j \in \mathbb{J}} < \xi, \xi_j >, < \xi_j, \xi >$$

converges in the norm of $\mathcal{A}$ to $< D_A \xi, \xi >$ by the Schwartz Inequality. Now a positive operator D_A is bounded and invertible if and only if $aI \le D_A \le bI$ for some constants a, $b > 0$. By [11, 2.1.3], this condition is equivalent to $a < \xi, \xi > \le < D_A \xi, \xi > \le b < \xi, \xi >$. Therefore, $\{\xi_j\}_{j \in \mathbb{J}}$ is a frame.

Conversely, assume that $\{\xi_j\}_{j \in \mathbb{J}}$ is a frame. Then $\sum_{j \in \mathbb{J}} < \theta_{\xi_j, \xi_j} \xi, \xi >$ converges in the norm of $\mathcal{A}$ to $< D_A \xi, \xi >$ for some positive operator $D_A \in B(\mathcal{H}_A)$. For any finite subset $\mathbb{F} \subset \mathbb{J}$, $< \sum_{j \in \mathbb{F}} \theta_{\xi_j, \xi_j} \xi, \xi > \le < D_A \xi, \xi >$, hence, again by [11, 2.1.3],

$$0 \le D_A - \sum_{j \in \mathbb{F}} \theta_{\xi_j, \xi_j} \le D_A.$$

But then, by (vi) in the above lemma,

$$\|(D_A - \sum_{j \in \mathbb{F}} \theta_{\xi_j, \xi_j}) \theta_{\xi, \eta}\| \le \|(D_A - \sum_{j \in \mathbb{F}} \theta_{\xi_j, \xi_j})^{1/2}\| \|\theta_{(D_A - \sum_{j \in \mathbb{F}} \theta_{\xi_j, \xi_j})^{1/2} \xi, \eta}\|$$
$$\le \|D_A\|^{1/2} \|(D_A - \sum_{j \in \mathbb{F}} \theta_{\xi_j, \xi_j})^{1/2} \xi\| \|\eta\|$$
$$= \|D_A\|^{1/2} \|\eta\| \| < (D_A - \sum_{j \in \mathbb{F}} \theta_{\xi_j, \xi_j}) \xi, \xi >\|^{1/2} \to 0.$$

Since the linear span of rank-one operators is by dense in $\mathcal{A} \otimes \mathcal{K}$, it follows that $\|(D_A - \sum_{j \in \mathbb{F}} \theta_{\xi_j, \xi_j}) S\| \to 0$ for all $S \in \mathcal{A} \otimes \mathcal{K}$. Since $D_A - \sum_{j \in \mathbb{F}} \theta_{\xi_j, \xi_j}$ is selfadjoint, this proves that the series $\sum_{j\in\mathbb{J}}\theta_{\xi_j,\xi_j}$ converges to D_A in the strict topology. The same argument as above shows that since D_A is bounded and invertible then

$$a<\xi,\xi>\ \leq\ <D_A\xi,\xi>=\sum_{j\in\mathbb{J}}<\xi,\xi_j,><\xi_j,\xi>\leq b<\xi,\xi>\quad\text{for all}\ \ \xi.$$

$\square$

Many of the results on frames in Hilbert C*-modules are obtained under the assumption that $\mathcal{A}$ is unital, which is of course the case for Hilbert space frames where $\mathcal{A}=\mathbb{C}$. When $\mathcal{A}$ is unital, in lieu of viewing frames as collections of vectors in $\mathcal{H}_\mathcal{A}$, we can view them as collections of rank-one operators on $\mathcal{H}_\mathcal{A}$ with range in a submodule $\mathcal{H}_0$. Indeed, if $\eta\in\mathcal{H}_\mathcal{A}$ is an arbitrary unital vector, i.e., $<\eta,\eta>=I$, then by Lemma 1.5 (vii), (i) and (ii), $E_0:=\theta_{\eta,\eta}\in\mathcal{A}\otimes K$ is a projection, actually the range projection of $\theta_{\eta,\xi}$ for every $\xi\in\mathcal{H}_\mathcal{A}$. Then $\mathcal{H}_0:=E_0\mathcal{H}_\mathcal{A}$ is a submodule of $\mathcal{H}_\mathcal{A}$ and we can identify $E_0M(\mathcal{A}\otimes\mathcal{K})$ with $B(\mathcal{H}_\mathcal{A},\mathcal{H}_0)$, the set of linear bounded adjointable operators from $\mathcal{H}_\mathcal{A}$ to the submodule $\mathcal{H}_0$. Notice that all this would hold also under the weaker hypothesis that $<\eta,\eta>$ is a projection. Then for every collection $\{\xi_j\}_{j\in\mathbb{J}}$ in $\mathcal{H}_\mathcal{A}$, define the rank-one operators $A_j:=\theta_{\eta,\xi_j}$. Since $E_0A_j=A_j$, $A_j\in B(\mathcal{H}_\mathcal{A},\mathcal{H}_0)$. Again, by Lemma 1.5, $A_j^*A_j=\theta_{\xi_j<\eta,\eta>,\xi_j}=\theta_{\xi_j,\xi_j}$. It follows from Theorem 1.4 that $\{\xi_j\}_{j\in\mathbb{J}}$ is a frame if and only if the series $\sum_{j\in\mathbb{J}}A_j^*A_j$ converges in the strict topology to a bounded invertible operator on $\mathcal{H}_\mathcal{A}$.

This leads naturally to the following definition.

1.6 Definition Let $\mathcal{A}$ be a σ-unital C*-algebra and $\mathbb{J}$ be a countable index set. Let E_0 be a projection in $M(\mathcal{A}\otimes\mathcal{K})$. Denote by $\mathcal{H}_0$ the submodule $E_0\mathcal{H}_\mathcal{A}$ and identify $B(\mathcal{H}_\mathcal{A},\mathcal{H}_0)$ with $E_0M(\mathcal{A}\otimes\mathcal{K})$. A collection $A_j\in B(\mathcal{H}_\mathcal{A},\mathcal{H}_0)$ for $j\in\mathbb{J}$ is called an operator-valued frame on $\mathcal{H}_\mathcal{A}$ with range in $\mathcal{H}_0$ if the sum $\sum_{j\in\mathbb{J}}A_j^*A_j$ converges in the strict topology to a bounded invertible operator on $\mathcal{H}_\mathcal{A}$, denoted by D_A. $\{A_j\}_{j\in\mathbb{J}}$ is called a tight operator-valued frame (resp., a Parseval operator-valued frame) if $D_A=\lambda I$ for a positive number λ (resp., $D_A=I$). If the set $\bigcup\{A_j\mathcal{H}_\mathcal{A}:j\in\mathbb{J}\}$ is dense in $\mathcal{H}_0$, then the frame is said to be non-degenerate.

From now on, by frame, we will mean an operator-valued frame on a Hilbert C*-module. Notice that if $\{A_j\}_{j\in\mathbb{J}}$ is a frame with range in $\mathcal{H}_0$ then it is also a frame with range in any larger submodule.

A minor difference with the definition given in [8] for operator-valued frames on a Hilbert space, is that here, in order to avoid introducing maps between different modules, we take directly $\mathcal{H}_0$ as a submodule of $\mathcal{H}_\mathcal{A}$. For operator-valued frames on a Hilbert space we do not assume in [8] that $\mathcal{H}_0\subset\mathcal{H}$ and hence we are left with the flexibility of considering $\dim\mathcal{H}_0>\dim\mathcal{H}$.

1.7 Example Let $\sum_{j\in\mathbb{J}}E_j=I$ be a decomposition of the identity of $M(\mathcal{A}\otimes\mathcal{K})$ into mutually orthogonal equivalent projections in $M(\mathcal{A}\otimes\mathcal{K})$, i.e., $L_jL_j^*=E_j$ and $L_j^*L_j=E_0$ for some collection of partial isometries $L_j\in M(\mathcal{A}\otimes\mathcal{K})$, and the series converges in the strict topology. Let T be a left-invertible element of $M(\mathcal{A}\otimes\mathcal{K})$ and let $A_j:=L_j^*T$. Then $A_j\in B(\mathcal{H}_\mathcal{A},E_0\mathcal{H}_\mathcal{A})$ for $j\in\mathbb{J}$, and $\sum_{j\in\mathbb{J}}A_j^*A_j=\sum_{j\in\mathbb{J}}T^*A_jT=T^*T$ is an invertible element of $M(\mathcal{A}\otimes\mathcal{K})$, where the convergence is in the strict topology. Thus $\{A_j\}_{j \in \mathbb{J}}$ is a frame with range in $E_0 \mathcal{H}_A$. The frame is Parseval precisely when T is an isometry. We will see in the next section that this example is generic.

2. Frame Transforms

2.1. Definition Assume that $\{A_j\}_{j \in \mathbb{J}}$ is a frame in $B(\mathcal{H}_A, E_0 \mathcal{H}_A)$ for the Hilbert C^*-module $\mathcal{H}_A$ and set $\mathcal{H}_0 := E_0 \mathcal{H}_A$. Decompose the identity of $M(\mathcal{A} \otimes \mathcal{K})$, into a strictly converging sum of mutually orthogonal projections $\{E_j\}_{j \in \mathbb{J}}$ in $M(\mathcal{A} \otimes \mathcal{K})$ with $E_j \sim E_{00} \geq E_0$. Let L_j be partial isometries in $M(\mathcal{A} \otimes \mathcal{K})$ such that $L_j L_j^* = E_j$ and $L_j^* L_j = E_{00}$. Define the frame transform θ_A of the frame $\{A_j\}_{j \in \mathbb{J}}$ as

$$\theta_A = \sum_{j \in \mathbb{J}} L_j A_j : \quad \mathcal{H}_A \longrightarrow \mathcal{H}_A.$$

2.2. Theorem Assume that $\{A_j\}_{j \in \mathbb{J}}$ is a frame in $B(\mathcal{H}_A, \mathcal{H}_0)$.
(a) The sum $\sum_{j \in \mathbb{J}} L_j A_j$ converges in the strict topology, and hence θ_A is an element of $M(\mathcal{A} \otimes \mathcal{K})$.
(b) $D_A = \theta_A^* \theta_A$, $\theta_A D_A^{-1/2}$ is an isometry, $P_A := \theta_A D_A^{-1} \theta_A^*$ is the range projection of θ_A, and all these three elements belong to $M(\mathcal{A} \otimes \mathcal{K})$.
(c) $\{A_j\}_{j \in \mathbb{J}}$ is a Parseval frame, if and only if θ_A is an isometry of $M(\mathcal{A} \otimes \mathcal{K})$, and again if and only if $\theta_A \theta_A^*$ is a projection.
(d) $A_j = L_j^* \theta_A$ for all $j \in \mathbb{J}$.

PROOF. **(a)** For every finite subset F of $\mathbb{J}$, let $S_F = \sum_{j \in F} L_j A_j$. We need to show that $\{S_F : F$ is a finite subset of $\mathbb{J}\}$ is a Cauchy net in the strict topology of $M(\mathcal{A} \otimes \mathcal{K})$, i.e., for every for any $a \in \mathcal{A} \otimes \mathcal{K}$, $\max\{\|(S_F - S_{F'})a\|, \|a(S_F - S_{F'})\| \longrightarrow 0$, in the sense that for every $\varepsilon > 0$ there is a finite set G such that

$$\max\{\|(S_F - S_{F'})a\|, \|a(S_F - S_{F'})\| < \varepsilon \quad \text{for any finite sets } F \supset G, \ F' \supset G.$$

Firstly, since the partial isometries L_j have mutually orthogonal ranges, one has

$$\begin{aligned}
\|(S_F - S_{F'})a\| &= \|a^*(S_F - S_{F'})^*(S_F - S_{F'})a\|^{\frac{1}{2}} \\
&= \|a^*\Big(\sum_{j \in (F \setminus F') \cup (F' \setminus F)} A_j^* L_j^* L_j A_j \Big)a\|^{\frac{1}{2}} \\
&= \|a^*\Big(\sum_{j \in (F \setminus F') \cup (F' \setminus F)} A_j^* E_{00} A_j \Big)a\|^{\frac{1}{2}} \\
&\leq \|a\|^{\frac{1}{2}} \| \sum_{j \in (F \setminus F') \cup (F' \setminus F)} A_j^* A_j a\|^{\frac{1}{2}} \longrightarrow 0,
\end{aligned}$$

where the last term above converges to 0 because of the assumption that $\sum_{j \in \mathbb{J}} A_j^* A_j$ converges in the strict topology. Secondly, for all $\xi \in \mathcal{H}_A$

$$\|(S_F - S_{F'})\xi\|^2 = \sum_{j \in (F \setminus F') \cup (F' \setminus F)} \|L_j A_j \xi\|^2$$

$$= \sum_{j \in (F \setminus F') \cup (F' \setminus F)} \|A_j \xi\|^2$$

$$= \sum_{j \in (F \setminus F') \cup (F' \setminus F)} < A_j^* A_j \xi, \xi >$$

$$= < \sum_{j \in (F \setminus F') \cup (F' \setminus F)} A_j^* A_j \xi, \xi >$$

$$\leq < D_A \xi, \xi >$$

$$= \|D_A^{1/2} \xi\|^2$$

Thus $\|S_F - S_{F'}\| \leq \|D_A\|^{1/2}$ for any finite sets F and F'. Moreover, $\sum_{j \in (F \setminus F') \cup (F' \setminus F)} E_j(S_F - S_{F'}) = S_F - S_{F'}$, hence for every $a \in \mathcal{A} \otimes \mathcal{K}$,

$$\|a(S_F - S_{F'})\| = \|a(S_F - S_{F'})(S_F - S_{F'})^* a^*\|^{\frac{1}{2}}$$

$$= \|a \sum_{j \in (F \setminus F') \cup (F' \setminus F)} E_j (S_F - S_{F'})(S_F - S_{F'})^* \sum_{j \in (F \setminus F') \cup (F' \setminus F)} E_j a^*\|^{\frac{1}{2}}$$

$$\leq \|S_F - S_{F'}\| \|a \sum_{j \in (F \setminus F') \cup (F' \setminus F)} E_j a^*\|^{\frac{1}{2}}$$

$$\leq \|D_A\|^{1/2} \|a\|^{1/2} \| \sum_{j \in (F \setminus F') \cup (F' \setminus F)} E_j a^*\|^{1/2} \longrightarrow 0$$

by the strict convergence of the series $\sum_{j \in \mathbb{J}} E_j$. The statements (b) and (c) are now obvious.

(d) Also obvious since the series $\sum_{i \in \mathbb{J}} L_i^* L_i A_i$ converges strictly to $L_j^* \theta_A$ and

$$L_j^* L_i A_i = \delta_{i,j} E_{00} A_i = \delta_{i,j} A_j.$$

$\square$

The projection P_A is called the *frame projection*.

2.3 Remark (i) Since θ_A is left-invertible as $(D_A^{-1}\theta_A^*)\theta_A = I$, Example 1.7 is indeed generic, i.e., every frame $\{A_j\}_{j \in \mathbb{J}}$ is obtained from partial isometries L_j with mutually orthogonal range projections summing to the identity and same first projection majorizing E_0. The relation with the by now familiar "dilation" point of view of the theory of frames is clarified in (ii) below.

(ii) For vector frames on a Hilbert space, the frame transform is generally defined as a map from the Hilbert space $\mathcal{H}$ into $\ell^2(\mathbb{J})$ - a dilation of $\mathcal{H}$. If $\mathcal{H}$ is infinite dimensional and separable and if $\mathbb{J}$ is infinite and countable, which are the most common assumptions, then $\mathcal{H}$ can be identified with $\ell^2(\mathbb{J})$, and hence, the frame transform can be seen as mapping of $\mathcal{H}$ onto a subspace. In the case of Hilbert C*-modules, it is convenient to choose the latter approach, so to identify the frame transform and the frame projection with elements of $M(\mathcal{A} \otimes \mathcal{K})$.

(iii) The range projections of elements of $M(\mathcal{A} \otimes \mathcal{K})$ and even of elements of $\mathcal{A} \otimes \mathcal{K}$ always belong to $(\mathcal{A} \otimes \mathcal{K})^{**}$, but may fail to belong to $M(\mathcal{A} \otimes \mathcal{K})$. As shown above, however, the frame projection P_A is always in $M(\mathcal{A} \otimes \mathcal{K})$ and $P_A \sim I$ since $\theta_A D_A^{1/2}$ is an isometry.

(iv) When $\mathcal{A}$ is not simple, given an arbitrary nonzero projection $E_0 \in M(\mathcal{A} \otimes \mathcal{K})$ there may be no decomposition of the identity in projections equivalent to E_0. Nevertheless, there is always a decomposition of the identity into a strictly convergent sum of mutually orthogonal projections that are all equivalent to a projection $E_{00} \geq E_0$, e.g., $E_{00} = I$.

(v) There seem to be no major advantage in considering only non-degenerate frames, i.e., seeking a "minimal" Hilbert module $\mathcal{H}_0$ that contains the ranges of all the operators A_j or similarly, choosing a frame transform with a minimal projection E_{00}. In fact, if we view the operators A_j as having their ranges in some $E_{00}\mathcal{H}_A$, the ensuing frame transform will, as seen in (d) above and in the next section, carry equally well all the "information" of the frame.

(vi) For the vector case, [**5**, 4.1] proves that the frame transform θ as a map from a finite or countably generated Hilbert C*- module $\mathcal{H}$ to the standard module $\mathcal{H}_A$ is adjointable. This is obvious in the case that we consider, where $\mathcal{H} = \mathcal{H}_A$ as then θ_A is in the C*-algebra $M(\mathcal{A} \otimes \mathcal{K})$.

(vii) A compact form of the *reconstruction formula* for a frame is simply

$$D_A^{-1} \sum_{j \in \mathbb{J}} A_j^* A_j = D_A^{-1} \theta_A^* \theta_A = I.$$

In the special case that $\mathcal{A}$ is unital and that $\{A_j\}_{j \in \mathbb{J}}$ is a vector frame, we have seen in the course of the proof of Theorem 1.4 that $\sum_{j \in \mathbb{J}} A_j^* A_j$ converges strongly and the same result was obtained in [**5**, 4.1]. Thus, assuming for the sake of simplicity that the frame is Parseval, the reconstruction formula has the more familiar form

$$\sum_{j \in \mathbb{J}} A_j^* A_j \xi = \sum_{j \in \mathbb{J}} \xi_j < \xi_j, \xi >= \xi \text{ for all } \xi \in \mathcal{H}_A$$

where the convergence is in the norm of $\mathcal{H}_A$.

3. Similarity of frames

3.1. Definition Two frames $\{A_j\}_{j \in \mathbb{J}}$ and $\{B_j\}_{j \in \mathbb{J}}$ in $B(\mathcal{H}_A, \mathcal{H}_0)$ are said to be right-similar (resp. right-unitarily equivalent) if there exists an invertible (resp. a unitary) element $T \in M(\mathcal{A} \otimes \mathcal{K})$ such that $B_j = A_j T$ for all $j \in \mathbb{J}$.

The following facts are immediate and their proofs are left to the reader.

3.2. Lemma
(i) If $\{A_j\}_{j \in \mathbb{J}}$ is a frame and T is an invertible element in $M(\mathcal{A} \otimes \mathcal{K})$, then $\{A_j T\}_{j \in \mathbb{J}}$ is also a frame.

(ii) If $\{A_j\}_{j \in \mathbb{J}}$ and $\{B_j\}_{j \in \mathbb{J}}$ are right-similar and T is an invertible element in $M(\mathcal{A} \otimes \mathcal{K})$ for which $B_j = A_j T$ for all $j \in \mathbb{J}$, then $\theta_B = \theta_A T$. Therefore $T = D_A^{-1} \theta_A^* \theta_B$, hence T is uniquely determined. Moreover, $P_A = P_B$ and $D_B = T^* D_A T$. Conversely, if $\theta_B = \theta_A T$ for some invertible element $T \in M(\mathcal{A} \otimes \mathcal{K})$, then $B_j = A_j T$ for all $j \in \mathbb{J}$.

(iii) Every frame is right-similar to a Parseval frame, i.e., $\{A_j\}_{j\in\mathbb{J}}$ is right-similar to $\{A_j\}_{j\in\mathbb{J}}D_A^{-1/2}$. Two Parseval frame are right-similar if and only if they are right-unitarily equivalent.

3.3. Theorem Let $\{A_j\}_{j\in\mathbb{J}}$ and $\{B_j\}_{j\in\mathbb{J}}$ be two frames in $B(\mathcal{H}_\mathcal{A}, \mathcal{H}_0)$. Then the following are equivalent:

(i) $\{A_j\}_{j\in\mathbb{J}}$ and $\{B_j\}_{j\in\mathbb{J}}$ are right-similar.
(ii) $\theta_B = \theta_A T$ for some invertible operator $T \in M(\mathcal{A} \otimes \mathcal{K})$.
(iii) $P_A = P_B$.

PROOF. The implications $(i) \Rightarrow (ii) \Rightarrow (iii)$ are given by Lemma 3.2. $(iii) \Rightarrow (i)$. Assume that $P_A = P_B$. Then by Theorem 2.2 (b)

$$\theta_A D_A^{-1}\theta_A^* = \theta_B D_B^{-1}\theta_B^*.$$

By Theorem 2.2. (d)

$$B_j = L_j^*\theta_B = L_j^* P_A \theta_B = L_j^* \theta_A D_A^{-1}\theta_A^*\theta_B = A_j D_A^{-1}\theta_A^*\theta_B.$$

Let $T := D_A^{-1}\theta_A^*\theta_B$, then $T \in M(\mathcal{A} \otimes \mathcal{K})$ and

$$(D_B^{-\frac{1}{2}}\theta_B^*\theta_A D_A^{-\frac{1}{2}})(D_A^{-\frac{1}{2}}\theta_A^*\theta_B D_B^{-\frac{1}{2}}) = D_B^{-\frac{1}{2}}\theta_B^* P_B \theta_B D_B^{-\frac{1}{2}} = I.$$

Interchanging A and B, one has also

$$(D_A^{-\frac{1}{2}}\theta_A^*\theta_B D_B^{-\frac{1}{2}})(D_B^{-\frac{1}{2}}\theta_B^*\theta_A D_A^{-\frac{1}{2}}) = I.$$

Thus, $D_A^{-\frac{1}{2}}\theta_A^*\theta_B D_B^{-\frac{1}{2}}$ is unitary, hence $\theta_A^*\theta_B$ is invertible, and thus so is $T = D_A^{-1}\theta_A^*\theta_B$. This concludes the proof. □

As is the case in $B(H)$, all the projections in $M(\mathcal{A} \otimes \mathcal{K})$ that are equivalent to the frame projection of a given frame, are also the frame projection of a frame, which by Theorem 3.3 is unique up to right similarity.

3.4. Theorem Let $\{A_j\}_{j\in\mathbb{J}}$ be a frame in $B(\mathcal{H}_\mathcal{A}, \mathcal{H}_0)$ and let P be a projection in $M(\mathcal{A}\otimes\mathcal{K})$. Then $P \sim P_A$ if and only if there exists a frame $\{B_j\}_{j\in\mathbb{J}}$ in $B(\mathcal{H}_\mathcal{A}, \mathcal{H}_0)$ such that $P = P_B$.

PROOF. If $P = P_B$, let $V = \theta_B D_B^{-\frac{1}{2}} D_A^{-\frac{1}{2}}\theta_A$. Then $V \in B(\mathcal{H}_\mathcal{A})$, $VV^* = P$, and $V^*V = P_A$, i.e., $P \sim P_A$. Conversely, if there exists $V \in M(\mathcal{A} \otimes \mathcal{K})$ such that $VV^* = P$ and $V^*V = P_A$, then set $B_j = L_j^* V \theta_A$. Then

$$\sum_{j\in\mathbb{J}} B_j^* B_j = \theta_A^* V^* (\sum_{j\in\mathbb{J}} L_j^* L_j) V \theta_A = \theta_A^* V^* V \theta_A = D_A.$$

Thus $\{B_j\}_{j\in\mathbb{J}}$ is a frame with $D_A = D_B$. Moreover,

$$\theta_B = \sum_{j\in\mathbb{J}} L_j^* L_j V \theta_A = V \theta_A.$$

It follows that $P_B = VV^* = P$. □

The proof of Theorem 3.4 actually yields a parametrization of all the operator-valued frames with range in $B(\mathcal{H}_A, \mathcal{H}_0)$. For simplicity's sake, we formulate it in terms of Parseval frames

3.5. Corollary Let $\{A_j\}_{j \in \mathbb{J}}$ be a Parseval frame in $B(\mathcal{H}_A, \mathcal{H}_0)$. Then $\{B_j\}_{j \in \mathbb{J}}$ is a Parseval frame in $B(\mathcal{H}_A, \mathcal{H}_0)$ if and only if $B_j = L_j^* V \theta_A$ for some partial isometry $V \in M(\mathcal{A} \otimes \mathcal{K})$ such that $VV^* = P_B$ and $V^*V = P_A$.

3.6. Remark For a given equivalence $P \sim P_A$ the choice of partial isometry $V \in M(\mathcal{A} \otimes \mathcal{K})$ with $VV^* = P$ and $V^*V = P_A$ is determined up to a unitary that commutes with P_A, i.e., $V_1 V_1^* = P$ and $V_1^* V_1 = P_A$ implies $V_1 = VU$ for some unitary U that commutes with P_A, or, equivalently, $V_1 = U_1 V$ for some unitary U_1 that commutes with P. Notice that if V and U are as above, then $\{L_j^* V \theta_A D_A^{-\frac{1}{2}}\}_{j \in \mathbb{J}}$ and $\{L_j^* V U \theta_A D_A^{-\frac{1}{2}}\}_{j \in \mathbb{J}}$ are two frames having the same frame projections P. It then follows from Theorem 3.3 that the two frames are right-similar, actually, right-unitarily equivalent, because $L_j^* V U \theta_A D_A^{-\frac{1}{2}} = L_j^* V \theta_A D_A^{-\frac{1}{2}} (D_A^{-\frac{1}{2}} \theta_A^* U \theta_A D_A^{-\frac{1}{2}})$ and $D_A^{-\frac{1}{2}} \theta_A^* U \theta_A D_A^{-\frac{1}{2}}$ is a unitary operator on $\mathcal{H}_A$ as $\theta_A D_A^{-\frac{1}{2}}$ is an isometry.

For operator valued frames it is natural to consider also the notion of left similarity.

3.7. Definition Two frames $\{A_j\}_{j \in \mathbb{J}}$ and $\{B_j\}_{j \in \mathbb{J}}$ in $B(\mathcal{H}_A, \mathcal{H}_0)$ are said to be left-similar (resp., left-unitarily equivalent) if there exists an invertible (resp., a unitary) element S in the corner algebra $E_0 M(\mathcal{A} \otimes \mathcal{K}) E_0$ such that $B_j = SA_j$ for all $j \in \mathbb{J}$, where E_0 is the projection in $M(\mathcal{A} \otimes \mathcal{K})$ such that $E_0 \mathcal{H}_A = \mathcal{H}_0$.

The following results are elementary.

3.8. Lemma If $\{A_j\}_{j \in \mathbb{J}}$ is a frame in $B(\mathcal{H}_A, \mathcal{H}_0)$ and S is an invertible element in $E_0 M(\mathcal{A} \otimes \mathcal{K}) E_0$, then $\{A_j\}_{j \in \mathbb{J}}$ is also a frame. Moreover, $\theta_B = \sum_{j \in \mathbb{J}} L_j SA_j$, hence $D_B = \sum_{j \in \mathbb{J}} A_j^* S^* SA_j$ and thus

$$\|S^{-1}\|^{-2} D_A \leq D_B \leq \|S\|^2 D_A.$$

In particular, if S is unitary, then $D_A = D_B$.

3.9. Proposition Given two left-unitarily equivalent frames $\{A_j\}_{j \in \mathbb{J}}$ and $\{B_j\}_{j \in \mathbb{J}}$ in $B(\mathcal{H}_A, \mathcal{H}_0)$ with $B_j = SA_j$ for some unitary element $S \in M(\mathcal{A} \otimes \mathcal{K})$, then the following are equivalent:

(i) S commutes with $A_j D_A^{-1} A_i^*$ for all $i, j \in \mathbb{J}$.
(ii) $\{A_j\}_{j \in \mathbb{J}}$ and $\{B_j\}_{j \in \mathbb{J}}$ are right-unitarily equivalent.

PROOF. $(i) \implies (ii)$ One has

$$
\begin{aligned}
B_j =& SA_j \\
=& SA_j(D_A^{-1}D_A) = \sum_{i \in \mathbb{J}} SA_j D_A^{-1} A_i^* A_i \\
=& \sum_{i \in \mathbb{J}} A_j D_A^{-1} A_i^* SA_i = A_j D_A^{-1} \sum_{i \in \mathbb{J}} A_i^* SA_i
\end{aligned}
$$

where the series here and below converge in the strict topology. Let

$$
U = D_A^{-\frac{1}{2}} \sum_{i \in \mathbb{J}} A_i^* SA_i D_A^{-\frac{1}{2}}.
$$

Then

$$
\begin{aligned}
UU^* =& D_A^{-\frac{1}{2}} \sum_{i,j \in \mathbb{J}} A_i^* SA_i D_A^{-1} A_j^* S^* A_j D_A^{-\frac{1}{2}} \\
=& D_A^{-\frac{1}{2}} \sum_{i,j \in \mathbb{J}} A_i^* A_i D_A^{-1} A_j^* SS^* A_j D_A^{-\frac{1}{2}} \\
=& D_A^{-\frac{1}{2}} (\sum_{i \in \mathbb{J}} A_i^* A_i) D_A^{-1} (\sum_{j \in \mathbb{J}} A_j^* A_j) D_A^{-\frac{1}{2}} \\
=& I.
\end{aligned}
$$

Similarly, $U^*U = I$. Notice that $B_j = A_j D_A^{-\frac{1}{2}} U D_A^{\frac{1}{2}}$, and that $D_A^{-\frac{1}{2}} U D_A^{\frac{1}{2}}$ is invertible. It follows that $\{A_j\}_{j \in \mathbb{J}}$ and $\{B_j\}_{j \in \mathbb{J}}$ are right-unitarily equivalent frames.

$(ii) \implies (i)$ $P_A = P_B$ by Theorem 3.3 and $D_A = D_B$ by Lemma 3.8. Then

$$
P_A = \theta_A D_A^{-1} \theta_A^* = \sum_{i,j \in \mathbb{J}} L_i A_i D_A^{-1} A_j^* L_j^*
$$

$$
= P_B = \theta_B D_B^{-1} \theta_B^* = \sum_{i,j \in \mathbb{J}} L_i SA_i D_A^{-1} A_j^* S^* L_j^*.
$$

Multiplying on the left by L_i^* and on the right by L_j, one has

$$
A_i D_A^{-1} A_j^* = SA_i D_A^{-1} A_j^* S^* \quad \forall\, i,j \in \mathbb{J}.
$$

$\square$

3.10. Composition of frames Let $\{A_j\}_{j \in \mathbb{J}}$ be a frame in $B(\mathcal{H}_\mathcal{A}, \mathcal{H}_0)$ and $\{B_i\}_{i \in \mathbb{I}}$ be a frame in $B(\mathcal{H}_0, \mathcal{H}_1)$. Then it is easy to check that $\{C_{i,j} := B_i A_j\}_{j \in \mathbb{J}, i \in \mathbb{I}}$ is a frame in $B(\mathcal{H}_\mathcal{A}, \mathcal{H}_1)$, called the composition of the frames $\{A_j\}_{j \in \mathbb{J}}$ and $B(\mathcal{H}_\mathcal{A}, \mathcal{H}_0)$. In symbols, $C = BA$

It is easy to see that the composition of two Parseval frames is also Parseval.

3.11. Remarks (i) If $BA = BA'$, then $A = A'$. Indeed, if for all $i \in \mathbb{I}$ and $j \in \mathbb{J}$ and $B_i A_j = B_i A_j'$, then $\sum_{i \in I} B_i^* B_i A_j = D_B A_j = D_B A_j'$. Then $A_j = A_j'$, since D_B is invertible.

(ii) If $A := \{A_j\}_{j \in \mathbb{J}}$ is non-degenerate (i.e., the closure of $\bigcup_{j \in \mathbb{J}} A_j \mathcal{H}_2$ is $\mathcal{H}_2$), then $BA = B'A$ implies $B = B'$.

3.12. Remark Let $\mathcal{A}$ be unital, then as shown in Theorem 1.4, we can identify a vector frame $\{\xi_i\}_{i\in\mathbb{I}}$ on on a submodule $\mathcal{H}_1 \subset \mathcal{H}_\mathcal{A}$, with the (rank-one) operator valued frame $\{\theta_{\eta,\xi_i}\}_{i\in\mathbb{I}}$ in $B(\mathcal{H}_1, \theta_{\eta,\eta}\mathcal{H}_\mathcal{A})$, where $\eta \in \mathcal{H}_\mathcal{A}$ is a vector for which $<\eta,\eta> = I$. But then, for any operator-valued frame $\{A_j\}_{\in\mathbb{J}}$ in $B(\mathcal{H}_\mathcal{A}, \mathcal{H}_0)$ and $\mathcal{H}_1 \subset \mathcal{H}_0$, the composition $\{\theta_{\eta,\xi_i}A_j = \theta_{\eta,A_j^*\xi_i}\}_{i\in\mathbb{I},i\in\mathbb{I}}$ is identified with the vector frame $\{A_j^*\xi_i\}_{i\in\mathbb{I}, j\in\mathbb{J}}$. We can view this as a decomposition of the operator valued frame $\{A_j\}_{\in\mathbb{J}}$ into the collection of (vector-valued) frames $\{A_j^*\xi_i\}_{j\in\mathbb{J}}$ indexed by $\mathbb{I}$, i.e., a "multiframe".

References

[1] C.A. Akemann, G.K. Pedersen and J. Tomiyama, Multipliers of C*- algebras, J. Funct. Anal., 13(1973), 277 - 301.

[2] B. Blackadar, K-theory for operator algebras, Springer-Verlag, New York, Berlin, Heidelberg, London, Paris, Tokyo 1987.

[3] L.G. Brown, Semicontinuity and multipliers of $C*$-algebras, Can. J. Math., 40(1989), 769 - 887.

[4] R. Busby, Double centralizers and extensions of C*-algebras, Trans. Amer. Math., 132(1968), 79 - 99.

[5] M. Frank and D. Larson, Frames in Hilbert C*-modules and C*-algebras, J. Operator Theory, 48(2002), 273 - 314.

[6] D, Han and D. Larson, Frames, bases and group representations, Memoirs Amer. Math. Soc., 147(2000), No. 697.

[7] D, Han, W. Jing, and R. M. Mohapatra, Structured Parseval frames in Hilbert C*-modules, (preprint).

[8] V. Kaftal, D. Larson, and S. Zhang, Operator valued frames (preprint).

[9] G. G. Kasparov, Hilbert C*-modules: theorems of Stinespring and Voiculescu, J. Operator Theory, 3 (1980), 133 - 150.

[10] G. G. Kasparov, The operator K-functor and extensions of C*-algebras, Math. USSR, Izv., 16(1981), 513 - 572.

[11] V.M. Manuilov and E.V. Troitsky, Hilbert C*-modules, Transl of Math Monogr. AMS, Providence, RI 226, 2001, 513 - 572.

[12] G.K. Pedersen, C*-algebras and their automorphism groups, Academic Press, London, New York, San Francisco, 1979.

[13] P. Wood, Wavelets and projective Hilbert modules (preprint).

[14] S. Zhang, K-theory and a bivariable Fredholm index, Contemporary Math., Amer. Math. Soc., 148(1993), 155 - 190.

[15] S. Zhang, A Riesz decomposition property and ideal structure of multiplier algebras, J. Operator Theory, 24 (1990), 209 - 225.

DEPARTMENT OF MATHEMATICAL SCIENCES, UNIVERSITY OF CINCINNATI, CINCINNATI, OHIO 45221-0025

E-mail address: `victor.kaftal@uc.edu`

DEPARTMENT OF MATHEMATICS, TEXAS A&M UNIVERSITY, COLLEGE STATION, TEXAS 77843, USA

E-mail address: `larson@math.tamu.edu`

DEPARTMENT OF MATHEMATICAL SCIENCES, UNIVERSITY OF CINCINNATI, CINCINNATI, OHIO 45221-0025

E-mail address: `zhangs@math.uc.edu`

Contemporary Mathematics
Volume **451**, 2008

Coxeter Groups and Wavelet Sets

David R. Larson and Peter Massopust

ABSTRACT. A traditional wavelet is a special case of a vector in a separable
Hilbert space that generates a basis under the action of a system of unitary operators defined in terms of translation and dilation operations. A
Coxeter/fractal-surface wavelet is obtained by defining fractal surfaces on foldable figures, which tesselate the embedding space by reflections in their bounding hyperplanes instead of by translations along a lattice. Although both
theories look different at their onset, there exist connections and communalities which are exhibited in this semi-expository paper. In particular, there is
a natural notion of a dilation-reflection wavelet set. We prove that dilation-reflection wavelet sets exist for arbitrary expansive matrix dilations, paralleling
the traditional dilation-translation wavelet theory. There are certain measurable sets which can serve simultaneously as dilation-translation wavelet sets
and dilation-reflection wavelet sets, although the orthonormal structures generated in the two theories are considerably different.

1. Introduction

This article is meant to mesh together two distinct approaches to wavelet theory: the traditional dilation-translation approach that was the subject of the memoir [**DL**] co-authored by the first author, and the Coxeter/fractal-surface approach
that was developed by the second author in [**GHM1, GHM2**] and the book [**M2**].
The two approaches seem distinctly different under initial scrutiny. However, it
turns out that both approaches carry a natural notion of "wavelet set", and while
these have different meanings in the two theories, we have recently discovered,
with some surprise, that there is more than a little connection between the two
notions of "wavelet set." Indeed, in the plane $\mathbb{R}^2$, with dilation (scale factor) 2,
the "dyadic case", some of the known wavelet sets in the dilation-translation theory, and in particular the "wedding cake set" (c.f. [**DL**], p. 59, and [**DLS2**])
and the "four-corners set" (c.f. [**DL**], p.57, and [**DLS2**]) are both also wavelet

---

2000 *Mathematics Subject Classification.* Primary 20F55, 28A80, 42C40, 51F15; Secondary
46E25, 65T60.

Key words and phrases. Coxeter groups, reflection groups, Weyl groups, root systems, fractal
functions, fractal surfaces, wavelet sets.

The first author was partially supported by grants from the National Science Foundation
and the second author was partially supported by the grant MEXT-CT-2004-013477, Acronym
MAMEBIA, of the European Commission. Both authors participated in NSF supported Workshops in Linear Analysis and Probability, Texas A&M University.

sets in the Coxeter/fractal-surface multiresolution-analysis theory. We remark that Coxeter/fractal-surface "wavelet sets", as such, were not formally defined in the book [**M2**] and were not part of the original Coxeter-MRA theory that was formally defined and developed in [**GHM1, GHM2**] and [**M2**]. However, in a project we began three years ago, while both of us were participants at an international conference on Abstract and Applied Analysis in Hanoi, Vietnam, we decided to pursue the possibility of a connection between the subjects of our two talks. This resulted in developing the [**GHM1, GHM2, M2**] theory a bit further, including a proper notion of "wavelet set" in that context.

The structure of this paper is as follows. In Section 2, we review some aspects of traditional wavelet theory, introduce the concept of wavelet set, translation and dilation congruence, and abstract dilation-translation pair, and summarize some of the results from the theory of dilation-translation wavelet sets. A class of fractal functions and fractal surfaces is introduced in Section 3 and some of their properties are mentioned. Section 4 deals with Coxeter groups, affine Weyl groups, and foldable figures to the extend that is necessary for this paper. Fractal surfaces on foldable figures are defined in Section 5 and are shown to generate multiresolution analyses in Section 6. The concept of dilation-reflection wavelet set is introduced in Section 7 and it is shown that these wavelet sets exist for arbitrary expansive matrix dilations. We exhibit two examples of measurable sets that serve simultaneously as dilation-translation as well as dilation-reflection wavelet sets, although the orthonormal structures they define are intrinsically different. In Section 8, we consider some questions of a general nature regarding dilation-reflection wavelet sets and pose two open problems.

2. Some Aspects of Traditional Wavelet Theory

A traditional wavelet is a special case of a vector in a separable Hilbert space that generates a basis under the action of a collection, or "system", of unitary operators defined in terms of translation and dilation operations. A traditional wavelet set is a measurable set whose characteristic function, scaled appropriately, is the Fourier Transform of a (single) orthonormal wavelet. Multi-versions have been studied, as well as frame analogues. In [**DLS1**], Dai and Speegle, together with the first author, found a proof of the existence of wavelet sets in the plane and in higher dimensions. The announcement of the existence of such wavelet sets stimulated the construction of examples of such sets by several authors, beginning with Soardi and Weiland [**SW**], and then Baggett, Medina and Merrill [**BMM**], and Benedetto and Leon [**BL**]. Wavelet sets have been a part of the wavelet literature for a number of years now.

A *dyadic orthonormal* wavelet in one dimension is a unit vector $\psi \in L^2(\mathbb{R}, \mu)$, with μ Lebesgue measure, with the property that the set

$$(2.1) \qquad \{2^{\frac{n}{2}} \psi(2^n t - \ell) \mid n, \ell \in \mathbb{Z}\}$$

of all integral translates of ψ followed by dilations by arbitrary integral powers of 2, is an orthonormal basis for $L^2(\mathbb{R}, \mu)$. The term *dyadic* refers to the dilation factor "2". The term *mother wavelet* is also used in the literature for ψ. Then the functions

$$\psi_{n,\ell}(t) := 2^{\frac{n}{2}} \psi(2^n t - \ell)$$

are called elements of the wavelet basis generated by the "mother". The functions $\psi_{n,\ell}$ will not themselves be mother wavelets unless $n = 0$. Let T and D be the translation (by 1) and dilation (by 2) unitary operators in $B(L^2(\mathbb{R}))$, the Banach space of bounded linear operators from $L^2(\mathbb{R})$ to itself, given by $(Tf)(t) = f(t-1)$ and $(Df)(t) = \sqrt{2}f(2t)$. Then

$$2^{\frac{n}{2}}\psi(2^n t - \ell) = (D^n T^\ell \psi)(t)$$

for all $n, \ell \in \mathbb{Z}$. Operator-theoretically, the operators T, D are *bilateral shifts* of *infinite multiplicity*. It is obvious that $L^2([0,1])$, considered as a subspace of $L^2(\mathbb{R})$, is a complete wandering subspace for T, and that $L^2([-2,-1] \cup [1,2])$ is a complete wandering subspace for D.

A *complete wandering subspace* E for a unitary operator D acting on a Hilbert space H is a closed subspace of H for which $\{D^n E \mid n \in \mathbb{Z}\}$ is an orthogonal decomposition of H. A unitary operator is called a *bilateral shift* if it has a complete wandering subspace. In that case its multiplicity is defined to be the dimension of the wandering subspace.

An abstract interpretation is that, since D is a bilateral shift it has (many) complete wandering subspaces, and a wavelet for the system is a vector ψ whose translation space (that is, the closed linear span of $\{T^k \mid k \in \mathbb{Z}\}$ is a complete wandering subspace for D. Hence ψ must generate an orthonormal basis for the entire Hilbert space under the action of the unitary system.

In one dimension, there are non-dyadic orthonormal wavelets: i.e. wavelets for all possible dilation factors besides 2 (the dyadic case). We said "possible", because the scales $\{0, 1, -1\}$ are excluded as scales because the dilation operators they would introduce are not bilateral shifts. All other real numbers for scales yield wavelet theories. In [**DL**], Example 4.5 (x), a family of examples is given of three-interval wavelet sets (and hence wavelets) for all scales $d \geq 2$, and it was noted there that such a family also exists for dilation factors $1 < d \leq 2$.

Let $1 \leq n < \infty$, and let A be an $n \times n$ real invertible matrix. The most tractable such matrices for dilations in wavelet theory are those that are *expansive*.

REMARK 2.1. There are at least six equivalent characterizations of the property "expansive" for an $n \times n$ real invertible matrix A. It may be good to give them here for sake of exposition (and it can make a good student exercise to verify these equivalences). Any of the first five can be (and sometimes is) taken as the definition, and the sixth is particularly useful for wavelet theory. The first characterization is that all (complex) eigenvalues of A have modulus > 1. A second is that $\bigcup\{A^\ell B_1 \mid \ell \in \mathbb{N}\} = \mathbb{R}^n$, where $B_1 := B_1(0)$ is the open unit ball of $\mathbb{R}^n$. A third is that for each nonzero vector $x \in \mathbb{R}^n$, the sequence of norms $\{\|A^\ell x\| \mid \ell \in \mathbb{N}\}$ is unbounded. A fourth characterization is that the sequence of norms $\{\|A^{-\ell}\| : \ell \in \mathbb{N}\}$ converges to 0. A fifth (which is a quantitative version of the second) is that for each neighborhood N of 0 and each $r > 0$, there exists an $\ell \in \mathbb{N}$ such that $B_0(r) \subseteq A^\ell N$. And the sixth is that for each open set F which is bounded away from 0, and each $r > 0$, there exists an $\ell \in \mathbb{N}$ such that $A^\ell F$ contains a ball $B_r(p)$ of radius r and some center p (where p depends on F and r).

By a *dilation - A regular–translation orthonormal wavelet* we mean a function $\psi \in L^2(\mathbb{R}^n)$ such that

(2.2) $$\{|\det(A)|^{\frac{n}{2}}\psi(A^n t - \ell) \mid n \in \mathbb{Z}, \ell \in \mathbb{Z}^n, i = 1, \ldots, n\}$$

where $\ell = (\ell_1, \ell_2, ..., \ell_n)^\top$, is an orthonormal basis for $L^2(\mathbb{R}^n; m)$. (Here m is product Lebesgue measure, and the superscript $^\top$ means transpose.)If $A \in M_n(\mathbb{R})$ is invertible (so in particular if A is expansive), then the operator defined by

$$(2.3) \qquad (D_A f)(t) = |\det A|^{\frac{1}{2}} f(At)$$

for $f \in L^2(\mathbb{R}^n)$, $t \in \mathbb{R}^n$, is *unitary*. For $1 \leq i \leq n$, let T_i be the unitary operator determined by translation by 1 in the ith coordinate direction. The set (5) above is then

$$(2.4) \qquad \{D_A^k T_1^{\ell_1} \cdots T_n^{\ell_n} \psi \mid k, \ell_i \in \mathbb{Z}\}$$

If the dilation matrix A is expansive, but the translations are along some oblique lattice, then there is an invertible real $n \times n$ matrix T such that conjugation with D_T takes the entire wavelet system to a regular-translation expansive-dilation matrix. This is easily worked out, and was shown in detail in [**ILP**] in the context of working out a complete theory of unitary equivalence of wavelet systems. Hence the wavelet theories are equivalent.

Much work has also been accomplished concerning the existence of wavelets for dilation matrices A which are not expansive. But there is no need to go into that here.

2.1. Fourier Transform. We will use the following form of the Fourier–Plancherel transform $\mathscr{F}$ on $\mathcal{H} = L^2(\mathbb{R})$, which is a form that is *normalized* so it is a unitary transformation, a property that is desirable for our treatment. If $f, g \in L^1(\mathbb{R}) \cap L^2(\mathbb{R})$ then

$$(2.5) \qquad (\mathscr{F}f)(s) := \frac{1}{\sqrt{2\pi}} \int_{\mathbb{R}} e^{-ist} f(t) dt := \hat{f}(s),$$

and

$$(2.6) \qquad (\mathscr{F}^{-1} g)(t) = \frac{1}{\sqrt{2\pi}} \int_{\mathbb{R}} e^{ist} g(s) ds.$$

We have

$$(\mathscr{F}T_\alpha f)(s) = \frac{1}{\sqrt{2\pi}} \int_{\mathbb{R}} e^{-ist} f(t - \alpha) dt = e^{-is\alpha} (\mathscr{F}f)(s).$$

So $\mathscr{F}T_\alpha \mathscr{F}^{-1} g = e^{-is\alpha} g$. For $A \in B(\mathcal{H})$ let $\hat{A}$ denote $\mathscr{F}A\mathscr{F}^{-1}$. Thus

$$(2.7) \qquad \hat{T}_\alpha = M_{e^{-i\alpha s}},$$

where for $h \in L^\infty$ we use M_h to denote the multiplication operator $f \mapsto hf$. Since $\{M_{e^{-i\alpha s}} \mid \alpha \in \mathbb{R}\}$ generates the m.a.s.a. (maximal abelian self adjoint operator algebra) $\mathcal{D}(\mathbb{R}) := \{M_h \mid h \in L^\infty(\mathbb{R})\}$ as a von Neumann algebra. Let $\mathcal{A}_T$ denote the von Neumann algebra generated by $\{T_\alpha : \alpha \in \mathbb{R}\}$. We then have

$$\mathscr{F}\mathcal{A}_T \mathscr{F}^{-1} = \mathcal{D}(\mathbb{R}).$$

Similarly,

$$
\begin{aligned}
(\mathscr{F}D^n f)(s) &= \frac{1}{\sqrt{2\pi}} \int_{\mathbb{R}} e^{-ist} (\sqrt{2})^n f(2^n t) dt \\
&= (\sqrt{2})^{-n} \cdot \frac{1}{\sqrt{2\pi}} \int_{\mathbb{R}} e^{-i2^{-n} st} f(t) dt \\
&= (\sqrt{2})^{-2} (\mathscr{F}f)(2^{2^{-n} s}) = (D^{-n} \mathscr{F}f)(s).
\end{aligned}
$$

So $\widehat{D}^n = D^{-n} = D^{*n}$. Therefore,

$$(2.8) \qquad\qquad \widehat{D} = D^{-1} = D^*.$$

Wavelet sets belong to the theory of wavelets via the Fourier transform. As mentioned earlier, we define a *wavelet set in* $\mathbb{R}$ to be a measurable subset E of $\mathbb{R}$ for which $\frac{1}{\sqrt{2\pi}}\chi_E$ is the Fourier transform of a wavelet. The wavelet $\widehat{\psi}_E := \frac{1}{\sqrt{2\pi}}\chi_E$ is called *s-elementary* in [**DL**]. The class of wavelet sets was also discovered and systematically explored completely independently, and in about the same time period, by Guido Weiss (Washington University), his colleague and former student E. Hernandez (U. Madrid), and his students X. Fang and X. Wang [**HWW, FW**]. In this theory the corresonding wavelets are are called MSF (minimally supported frequency) wavelets.

2.2. Shannon Wavelet. The two most elementary dyadic orthonormal wavelets are the well-known *Haar wavelet* and *Shannon's wavelet* (also called the Littlewood–Paley wavelet). The Haar wavelet is the prototype of a large class of wavelets, and is the function given in the time domain by $\psi = \chi_{[0,1/2)} - \chi_{[1/2,1)}$. The Shannon set (2.9) is the prototype of the class of wavelet sets.

Shannon's wavelet is the $L^2(\mathbb{R})$-function ψ_S with Fourier transform $\widehat{\psi}_S = \frac{1}{\sqrt{2\pi}}\chi_{E_0}$ where

$$(2.9) \qquad\qquad E_0 = [-2\pi, -\pi) \cup [\pi, 2\pi).$$

The argument that $\widehat{\psi}_S$ is a wavelet is in a way even more transparent than for the Haar wavelet. And it has the advantage of generalizing nicely. For a simple argument, start from the fact that the set of exponentials

$$\{e^{i\ell s} \mid \ell \in \mathbb{Z}\}$$

restricted to $[0, 2\pi]$ and normalized by $\frac{1}{\sqrt{2\pi}}$ is an orthonormal basis for $L^2[0, 2\pi]$. Write $E_0 = E_- \cup E_+$ where $E_- = [-2\pi, -\pi)$, $E_+ = [\pi, 2\pi)$. Since $\{E_- + 2\pi, E_+\}$ is a partition of $[0, 2\pi)$ and since the exponentials $e^{i\ell s}$ are invariant under translation by 2π, it follows that

$$(2.10) \qquad\qquad \left\{ \frac{e^{i\ell s}}{\sqrt{2\pi}}\Big|_{E_0} \;\Big|\; \ell \in \mathbb{Z} \right\}$$

is an orthonormal basis for $L^2(E_0)$. Since $\widehat{T} = M_{e^{-is}}$, this set can be written

$$(2.11) \qquad\qquad \{\widehat{T}^\ell \widehat{\psi}_s \mid \ell \in \mathbb{Z}\}.$$

Next, note that any "dyadic interval" of the form $J = [b, 2b)$, for some $b > 0$ has the property that $\{2^n J \mid n \in \mathbb{Z}\}$, is a partition of $(0, \infty)$. Similarly, any set of the form

$$(2.12) \qquad\qquad \mathcal{K} = [-2a, -a) \cup [b, 2b)$$

for $a, b > 0$, has the property that

$$\{2^n \mathcal{K} \mid n \in \mathbb{Z}\}$$

is a partition of $\mathbb{R}\backslash\{0\}$. It follows that the space $L^2(\mathcal{K})$, considered as a subspace of $L^2(\mathbb{R})$, is a complete wandering subspace for the dilation unitary $(Df)(s) = \sqrt{2}\, f(2s)$. For each $n \in \mathbb{Z}$,

$$(2.13) \qquad\qquad D^n(L^2(\mathcal{K})) = L^2(2^{-n}\mathcal{K}).$$

So $\bigoplus_{n \in \mathbb{Z}} D^n(L^2(\mathcal{K}))$ is a direct sum decomposition of $L^2(\mathbb{R})$. In particular E_0 has this property. So

$$(2.14) \qquad D^n \left\{ \frac{e^{i\ell s}}{\sqrt{2\pi}} \Big|_{E_0} \mid \ell \in \mathbb{Z} \right\} = \left\{ \frac{e^{2^n i \ell s}}{\sqrt{2\pi}} \Big|_{2^{-n} E_0} \mid \ell \in \mathbb{Z} \right\}$$

is an orthonormal basis for $L^2(2^{-n} E_0)$ for each n. It follows that

$$\{ D^n \widehat{T}^\ell \widehat{\psi}_s \mid n, \ell \in \mathbb{Z} \}$$

is an orthonormal basis for $L^2(\mathbb{R})$. Hence $\{ D^n T^\ell \psi_s \mid n, \ell \in \mathbb{Z} \}$ is an orthonormal basis for $L^2(\mathbb{R})$, as required.

2.3. Spectral Set Condition.
From the argument above describing why Shannon's wavelet is, indeed, an orthonormal basis generator, it is clear that *sufficient* conditions for E to be a wavelet set are

> (i) the normalized exponential $\frac{1}{\sqrt{2\pi}} e^{i\ell s}$, $\ell \in \mathbb{Z}$, when restricted to E should constitute an orthonormal basis for $L^2(E)$ (in other words E is a *spectral set* for the integer lattice $\mathbb{Z}$),

and

> (ii) The family $\{ 2^n E \mid n \in \mathbb{Z} \}$ of dilates of E by integral powers of 2 should constitute a measurable partition (i.e. a partition modulo null sets) of $\mathbb{R}$.

These conditions are also necessary. In fact if a set E satisfies (i), then for it to be a wavelet set it is obvious that (ii) must be satisfied. To show that (i) must be satisfied by a wavelet set E, consider the vectors

$$\widehat{D}^n \widehat{\psi}_E = \frac{1}{\sqrt{2\pi}} \chi_{2^{-n} E}, \qquad n \in \mathbb{Z}.$$

Since $\widehat{\psi}_E$ is a wavelet these must be orthogonal, and so the sets $\{ 2^n E \mid n \in \mathbb{Z} \}$ must be disjoint modulo null sets. It follows that $\{ \frac{1}{\sqrt{2\pi}} e^{i\ell s}|_E \mid \ell \in \mathbb{Z} \}$ is not only an orthonormal set of vectors in $L^2(E)$, it must also *span* $L^2(E)$. It is known from the theory of *spectral sets* (as an elementary special case) that a measurable set E satisfies (i) if and only if it is a generator of a measurable partition of $\mathbb{R}$ under translation by 2π (i.e. iff $\{ E + 2\pi n \mid n \in \mathbb{Z} \}$ is a measurable partition of $\mathbb{R}$). This result generalizes to spectral sets for the integral lattice in $\mathbb{R}^n$. For this elementary special case a direct proof is not hard.

2.4. Translation and Dilation Congruence.
We say that measurable sets E, F are *translation congruent modulo* 2π if there is a measurable bijection $\phi \mid E \to F$ such that $\phi(s) - s$ is an integral multiple of 2π for each $s \in E$; or equivalently, if there is a measurable partition $\{ E_n \mid n \in \mathbb{Z} \}$ of E such that

$$(2.15) \qquad \{ E_n + 2n\pi \mid n \in \mathbb{Z} \}$$

is a measurable partition of F. Analogously, define measurable sets G and H to be *dilation congruent modulo* 2 if there is a measurable bijection $\tau \mid G \to H$ such that for each $s \in G$ there is an integer n, depending on s, such that $\tau(s) = 2^n s$; or equivalently, if there is a measurable partition $\{ G_n \}_{n=-\infty}^{+\infty}$ of G such that

$$(2.16) \qquad \{ 2^n G \}_{n=-\infty}^{+\infty}$$

is a measurable partition of H. (Translation and dilation congruency modulo other positive numbers of course make sense as well.) The following lemma is useful. It is Lemma 4.1 of [**DL**].

LEMMA 2.2. *Let $f \in L^2(\mathbb{R})$, and let $E = \text{supp}(f)$. Then f has the property that*

$$\{e^{i\ell s} f \mid \ell \in \mathbb{Z}\}$$

is an orthonormal basis for $L^2(E)$ if and only if

  (i) *E is congruent to $[0, 2\pi)$ modulo 2π, and*
  (ii) *$|f(s)| = \frac{1}{\sqrt{2\pi}}$ a.e. on E.*

We include a sketch of the proof of Lemma 2.2 for the case $f = \chi_E$, because the ideas in it are relevent to the sequel. For general f the proof is a slight modification of this. If E is a measurable set which is 2π–translation congruent to $[0, 2\pi)$, then since

$$\left\{ \frac{e^{i\ell s}}{\sqrt{2\pi}} \Big|_{[0,2\pi)} \Big| n \in \mathbb{Z} \right\}$$

is an orthonormal basis for $L^2[0, 2\pi]$ and the exponentials $e^{i\ell s}$ are 2π–invariant, as in the case of Shannon's wavelet, it follows that

$$\left\{ \frac{e^{i\ell s}}{\sqrt{2\pi}} \Big|_E \Big| \ell \in \mathbb{Z} \right\}$$

is an orthonormal basis for $L^2(E)$.

Conversely, if E is not 2π–translation congruent to $[0, 2\pi)$, then either E is congruent to a proper subset Ω of $[0, 2\pi)$, which is not of full measure, or there exists an integer k such that $E \cap (E + 2\pi k)$ has positive measure. In the first case, since the exponentials $\frac{e^{i\ell s}}{\sqrt{2\pi}}$ restricted to $[0.2\pi)$ do not form an orthonormal basis for $L^2([0, 2\pi))$, the same exponentials restricted to E cannot form an orthonormal basis for $L^2(E)$. In the second case, let $E_1 = E \cap (E + 2\pi k)$, and $E_2 = E - 2\pi k = (E - 2\pi k) \cap E$, and let $h = \chi_{E_1} - \chi_{E_2}$. Then $h \in L^2(E)$, and

$$h \perp \frac{e^{i\ell s}}{\sqrt{2\pi}} \Big|_E,$$

for all $\ell \in \mathbb{Z}$. Thus the exponentials $\frac{e^{i\ell s}}{\sqrt{2\pi}} \Big|_E$ cannot even span $L^2(E)$. This completes the proof sketch.

Next, observe that if E is 2π–translation congruent to $[0, 2\pi)$, then since

$$\{[0, 2\pi) + 2\pi n \mid n \in \mathbb{Z}\}$$

is a measurable partition of $\mathbb{R}$, so is

$$\{E + 2\pi n \mid n \in \mathbb{Z}\}.$$

Similarly, if F is 2–dilation congruent to the Shannon set $E_0 = [-2\pi, -\pi) \cup [\pi, 2\pi)$, then since $\{2^n E_0 : n \in \mathbb{Z}\}$ is a measurable partition of $\mathbb{R}$, so is $\{2^n F : n \in \mathbb{Z}\}$.

These arguments can be reversed. We say that a measurable subset $G \subset \mathbb{R}$ is a 2–*dilation generator* of a *partition* of $\mathbb{R}$ if the sets

(2.17) $2^n G := \{2^n s \mid s \in G\}, \qquad n \in \mathbb{Z}$

are disjoint and $\mathbb{R}\backslash\cup_n 2^n G$ is a null set. Also, we say that $E \subset \mathbb{R}$ is a 2π–*translation generator of a partition* of $\mathbb{R}$ if the sets

$$(2.18) \qquad\qquad E + 2n\pi := \{s + 2n\pi \mid s \in E\}, \qquad n \in \mathbb{Z},$$

are disjoint and $\mathbb{R}\backslash\cup_n (E + 2n\pi)$ is a null set.

The following is Lemma 4.2 of [**DL**].

LEMMA 2.3. *A measurable set $E \subseteq \mathbb{R}$ is a 2π–translation generator of a partition of $\mathbb{R}$ if and only if, modulo a null set, E is translation congruent to $[0, 2\pi)$ modulo 2π. Also, a measurable set $G \subseteq \mathbb{R}$ is a 2-dilation generator of a partition of $\mathbb{R}$ if and only if, modulo a null set, G is a dilation congruent modulo 2 to the set $[-2\pi, -\pi) \cup [\pi, 2\pi)$.*

DEFINITION 2.4. By a fundamental domain for a group of (measurable) transformations $\mathcal{G}$ on a measure space (Ω, μ) we will mean a measurable set C with the property that $\{g(C) : g \in \mathcal{G}\}$ is a measurable partition (tessellation) of Ω; that is,

$$\Omega \setminus \left(\bigcup_{g \in \mathcal{G}} g(C) \right) \text{ is a μ-null set and } g_1(C) \cap g_2(C) \text{ is a μ-null set for } g_1 \neq g_2.$$

Thus the sets E and F in Lemma 2.3 are fundamental domains for the groups generated by 2π and dilation by 2, respectively. Moreover, Lemma 2.3 actually characterizes the fundamental domains for those groups.

2.5. A Criterion. The following is a useful criterion for wavelet sets. It was published independently by Dai–Larson in [**DL**] and by Fang and Wang in [**FW**] at about the same time (December 1994).

PROPOSITION 2.5. *Let $E \subseteq \mathbb{R}$ be a measurable set. Then E is a wavelet set if and only if E is both a 2–dilation generator of a partition (modulo null sets) of $\mathbb{R}$ and a 2π–translation generator of a partition (modulo null sets) of $\mathbb{R}$. Equivalently, E is a wavelet set if and only if E is both translation congruent to $[0, 2\pi)$ modulo 2π and dilation congruent to $[-2\pi, -\pi) \cup [\pi, 2\pi)$ modulo 2. In the terminology of Definition 2.4, E is a wavelet set if and only if E is a fundamental domain for this dilation group and at the same time a fundamental domain for this translation group.*

Note that a set is 2π–translation congruent to $[0, 2\pi)$ iff it is 2π–translation congruent to $[-2\pi, \pi) \cup [\pi, 2\pi)$. So the last sentence of Proposition 2.5 can be stated: A measurable set E is a wavelet set if and only if it is both 2π–translation and 2–dilation congruent to the Littlewood–Paley set $[-2\pi, -\pi) \cup [\pi, 2\pi)$.

For our later purposes, we need the generalization of the above results to $\mathbb{R}^n$. To this end, a few definitions are necessary.

Let X be a metric space and m a σ-finite non-atomic Borel measure on X for which the measure of every open set is positive and for which bounded sets have finite measure. Let $\mathcal{T}$ and $\mathcal{D}$ be countable groups of homeomorphisms of X that map bounded sets to bounded sets and which are absolutely continuously in the sense that they map m-null sets to m-null sets. Furthermore, let $\mathcal{G}$ be a countable group of absolutely continuous Borel isomorphisms of X. Denote by $\mathcal{B}$ the family of Borel sets of X.

The following definition completely generalizes our definitions of 2π–translation congruence and 2–dilation congruence given in the beginning of subsection 2.4.

DEFINITION 2.6. Let $E, F \in \mathcal{B}$. We call E and F $\mathcal{G}$–*congruent* and write $E \sim_{\mathcal{G}} F$, if there exist measurable partitions $\{E_g : g \in \mathcal{G}\}$ and $\{F_g : g \in \mathcal{G}\}$ of E and F, respectively, such that $F_g = g(E_g)$, for all $g \in \mathcal{G}$, modulo m-null sets.

PROPOSITION 2.7.

(1) $\mathcal{G}$–congruence is an equivalence relation on the family of m-measurable sets.

(2) If E is a fundamental domains for $\mathcal{G}$, then F is a fundamental domain for $\mathcal{G}$ iff $F \sim_{\mathcal{G}} E$.

PROOF. See [**DL**]. □

DEFINITION 2.8. We call $(\mathcal{D}, \mathcal{T})$ an *abstract dilation–translation pair* if

(1) For each bounded set E and each open set F there exist elements $\delta \in \mathcal{D}$ and $\tau \in \mathcal{T}$ such that $\tau(F) \subset \delta(E)$.

(2) There exists a fixed point $\theta \in X$ for $\mathcal{D}$ with the property that if N is any neighborhood of θ and E any bounded set, there is an element $\delta \in \mathcal{D}$ such that $\delta(E) \subset N$.

The following result and its proof can be found in [**DLS1**].

THEOREM 2.9. *Let X, $\mathcal{B}$, m, $\mathcal{D}$, and $\mathcal{T}$ as above. Let $(\mathcal{D}, \mathcal{T})$ be an abstract dilation–translation pair with θ being the $\mathcal{D}$ fixed point. Assume that E and F are bounded measurable sets in X such that E contains a neighborhood of θ, and F has non-empty interior and is bounded away from θ. Then there exists a measurable set $G \subset X$, contained in $\bigcup_{\delta \in \mathcal{D}} \delta(F)$, which is both $\mathcal{D}$–congruent to F and $\mathcal{T}$–congruent to E.*

The following is a consequence of Proposition 2.7 and Theorem 2.9 and is the key to obtaining wavelet sets.

COROLLARY 2.10. *With the terminology of Theorem 2.9, if in addition F is a fundamental domain for $\mathcal{D}$ and E is a fundamental domain for $\mathcal{T}$, then there exists a set G which is a common fundamental domain for both $\mathcal{D}$ and $\mathcal{T}$.*

In order to apply the above result to wavelet sets in $\mathbb{R}^n$, we make the following definition.

DEFINITION 2.11. *A dilation A–wavelet set is a measurable subset E of $\mathbb{R}^n$ for which the inverse Fourier transform of $(m(E))^{-1/2} \chi_E$ is an orthonormal dilation A–wavelet.*

Two measurable subsets H and K of $\mathbb{R}^n$ are called A–*dilation congruent*, in symbols $H \sim_{\delta_A} K$, if there exist measurable partitions $\{H_\ell \,|\, \ell \in \mathbb{Z}\}$ of H and $\{K_\ell \,|\, \ell \in \mathbb{Z}\}$ of K such that $K_\ell = A^\ell H_\ell$ modulo Lebesgue null sets. Moreover, two measurable sets E and F of $\mathbb{R}^n$ are called 2π–*translation congruent*, written $E \sim_{\tau_{2\pi}} F$, if there exists measurable partitions $\{E_\ell \,|\, \ell \in \mathbb{Z}^n\}$ of E and $\{F_\ell \,|\, \ell \in \mathbb{Z}^n\}$ of F such that $F_\ell = E_\ell + 2\pi\ell$ modulo Lebesgue null sets.

Note that this generalizes to $\mathbb{R}^n$ our definition of 2π–translation congruence for subsets of $\mathbb{R}$. Observe that A–dilation by an expansive matrix together with 2π–translation congruence is a special case (in fact, it is really the *prototype* case) of an abstract dilation-translation pair (Definition 2.8). Let $\mathcal{D}$ be the group of dilations by powers of A, $\{A^\ell \,|\, \ell \in \mathbb{Z}\}$ on $\mathbb{R}^n$, and let $\mathcal{T}$ be the group of translations by the vectors $\{2\pi k \,|\, k \in \mathbb{Z}^n\}$. Let E be any bounded set, and let F be any open set that is bounded away from 0. Let $r > 0$ be such that $E \subseteq B_r(0)$. Since A is expansive there is an $\ell \in \mathbb{N}$ such that $A^\ell F$ contains a ball B of radius large enough so that B contains some lattice point $2k\pi$ together with the ball $B_R(2k\pi)$ of radius $R > 0$ centered at the lattice point. Then $E + 2k\pi \subseteq A^\ell F$. That is, the $2k\pi$–translate of E is contained in the A^ℓ–dilate of F, as required in (1) of Definition 2.8. For (2) of Definition 2.8, let $\theta = 0$, and let N be a neighborhood of 0, and let E be any bounded set. As above, choose $r > 0$ with $E \subseteq B_r(0)$. Let $\ell \in \mathbb{N}$ be such that $A^\ell N$ contains $B_r(0)$. Then $A^{-\ell}$ is the required dilation such that $A^{-\ell} E \subseteq N$.

Note that if W is a measurable subset of $\mathbb{R}^n$ that is 2π–translation congruent to the n-cube $E := \mathsf{X}_{k=1}^n [-\pi, \pi)$, it follows from the exponential form of $\widehat{T}_j$ that $\left\{ \widehat{T}_1^{\ell_1} \widehat{T}_2^{\ell_2} \cdots \widehat{T}_n^{\ell_n} (m(W))^{-1/2} \chi_W \,|\, \ell = (\ell_1, \ell_2, \ldots, \ell_n) \in \mathbb{Z}^n \right\}$ is an orthonormal basis for $L^2(W)$. Furthermore, if A is an expansive matrix, i.e., A is similar to a strict dilation, and B the unit ball of $\mathbb{R}^n$ then with $F_A := A(B) \setminus B$ the collection $\{A^k F_A : k \in \mathbb{Z}\}$ is a partition of $\mathbb{R}^n \setminus \{0\}$. As a consequence, $L^2(F_A)$, considered as a subspace of $L^2(\mathbb{R}^n)$, is a complete wandering subspace for D_A. Hence, $L^2(\mathbb{R}^n)$ is a direct sum decomposition of the subspaces $\{D_A^k L^2(F_A) \,|\, k \in \mathbb{Z}\}$. Clearly, any other measurable set $F' \sim_{\delta_A} F_A$ has this same property.

The above theorem not only gives the existence of wavelet sets in $\mathbb{R}^n$, but also shows that there are sufficiently many to generate the Borel structure of $\mathbb{R}^n$. For details, we refer the reader to [**DLS1**]. For our purposes, we only quote Corollary 1 in [**DLS1**], as a theorem.

THEOREM 2.12. *Let $n \in \mathbb{N}$ and let A be an expansive $n \times n$ matrix. Then there exist dilation–A wavelet sets.*

Some concrete examples of wavelet sets in the plane were subsequently obtained by Soardi and Weiland, and others were obtained by Gu and by Speegle in their thesis work at Texas A&M University. Two additional examples were constructed by Dai for inclusion in the revised concluding remarks section of [**DL**].

3. Iterated function systems and fractal functions

Fractal (interpolation) functions were first systematically introduced in [**B**]. Their construction is based on iterated function systems and their properties [**BD, Hu**].

To this end, recall that a *contraction* on a metrizable space (M, d) is a mapping $f : M \to M$ such that there exists a $0 \le k < 1$, called the *contractivity constant*, so that for all $x, y \in M$
$$d(f(x), f(y)) \le k \, d(x, y).$$

DEFINITION 3.1. Let (X, d) be a complete metrizable space with metric d and let $\{T_i \,|\, i = 1, \ldots, N\}$ be a finite set of contractions on X. The pair $((X, d), \{T_i\})$ is called an iterated function system (IFS) on X.

With the finite set of contractions, one can associate a set-valued operator $\mathscr{T}$, called the Hutchinson operator, defined on the hyperspace $H(X)$ of nonempty compact subsets of X endowed with the Hausdorff metric d_H:
$$\mathscr{T}(E) := \bigcup_{i=1}^{N} T_i(E).$$

It is easy to show that the Hutchinson operator is contractive on the complete metric space $(H(X), d_H)$ with contractivity constant $\max_{1 \leq i \leq N} s_i$, where s_i is the contractivity constant of T_i. By the Banach Fixed Point Theorem, $\mathscr{T}$ has a unique fixed point, called the *fractal F associated with the IFS* $((X, d), \{T_i\})$. The fractal F satisfies

$$(3.1) \qquad F = \mathscr{T}(F) = \bigcup_{i=1}^{N} T_i(F),$$

i.e., F is made up of a finite number of images of itself. The proof of the Banach Fixed Point Theorem shows that the fractal can be iteratively obtained via the following procedure. Choose $F_0 \in H(X)$ arbitrary. Define

$$F_n := \mathscr{T}(F_{n-1}), \qquad n \in \mathbb{N}.$$

Then $F = \lim_{n \to \infty} F_n$, where the limit is taken in the Hausdorff metric.

A special situation occurs when $X := [a, b] \times \mathbb{R} \subset \mathbb{R}^2$, $a < b$, and d is the Euclidean metric. Let $\{(x_j, y_j) \mid x_0 := a < x_1 < \ldots x_N := b, \ y_j \in \mathbb{R}, \ j = 0, 1, \ldots, N\}$ be a given set of interpolation points. For $i = 1, \ldots, N$, let $s_i \in (-1, 1)$ and let $F_0 := [a, b] \times [a, b]$. Define images $T_i F_0$, $i = 1, \ldots, N$, of F_0 as follows. $T_i F_0$ is the unique parallelogram with vertices at (x_{i-1}, y_{i-1}), (x_i, y_i), $(x_i, y_i + s_i(b - a))$, and $(x_{i-1}, y_{i-1} + s_i(b - a))$. There exists a unique affine mapping $T_i : X \to X$ such that $T_i F_0 = T_i(F_0)$, namely

$$T_i \begin{pmatrix} x \\ y \end{pmatrix} = \begin{pmatrix} a_i & 0 \\ c_i & s_i \end{pmatrix} \begin{pmatrix} x \\ y \end{pmatrix} + \begin{pmatrix} \alpha_i \\ \beta_i \end{pmatrix},$$

where

$$a_i := \frac{x_i - x_{i-1}}{b - a}, \qquad\qquad c_i := \frac{y_i - y_{i-1} - s_i(y_N - y_0)}{b - a},$$

$$\alpha_i := \frac{b x_{i-1} - a x_i}{b - a}, \qquad\qquad \beta_i := \frac{b y_{i-1} - a y_i - s_i(b y_0 - a y_N)}{b - a}.$$

Since the scaling factors s_i are in modulus less than one, the affine mappings T_i are contractive on X and thus $((X, d), \{T_i\})$ is an IFS. As such, it has a unique fixed point F, which turns out to be the graph of a continuous function $f : [a, b] \to \mathbb{R}$ satisfying $f(x_j) = y_j$, for $j = 0, 1, \ldots, N$. (See [**B**]) The graph of f is in general a fractal set in the above sense and contains the given set of interpolation points.

EXAMPLE 3.2. Let $a = 0$, $b = 1$, $N = 2$, and choose interpolation points $\{(0, 0), (0.5, 0.7), (1, 0)\}$ and scaling factors $s_1 = 0.6$ and $s_2 = 0.4$. The sequence of graphs in Figure 1 shows the geometric construction of a fractal function with these parameters as outlined above.

Writing the fixed point equation (3.1) using the affine mappings T_i for a point $(x, f(x))$ on the graph of a continuous fractal function f yields

$$\begin{pmatrix} x \\ f(x) \end{pmatrix} \Bigg|_{x \in [x_{i-1}, x_i]} = \begin{pmatrix} a_i & 0 \\ c_i & s_i \end{pmatrix} \begin{pmatrix} x \\ f(x) \end{pmatrix} \Bigg|_{x \in [a, b]} + \begin{pmatrix} \alpha_i \\ \beta_i \end{pmatrix},$$

Setting $u_i(x) := a_i x + \alpha_i$ and $p_i(x) := c_i x + \beta_i$, $i = 1, \ldots, N$, one can rewrite the second component of the above equation as

$$f(x) = p_i(u_i^{-1}(x)) + s_i f(u_i^{-1}(x)), \qquad x \in [x_{i-1}, x_i],$$

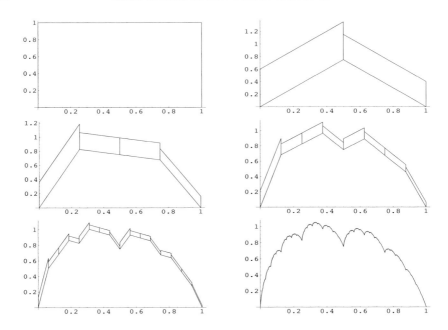

FIGURE 1. The geometric construction of a fractal function

or, equivalently,

(3.2)
$$f(x) = \sum_{i=1}^{N} \left[p_i(u_i^{-1}(x)) + s_i f(u_i^{-1}(x)) \right] \chi_{[x_{i-1}, x_i]}$$
$$=: p(x) + \sum_{i=1}^{N} s_i f(u_i^{-1}(x)) \chi_{[x_{i-1}, x_i]}, \qquad x \in [a, b],$$

where χ is the indicator function and p the linear spline whose restriction to any of the intervals $[x_{i-1}, x_i]$ equals $p_i \circ u_i^{-1}$. Such a fractal function may be considered as a *linear spline parametrized by the row vector* $\boldsymbol{s} := (s_1, \ldots, s_n)$.

Note that the functions p_i are uniquely determined by the interpolation points and that the fractal functions is uniquely determined by the row vector $\boldsymbol{p} := (p_1, \ldots, p_N)$ and the scaling factors $(s_1, \ldots, s_n)$. We suppress the dependence of a fractal function f on $(s_1, \ldots, s_n)$ but write $f = f_{\boldsymbol{p}}$ when necessary.

Denote by $\Pi^1[a, b]$ the linear space of real polynomials of degree at most one on $[a, b]$ and by $C[a, b]$ the space of continuous functions on $[a, b]$.

THEOREM 3.3. *The mapping* $\Theta : \Pi^1[a, b]^N \cap C[a, b] \ni \boldsymbol{p} \mapsto f_{\boldsymbol{p}}$ *is a linear isomorphism.*

PROOF. The result follows from the above observations and the uniqueness of the fixed point: $f_{\alpha \boldsymbol{p} + \boldsymbol{q}} = \alpha f_{\boldsymbol{p}} + f_{\boldsymbol{q}}$, $\alpha \in \mathbb{R}$ and $\boldsymbol{p}, \boldsymbol{q} \in \Pi^1[a, b]^N$. (See also [**M2, M3**].) $\square$

Since the space $\Pi^1[a, b]^N \cap C[a, b]$ is $(N+1)$-dimensional ($2N$ free parameters plus $N-1$ join-up conditions at the interior interpolation points), the space $\mathfrak{F}^1[a, b]$ of all continuous fractal functions on $[a, b]$ generated by functions in $\Pi^1[a, b]^N$ has also dimension $N + 1$. An basis for $\mathfrak{F}^1[a, b]$ can be found by choosing as the ith basis function the continuous fractal function e_i that interpolates according to

$$e_i(x_j) = \delta_{ij}, \quad i, j = 0, 1, \ldots, N.$$

It follows immediately from the uniqueness of the fixed point, that every continuous fractal function $f \in \mathfrak{F}[a, b]$ can be written as a linear combination of the form

$$(3.3) \qquad f(x) = \sum_{j=0}^{N} y_j \, e_j(x).$$

Using the Gram-Schmidt Orthonormalization procedure and the fact that the L^2-inner product of two fractal functions in $\mathfrak{F}^1[a, b]$ over the same knot set $\{x_j \,|\, j = 0, 1, \ldots, n\}$ can be explicitly computed in terms of the parameters $(s_1, \ldots, s_N)$ and the functions in $\Pi^1[a, b]^N \cap C[a, b]$ (cf. [**M2**]), the basis $\{e_j\}$ can also be orthonormalized.

Equation (3.2) can be interpreted as the fixed point equation for a function-valued operator $\mathscr{B}$ and this interpretation leads to a more abstract definition of fractal functions. The following theorem gives a general construction of fractal functions in terms of so-called Read-Bajraktarević operators.

THEOREM 3.4. *Let $\Omega \subset \mathbb{R}$ be compact and $1 < N \in \mathbb{N}$. Assume that $u_i : \Omega \to \Omega$ are contractive homeomorphisms inducing a partition on Ω, $\lambda_i : \mathbb{R} \to \mathbb{R}$ are bounded functions and s_i real numbers, $i = 1, \ldots, N$. Let*

$$(3.4) \qquad \mathscr{B}(f) := \sum_{i=1}^{N} \left[\lambda_i \circ u_i^{-1} + s_i f \circ u_i^{-1} \right] \chi_{u_i(\Omega)}$$

If $\max\{|s_i|\} < 1$, then the operator $\mathscr{B}$ is contractive on $L^\infty(\Omega)$ and its unique fixed point $f : \Omega \to \mathbb{R}$ satisfies

$$f = \sum_{i=1}^{N} \left[\lambda_i \circ u_i^{-1} + s_i f \circ u_i^{-1} \right] \chi_{u_i(\Omega)}$$

PROOF. Apply the Banach Fixed Point Theorem. $\qquad\qquad\qquad\qquad\qquad$ $\square$

The fixed point of such an operator is called a *fractal function*. Note again that f depends on the row vector of functions $\boldsymbol{\lambda} := (\lambda_1, \ldots, \lambda_N)$.

If $\mathscr{B}$ acts on a normeable or metrizable function space $\mathcal{F}$, then its fixed point, under appropriate conditions on $\boldsymbol{\lambda}$ and the scaling factors $(s_1, \ldots, s_n)$, is also an element of $\mathcal{F}$. In this manner, one can construct fractal functions with prescribed regularity or approximation properties [**M4**].

In case the contractive homeomorphisms u_i, $i = 1, \ldots, N$, induce a uniform partition of Ω, the above expressions become more transparent. In addition, w.l.o.g., suppose that $\Omega = [0, N]$. A natural uniform partition in this case is $[0, N] = \bigcup_{i=1}^{N-1} [i - 1, i) \cup [N - 1, N]$ and the mappings u_i are given by

$$u_i(x) = \frac{x}{N} + i - 1, \qquad i = 1, \ldots, N.$$

Hence, it suffices to define $u_1 := x/N$ and then all other mappings are given by translating u_1: $u_i = u_1 + (i - 1) = u_{i-1} + 1$, $i = 2, \ldots, N$. The fixed point equation for a continuous fractal function than reads

$$f(x) = \sum_{i=1}^{N} \left[\lambda_i(N(x-i+1)) + s_i \, f(N(x-i+1)) \right] \chi_{[i-1,i]}, \quad x \in [0, N].$$

Instead of using translations to obtain $u_2, \ldots, u_N$ from u_1, one may choose reflections about the partition points i of $[0, N]$, $i = 1, \ldots, N-1$, instead. The reflection R_i about the point $(i, 0)$ on the x-axis is given by $R_i(x) = 2i - x$. As above, let $u_1(x) = x/N$ and set $u_i = R_{i-1} \circ u_{i-1}$, $i = 2, \ldots, N$. This also generates a uniform partition of $[0, N]$.

EXAMPLE 3.5. As an example of the two types of continuous fractal functions, we consider $N := 3$ and $s_1 = s_2 = s_3 := 0.5$. The continuous fractal function f is generated by translations in the above sense with $\boldsymbol{\lambda} = ((1/3 - s_1/2)x, (-1/6 - s_2/2)x + 1, (1/3 - s_3/2)x + 1/2)$, whereas the continuous fractal function g is generated using reflections and with $\boldsymbol{\lambda} = ((1/3 - s_1/2)x, (1/6 - s_2/2)x + 1, (1/3 - s_3/2)x + 1/2)$. The values of both functions are the partition points $(i, 0)$, $i = 1, \ldots, 4$, are $0, 1, 1/2$, and $3/2$. The graphs of these two fractal functions are displayed in Figure 2.

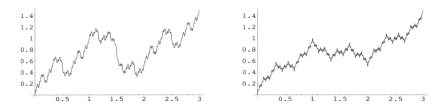

FIGURE 2. A fractal function generated via translations (left) and reflections (right).

Both fractal functions belong to the four-dimensional linear space $\Pi^1[a, b]^3 \cap C[a, b]$ and the four basis functions are depicted in Figures 3 and 4.

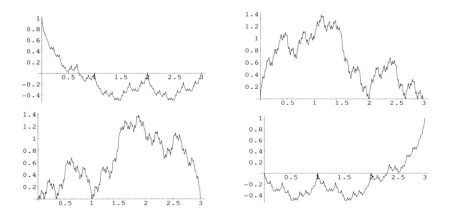

FIGURE 3. The four basis functions for the fractal function f.

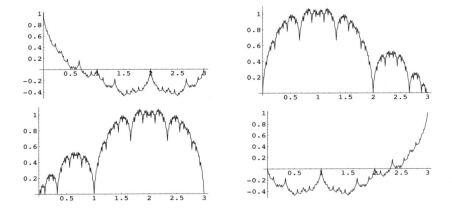

FIGURE 4. The four basis functions for the fractal function g.

At this point the question arises whether there is a generalization of the above procedures to higher dimensions. It is possible to define fractal surfaces in $\mathbb{R}^n$, $1 < n \in \mathbb{N}$, using reflections instead of translations in the definition of the contractive homeomorphisms u_i. We will see that this leads to a natural way of tessellating the embedding space and to the construction of a multiresolution analysis of $L^2(\mathbb{R}^n)$.

4. Coxeter groups and foldable figures

In order to carry out the construction of fractal surfaces using reflections, a short excursion into the theory of Coxeter groups and foldable figures is necessary. The interested reader is referred to [**Bo, C, G, Gu, H, HW**] for more details and proofs.

4.1. Coxeter groups.

DEFINITION 4.1. A Coxeter group $\mathcal{C}$ is a discrete groups with a finite number of generators $\{r_i \mid i = 1, \ldots, k\}$ satisfying

$$\mathcal{C} = \left\langle r_1, \ldots, r_k \mid (r_i r_j)^{m_{ij}} = 1, \ 1 \leq i, j \leq k \right\rangle$$

where $m_{ii} = 1$, for all i, and $m_{ij} \geq 2$, for all $i \neq j$. ($m_{ij} = \infty$ is used to indicate that no relation exists.)

A geometric representation of a Coxeter group is given by considering it as a subgroup of $GL(V)$, where V is a k-dimensional real vector space, which we take to be $\mathbb{R}^k$ endowed with its usual positive definite symmetric bilinear form $\langle \cdot, \cdot \rangle$. In this representation, the generators are interpreted in the following way.

A reflection about a linear hyperplane H is defined as a linear mapping $\rho : V \to V$ such that $\rho|_H = \mathrm{id}_H$ and $\rho(x) = -x$, if $x \in H^\perp$. In other words, ρ is an isometric isomorphism of V.

Now suppose that $0 \neq r \in H^\perp$, then an easy computation shows that

$$\rho_r(x) = x - \frac{2\langle x, r \rangle}{\langle r, r \rangle} r$$

is the reflection about the hyperplane H perpendicular to r. One can show that the linear mappings ρ_{r_i}, where $\{r_i \mid i = 1, \ldots, k\}$ are the generators of a Coxeter group $\mathcal{C}$, satisfy $(\rho_{r_i}\rho_{r_j})^{m_{ij}} = \mathrm{id}_V$. It is known that the map $r_i \mapsto \rho_{r_i}$ extends to a faithful representation of $\mathcal{C}$ into $GL(V)$.

If one considers the group generated by real reflections about linear hyperplanes in $\mathbb{R}^k$, then this group is isomorphic to a finite Coxeter group whose k generators correspond to the (unit) normal vectors of the set of hyperplanes.

4.2. Roots systems and Weyl groups. The normal vectors to a set of hyperplanes play an important role in the representation theory for Coxeter groups. We have seen above that they correspond to the generators of such groups. Two such normal vectors, $\pm r$, that are orthogonal to a hyperplane are called *roots*.

DEFINITION 4.2. A *root system* $\mathcal{R}$ is a finite set of nonzero vectors $r_1, \ldots, r_k \in \mathbb{R}^n$ satisfying

(1) $\mathbb{R}^n = \mathrm{span}\,\{r_1, \ldots, r_k\}$

(2) $r, \alpha r \in \mathcal{R}$ iff $\alpha = \pm 1$

(3) $\forall r, s \in \mathcal{R}$: $s - \dfrac{2\langle s, r \rangle}{\langle r, r \rangle}\, r \in \mathcal{R}$, i.e., the root system $\mathcal{R}$ is closed with respect to the reflection through the hyperplane orthogonal to r.

(4) $\forall r, s \in \mathcal{R}$: $\dfrac{2\langle s, r \rangle}{\langle r, r \rangle} \in \mathbb{Z}$, i.e., $\rho_r(s) - s \in \mathbb{Z}$

A subset $\mathcal{R}^+ \subset \mathcal{R}$ is called a set of *positive* roots if there exists a vector $v \in \mathbb{R}^n$ such that $\langle r, v \rangle > 0$ if $r \in \mathcal{R}^+$, and $\langle r, v \rangle > 0$ if $r \in \mathcal{R} \setminus \mathcal{R}^+$. Roots that are not positive are called *negative*. Since r is negative iff $-r$ is positive, there are exactly as many positive as there are negative roots.

The group generated by the set of reflections $\{\rho_r \mid r \in \mathcal{R}\}$ is called the *Weyl Group* $\mathcal{W}$ of $\mathcal{R}$.

It follows from the definition of root system, that the Weyl group $\mathcal{W}$ has finite order, indeed it is a finite Coxeter group.

EXAMPLE 4.3. A simple example of a Weyl group in $\mathbb{R}^2$ is given by the root system depicted in Figure 5. The roots are $r_1 = -r_3 = (1, 0)^\top$ and $r_2 = -r_4 = (0, 1)^\top$. The positive roots are r_1 and r_2. The group of reflections generated by these four roots is given by

$$V_4 := \langle \rho_1, \rho_2 \mid \rho_1^2 = \rho_2^2 = 1,\ (\rho_1\rho_2)^2 = 1 \rangle,$$

where ρ_1 and ρ_2 denotes the reflection about the y-, respectively, x-axis. This group is commutative and called *Klein's four-group* or the *group of order four*. In the classification scheme of Weyl groups V_4 is referred to as $A_1 \times A_1$ since it is the direct product of the group $A_1 := \langle \rho_1 \mid \rho_1^2 = 1 \rangle$ whose root system is $\mathcal{R} = \{r_1, r_3\}$ with itself.

FIGURE 5. The root system for Klein's four-group.

For the following, we need some properties of roots systems and Weyl groups, which we state in a theorem.

THEOREM 4.4. *Let $\mathcal{R}$ be a root system and $\mathcal{W}$ the associated Weyl group. Then the following hold.*

(1) *Every root system $\mathcal{R}$ has a basis $\mathcal{B} = \{b_i\}$ consisting of positive (negative) roots.*

(2) *Let $C_i := \{x \in \mathbb{R}^n \mid \langle x, b_i \rangle > 0\}$ be the* Weyl chamber *corresponding to the basis $\mathcal{B}$. Then the Weyl group $\mathcal{W}$ acts simply transitively on the Weyl chambers.*

(3) *The set $C := \overline{\bigcap_i C_i}$ is a noncompact fundamental domain for the Weyl group $\mathcal{W}$. It is a simplicial cone, hence convex and connected.*

In order to introduce foldable figures below, we need to consider reflections about affine hyperplanes. For this purpose, let $\mathcal{R}$ be a root system. An *affine hyperplane* with respect to $\mathcal{R}$ is given by

$$(4.1) \qquad H_{r,k} := \{x \in \mathbb{R}^n \mid \langle x, r \rangle = k\}, \qquad k \in \mathbb{Z}.$$

It is easy to show that reflections about affine hyperplanes have the form

$$(4.2) \qquad \rho_{r,k}(x) = x - \frac{2(\langle x, r \rangle - k)}{\langle r, r \rangle} \, r =: \rho_r(x) + k \, r^\vee,$$

where $r^\vee := 2 \, r / \langle r, r \rangle$ is the *coroot* of r.

DEFINITION 4.5. *The affine Weyl group $\widetilde{\mathcal{W}}$ for a root system $\mathcal{R}$ is the (infinite) group generated by the reflections $\rho_{r,k}$ about the affine hyperplanes $H_{r,k}$:*

$$\widetilde{\mathcal{W}} := \langle \rho_{r,k} \mid r \in \mathcal{R}, k \in \mathbb{Z} \rangle$$

We sometimes will refer to the concatenation of elements from $\widetilde{\mathcal{W}}$ as *words*.

THEOREM 4.6. *The affine Weyl group $\widetilde{\mathcal{W}}$ of a root system $\mathcal{R}$ is the semi-direct product $\mathcal{W} \ltimes \Gamma$, where Γ is the abelian group generated by the coroots $r^\vee$. Moreover, Γ is the subgroup of translations of $\widetilde{\mathcal{W}}$ and $\mathcal{W}$ the isotropy group (stabilizer) of the origin. The group $\mathcal{W}$ is finite and Γ infinite.*

REMARK 4.7. There exists a complete classification of all irreducible affine Weyl groups and their associated fundamental domains. These groups are given as types A_n $(n \geq 1)$, B_n $(n \geq 2)$, C_n $(n \geq 3)$, and D_n, $(n \geq 4)$, as well as E_n, $n = 6, 7, 8$, F_4, and G_2. (For more details, we refer the reader to [**Bo**] or [**H**].)

We need a few more definitions and related results. By a *reflection group* we mean a group of transformations generated by the reflections about a finite family of affine hyperplanes. Coxeter groups and affine Weyl groups are examples of reflections groups.

Let $\mathcal{G}$ be a reflection group and $\mathcal{O}_n$ the group of linear isometries of $\mathbb{R}^n$. Then there exists a homomorphism $\phi : \mathcal{G} \to \mathcal{O}_n$ given by

$$\phi(g)(x) = g(x) - g(0), \quad g \in \mathcal{G}, \, x \in \mathbb{R}^n.$$

The group $\mathcal{G}$ is called essential if $\phi(\mathcal{G})$ only fixes $0 \in \mathbb{R}^n$. The elements of $\ker \phi$ are called translations.

4.3. Foldable figures. In this subsection, we define for our later purposes the important concept of a foldable figure [**HW**].

DEFINITION 4.8. A compact connected subset F of $\mathbb{R}^n$ is called a *foldable figure* iff there exists a finite set $\mathcal{S}$ of affine hyperplanes that cuts F into finitely many congruent subfigures $F_1, \ldots, F_m$, each similar to F, so that reflection in any of the cutting hyperplanes in $\mathcal{S}$ bounding F_k takes it into some F_ℓ.

In Figure 6 are two examples of foldable figures shown. Properties of foldable

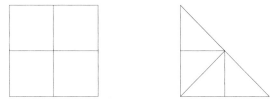

FIGURE 6. Examples of foldable figures.

figures are summarized in the theorem below. The statements and their proofs can be found in [**Bo**] and [**HW**].

THEOREM 4.9.

(1) *The reflection group generated by the reflections about the bounding hyperplanes of a foldable figure F is the affine Weyl group $\widetilde{W}$ of some root system. Moreover, $\widetilde{W}$ has F as a fundamental domain.*

(2) *Let $\mathcal{G}$ be a reflection group that is essential and without fixed points. Then $\mathcal{G}$ has a compact fundamental domain.*

(3) *There exists a one-to-one correspondence between foldable figures and reflection groups that are essential and without fixed points.*

5. Fractal surfaces on foldable figures

Affine fractal surfaces were first systematically introduced in [**M1**] and slightly generalized in [**GH**]. A further generalization was presented in [**HM**]. The construction is again based on IFS's now defined on simplicial regions $\Delta \subset \mathbb{R}^n$ such that the contractive homeomorphisms u_i are affine mappings from Δ into itself. In the present setting, we assume that the domain Δ is a foldable figure in the sense of the previous section and therefore partitioned into N congruent subsimplices Δ_i, $i = 1, \ldots, N$:

$$\Delta = \bigcup_{i=1}^{N} \Delta_i, \quad \text{and} \quad \overset{\circ}{\Delta}_i \cap \overset{\circ}{\Delta}_j = \varnothing, \; i \neq j,$$

Thus, one can find N similitudes $u_i : \Delta \to \Delta_i$ of the form

$$u_i = \sigma \, O_i + b_i$$

where $O_i \in E(n)$, the Euclidean group of $\mathbb{R}^n$, $b_i \in \mathbb{R}^n$, $i = 1, \ldots, N$, and $\sigma \in (0, 1)$ is the similarity ratio between Δ_i and Δ. As in Section 2, let $s \in (-1, 1)$ be an arbitrary scaling factor and $\{\lambda_i : \mathbb{R}^n \to \mathbb{R} \mid i = 1, \ldots, N\}$ a finite collection of continuous affine functions satisfying the following condition.

$(*)$ $\begin{cases} \text{Let } e_{ij} \text{ be the common face of } u_i(\Delta) \text{ and } u_j(\Delta). \text{ Then } \lambda_i(x) = \lambda_j(x) \\ \text{for all } x \in u_i^{-1}(e_{ij}) = u_j^{-1}(e_{ij}), \, i, j = 1, \ldots, N. \end{cases}$

Define an operator $\mathscr{B} : C(\Delta) \to L^\infty(\Delta)$ by

$$(5.1) \qquad \mathscr{B}f(x) := \sum_{i=1}^N \left[\lambda_i \circ u_i^{-1}(x) + s \, f \circ u_i^{-1}(x) \right] \chi_{\Delta_i}.$$

It can be shown [**HM, GH, M1, M2**] that $\mathscr{B}$ maps $C(\Delta)$ into itself and is a contractive operator in the sup-norm with contractivity constant $|s|$. Hence, $\mathscr{B}$ has a unique fixed point $f \in C(\Delta)$, called a *fractal surface over the foldable figure* Δ. As before, there exists a linear isomorphism $\boldsymbol{\lambda} := (\lambda_1, \ldots, \lambda_N) \mapsto f_{\boldsymbol{\lambda}}$ expressing the fact that the fractal surface f is uniquely determined by the vector of mappings λ_i.

EXAMPLE 5.1. Take the foldable figure on the right-hand side of Figure 6 as the domain Δ for an affine fractal function. The four subsimplices $\Delta_1, \ldots, \Delta_4$ induce four similitudes

$$u_1(x,y) = \frac{1}{2}\begin{pmatrix} 1 & 0 \\ 0 & 1 \end{pmatrix}\begin{pmatrix} x \\ y \end{pmatrix} + \begin{pmatrix} \frac{1}{2} \\ 0 \end{pmatrix}, \qquad u_2(x,y) = \frac{1}{2}\begin{pmatrix} -1 & 0 \\ 0 & 1 \end{pmatrix}\begin{pmatrix} x \\ y \end{pmatrix} + \begin{pmatrix} \frac{1}{2} \\ 0 \end{pmatrix}$$

$$u_3(x,y) = \frac{1}{2}\begin{pmatrix} 1 & 0 \\ 0 & -1 \end{pmatrix}\begin{pmatrix} x \\ y \end{pmatrix} + \begin{pmatrix} 0 \\ \frac{1}{2} \end{pmatrix}, \qquad u_4(x,y) = \frac{1}{2}\begin{pmatrix} 1 & 0 \\ 0 & 1 \end{pmatrix}\begin{pmatrix} x \\ y \end{pmatrix} + \begin{pmatrix} 0 \\ \frac{1}{2} \end{pmatrix}.$$

The similarity ratio σ equals $1/2$. Choose $s := 3/5$ and as functions $\lambda_1, \ldots, \lambda_4$:

$$\lambda_1(x,y) := -\frac{1}{5}x + \frac{3}{10}y + \frac{1}{5} =: \lambda_2(x,y),$$

$$\lambda_3(x,y) := \frac{1}{5}x - \frac{3}{10}y + \frac{3}{10} =: \lambda_4(x,y).$$

A short computation shows that this collection of functions satisfies condition $(*)$. Figure 7 shows the graph of the affine fractal surface generated by these maps.

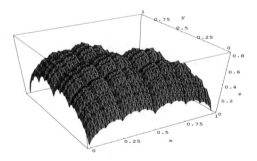

FIGURE 7. An affine fractal surface.

Notice that the affine fractal surface has the value $z = 0$ at the outer vertices $(0,0)$, $(1,0)$, and $(0,1)$ of Δ, and the values $z = 1/2$, $z = 1/2$, and $z = 3/10$ at the inner vertices $(1/2,0)$, $(1/2,1/2)$, and $(0,1/2)$ of Δ. It is not hard to see that the space of affine fractal surfaces over Δ is six-dimensional; there is one basis fractal surface for

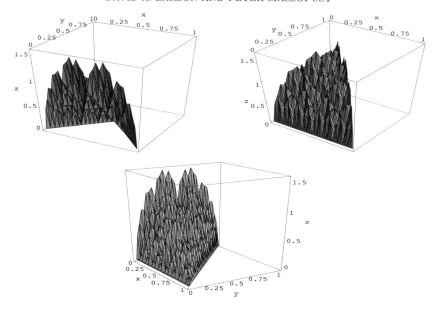

FIGURE 8. Three basis fractal surfaces.

each outer and inner vertex of Δ. If we denote the set of inner and outer vertices of Δ by $\{(x_i, y_j) \mid 2 \leq i + j \leq 4\}$ and by z_{ij} the associated z value of f, then

$$f = \sum_{i,j} z_{ij} \, \varphi_{ij},$$

where $\varphi_{ij}(x_k, y_\ell) = \delta_{ik} \delta_{j\ell}$ is a basis fractal surface. Three of these basis fractal surfaces, namely φ_{12}, φ_{21}, and φ_{22} are displayed in Figure 8.

REMARK 5.2. Let $\mathbf{i} := (i_1, \ldots, i_k) \in \{1, \ldots, N\}^k$. Then, if $\{\varphi_\nu\}$ is a fractal function basis for a fractal surface f defined on a foldable figure $\Delta \in \mathbb{R}^n$, then $\{u_{\mathbf{i}}^\sharp \varphi_\nu\}$, where $(u_{\mathbf{i}}^\sharp \varphi_\nu)(x) := \lambda_\nu \circ u_{\mathbf{i}}^{-1}(x) + s_\nu \varphi_\nu \circ u_{\mathbf{i}}^{-1}(x)$, $x \in u_{\mathbf{i}}(\Delta)$, is a basis for $f|_{u_{\mathbf{i}}(\Delta)}$.

REMARK 5.3. A more general class of fractal surfaces can be defined by taking the λ_i's to be continuous functions satisfying condition (*) and not merely affine functions. ([**GHM1, GHM2, M2**])

In order to achieve out goal, namely to construct a multiresolution analysis on $L^2(\mathbb{R}^n)$ with an orthonormal basis, we need to take into account the algebraic structure of a foldable figure, i.e., its associated affine Weyl group $\widetilde{\mathcal{W}}$, when constructing affine fractal surfaces on all of $\mathbb{R}^n$.

To this end, let $F \subset \mathbb{R}^n$ be a foldable figure with $0 \in \mathbb{R}^n$ as one of its vertices. Denote by $\mathcal{H}$ be the set of hyperplanes associated with F and by Σ be the tessellation of F induced by $\mathcal{H}$. The affine Weyl group of the foldable figure F, $\widetilde{\mathcal{W}}$, is then the group generated by $\mathcal{H}$. The following theorem summarizes some of the properties of F, $\mathcal{H}$, and $\widetilde{\mathcal{W}}$.

THEOREM 5.4. *Let F be a foldable figure with associated set of hyperplanes $\mathcal{H}$ and affine Weyl group $\widetilde{\mathcal{W}}$. Then*

(1) $\mathcal{H}$ consists of the translates of a finite set of linear hyperplanes.
(2) $\widetilde{\mathcal{W}}$ is simply-transitive on Σ, i.e., for all $\sigma, \tau \in \Sigma$ there exists a unique $r \in \widetilde{\mathcal{W}}$ such that $\tau = r\sigma$.
(3) $\forall \varkappa \in \mathbb{N} : \varkappa \mathcal{H} \subset \mathcal{H}$. [Here $\varkappa \mathcal{H} := \{\varkappa H \mid H \in \mathcal{H}\}$]

Now, fix $1 < \varkappa \in \mathbb{N}$ and define $\Delta := \varkappa F$. Then Δ is also a foldable figure, whose $N := \varkappa^n$ subfigures $\Delta_i \in \Sigma$. Assume w.l.o.g that $\Delta_1 = F$. The tessellation and set of hyperplanes for Δ are $\varkappa \Sigma$ and $\varkappa \mathcal{H}$, respectively. Moreover, the affine reflection group generated by $\varkappa \mathcal{H}$ is an isomorphic subgroup of $\widetilde{\mathcal{W}}$. Note that the similarity ratio $\sigma = 1/\varkappa$. By simple transitivity of $\widetilde{\mathcal{W}}$, define similitudes $u_i : \Delta \to \Delta_i$ by:

$$u_1 := (1/\varkappa)(\cdot) \quad \text{and} \quad \forall j = 2, \ldots, N : \quad u_j := r_{j,1} \circ u_1.$$

Denote by $\Pi^d = \Pi^d(\mathbb{R}^n)$ the linear space of real polynomials of degree at most d and choose functions $\lambda_1, \ldots, \lambda_N \in \Pi^d$ satisfying condition $(*)$.[1] We denote the linear space of all such polynomials by J^d. If we define an operator $\mathscr{B}$ as in Equation (5.1) but with $\boldsymbol{\lambda} = (\lambda_1, \ldots, \lambda_N) \in (J^d)^N$, then $\mathscr{B}$ is again contractive in the sup-norm on $C(\Delta)$ with contractivity $|s|$. Hence, it has a unique fixed point, which is the by $\boldsymbol{\lambda} \in (J^d)^N$ generated fractal surface $f = f_{\boldsymbol{\lambda}}$ on Δ.

To extend f to all of $\mathbb{R}^n$ we use that fact that the foldable figure F is a fundamental domain for its associated affine Weyl group and that it tessellates $\mathbb{R}^n$ by reflections in its bounding hyperplanes, i.e., under the action of $\widetilde{\mathcal{W}}$. To this end, let

$$J^{\widetilde{\mathcal{W}}} := \prod \left\{ (J^d)^N \mid r \in \widetilde{\mathcal{W}} \right\}.$$

For $\boldsymbol{\Lambda} \in J^{\widetilde{\mathcal{W}}}$, define $f_{\boldsymbol{\Lambda}}$ by

$$f_{\boldsymbol{\Lambda}}\big|_{r(\overset{\circ}{\Delta})} := f_{\boldsymbol{\Lambda}(r)} \circ r^{-1}, \quad r \in \widetilde{\mathcal{W}},$$

where $\boldsymbol{\Lambda}(r) = (\boldsymbol{\Lambda}(r)_1, \ldots, \boldsymbol{\Lambda}(r)_N)$ is the r-th coordinate of $\boldsymbol{\Lambda}$.

REMARK 5.5. The values of $f_{\boldsymbol{\Lambda}}$ are left unspecified on the hyperplanes $\varkappa \mathcal{H}$, a set of Lebesgue measure zero in $\mathbb{R}^n$, and thus $f_{\boldsymbol{\Lambda}}$ actually represents an equivalence class of functions.

6. Dilation- and $\widetilde{\mathcal{W}}$-invariant spaces

Let V be a linear space of functions $f : \mathbb{R}^n \to \mathbb{R}$, D an expansive unitary operator on V, and $\widetilde{\mathcal{W}}$ an affine Weyl group. (By an expansive unitary operator on V we mean a dilation operator of the form given in (2.3) for an expansive matrix on $\mathbb{R}^n$.)

DEFINITION 6.1. The linear space V is called *dilation–invariant* if

$$D^{-1}V \subset V$$

and $\widetilde{\mathcal{W}}$–invariant if

$$f \in V \Longrightarrow f \circ r \in V, \quad \forall r \in \widetilde{\mathcal{W}}.$$

Dilation–invariance of a global fractal function $f_{\boldsymbol{\Lambda}}$ can be expressed in terms of an associated dilation–invariance of $\boldsymbol{\Lambda} \in J^{\widetilde{\mathcal{W}}}$.

---

[1]More general functions can be used, but for the purposes of this paper the restriction to polynomials provides a large and important subclass.

THEOREM 6.2. *Let $1 < \varkappa \in \mathbb{N}$, $D_\varkappa := \varkappa \, id_{\mathbb{R}^n}$ and f_Λ a global fractal function generated by $\Lambda \in J^{\widetilde{\mathcal{W}}}$. Then*

(1) *f_Λ is $D_\varkappa$-invariant on $\mathbb{R}^n$ iff Λ is $\delta_\varkappa$-invariant on $J^{\widetilde{\mathcal{W}}}$, in other words, iff*

$$D_\varkappa^{-1} f_\Lambda = f_{\delta_\varkappa \Lambda},$$

where $\delta_\varkappa : J^{\widetilde{\mathcal{W}}} \to J^{\widetilde{\mathcal{W}}}$, is given by

$$\delta_\varkappa \Lambda(\varkappa r \, u_j)_i = \Lambda(r)_j \circ u_i + s \left[\Lambda(r)_i - \Lambda(r)_j \right], \quad r \in \widetilde{\mathcal{W}}, \, i, j = 1, \ldots, N.$$

(2) *$J^{\widetilde{\mathcal{W}}}$ is $\delta_\varkappa$-invariant.*

PROOF. For part (1) see [**GHM2, M2**], part (2) is a simple calculation. $\square$

The above theorem now allows us to generate a multiresolution analysis on $L^2(\mathbb{R}^n)$ and define an orthonormal basis for it consisting of fractal surfaces generated by functions in $J^{\widetilde{\mathcal{W}}}$.

DEFINITION 6.3. A multiresolution analysis (MRA) of $L^2(\mathbb{R}^n)$ with respect to dilation $D_\varkappa$ and affine Weyl group $\widetilde{\mathcal{W}}$ consists of a sequence of spaces $\{V_k \, | \, k \in \mathbb{Z}\} \subset L^2(\mathbb{R}^n)$ satisfying

(1) $V_k \subset V_{k+1}$, for all $k \in \mathbb{Z}$.
(2) $\bigcup \{V_k \, | \, k \in \mathbb{Z}\}$ is dense in $L^2(\mathbb{R}^n)$.
(3) There exists a finite set of generators $\{\phi^a \, | \, a \in A\} \subset L^2(\mathbb{R}^n)$ such that

$$\mathcal{B}_\phi := \{\phi^a \circ r \, | \, a \in A, \, r \in \widetilde{\mathcal{W}}\}$$

is a Riesz basis for V_0.
(4) $f \in V_k \implies D_\varkappa f \in V_{k+1}$, $\forall k \in \mathbb{Z}$.

We know that $J^{\widetilde{\mathcal{W}}}$ is $\delta_\varkappa$-invariant and thus we define

$$V_0 := \left\{ f_\Lambda \, | \, \Lambda \in J^{\widetilde{\mathcal{W}}} \right\}$$

and

$$V_k := D_\varkappa^k V_0, \qquad \forall k \in \mathbb{N}.$$

Notice that $D_\varkappa$ can be represented by the expansive matrix $\varkappa I$, where I is the $n \times n$-identity matrix. In the following, we will not distinguish between the operator $D_\varkappa$ and its matrix representation and denote both by $D_\varkappa$.

Since the dimension of $(J^d)^N = (d+1)N$, the dimension of $\dim V_0|_\Delta$ is also $(d+1)N$, and thus $|A| = (d+1)N$. We can take as a basis, a fractal surface basis of the type considered in Example 5.1, and apply the Gram-Schmidt orthogonalization procedure to it, to obtain an orthonormal basis $\{\phi^a \, | \, a = 1, \ldots, |A|\}$ for $V_0|_\Delta$.

Let $\Phi := (\phi^1, \ldots, \varphi^{|A|})^\top$. Then $V_1 \subset V_0$ implies the existence of sequence of $|A| \times |A|$-matrices $\{P(r) \, | \, r \in \widetilde{\mathcal{W}}\}$, only a finite number of which are nonzero, such that

$$(6.1) \qquad (D_\varkappa^{-1} \Phi)(x) = \Phi(x/\varkappa) = \sum_{r \in \widetilde{\mathcal{W}}} P(r) \, (\Phi \circ r)(x).$$

Equation (6.1) is the refinement equation for the scaling vector Φ in the current situation.

THEOREM 6.4. *The ladder of spaces $\{V_k \, | \, k \in \mathbb{Z}\}$ defines an MRA of $L^2(\mathbb{R}^n)$ with respect to dilation $D_\varkappa$ and affine Weyl group $\widetilde{\mathcal{W}}$.*

PROOF. See [**GHM1, GHM2, M2**].   □

For $k \in \mathbb{N}$, define the wavelet spaces $W_k := V_{k+1} \ominus V_k$. Since $\dim W_0|_\Delta = \dim V_1|_\Delta - \dim V_0|_\Delta = (\varkappa^n - 1)(d + 1)N$, we can again use the Gram-Schmidt orthonormalization procedure to construct an orthonormal basis $\{\psi^b \mid b \in B\}$ for $W_0|_\Delta$, where B has cardinality $|B| = (\varkappa^n - 1)(d + 1)N$.

Let $\Psi := (\psi^1, \ldots, \psi^{|B|})^\top$. As $W_0 \subset V_1$ there exists a sequence of $|B| \times |A|$-matrices $\{Q(r) \mid r \in \widetilde{\mathcal{W}}\}$, only a finite number of which are nonzero, such that

$$(6.2) \qquad (D_\varkappa^{-1}\Psi)(x) = \Psi(x/\varkappa) = \sum_{r \in \widetilde{\mathcal{W}}} Q(r)\,(\Phi \circ r)(x).$$

In addition, one can write down finite decomposition and reconstruction algorithms for these multigenerators Φ and Ψ. The interested reader is referred to [**GHM1, GHM2, M2**] for further details.

7. Wavelet sets constructed via Coxeter groups

In this section, a new type of wavelet set is introduced which we will call a *dilation-reflection wavelet set*. It belongs to the Coxeter/fractal-surface multiresolution analysis theory. The idea is to adapt Definition 2.8, replacing the group of translations $\mathcal{T}$ in the traditional wavelet theory by an affine Weyl group whose fundamental domain is a foldable figure C, and to use the orthonormal basis of fractal surfaces constructed in the previous section.

REMARK 7.1. In sections 3 - 6, the requirement that 0 is a vertex of C is not necessary. We chose it for the construction of fractal surfaces since the maps are easier to define (otherwise a shift is to be added) and this was consistent with earlier treatments of the subject [**GHM1, GHM2, M2**]. In the present section, however, where we define a dilation-reflection wavelet set, there is a disadvantage in requiring that 0 is a vertex of C. It agrees more with the dilation theory if 0 is an interior point of C to simplify the application of Definition 2.8 and Theorem 2.9 to this setting to produce the dilation-reflection wavelet sets. This requires an affine shift (as mentioned above) in the Coxeter/fractal surface multiresolution theory discussed in the previous sections. The expositional stance we take is to give a formal definition (Definition 7.6) of dilation-reflection wavelet set in more general terms involving an affine shift in both the Weyl (abstract translation) group and the matricial (abstract dilation) group. We simply take the dilation group fixed point in Definition 2.8 to be any point θ in the (nonempty) interior of C. We leave the details involving the affine shift in the Weyl group to the reader because they are straight-forward, and give explicit details on how the dilation group needs to be affinely shifted to agree with Definition 2.8 and Theorem 2.9. For the concrete examples 7.7 and 5.1, the theory is the simplest from the dilation group viewpoint in the case where 0 is in the interior of C and we therefore take $\theta = 0$. Hopefully, these concrete examples will clarify our treatment of the theory.

In Definition 2.8, we take $X := \mathbb{R}^n$ endowed with the Euclidean affine structure and distance, and for the abstract translation group $\mathcal{T}$ we take the affine Weyl group $\widetilde{\mathcal{W}}$ generated by a group of affine reflections arising from a locally finite collection of affine hyperplanes of X. Let C denote a fundamental domain for $\widetilde{\mathcal{W}}$ which is also a foldable figure. Recall that C is a simplex[2], i.e., a convex connected polytope (here we do not assume it has $n + 1$ vectors), which tessellates $\mathbb{R}^n$ by reflections about its bounding hyperplanes. Let θ be any fixed interior point of C. Let A be any real expansive matrix in $M_n(\mathbb{R})$ acting as a linear transformation on $\mathbb{R}^n$. In the case where θ is the orgin 0 in $\mathbb{R}^n$ we simply take D to be the usual dilation by A and the abstract dilation group to be $\mathcal{D} = \{D^k \,|\, k \in \mathbb{Z}\}$. For a general θ, define D to be the affine mapping $D(x) := A(x - \theta) + \theta, x \in \mathbb{R}^n$ and $\mathcal{D}_\theta = \{D^k \,|\, k \in \mathbb{Z}\}$.

PROPOSITION 7.2. $(\mathcal{D}_\theta, \widetilde{\mathcal{W}})$ is an abstract dilation-translation pair in the sense of Definition 2.8.

PROOF. By the definition of D, θ is a fixed point for $\mathcal{D}_\theta$. By a change of coordinates we may assume without loss of generality that $\theta = 0$ and consequently that D is multiplication by A on $\mathbb{R}^n$.

Let $B_r(0)$ be an open ball centered at 0 with radius $r > 0$ containing both E and C. Since F is open and A is expansive, there exists a $k \in \mathbb{N}$ sufficiently large so that $D^k F$ contains an open ball $B_{3r}(p)$ of radius $3r$ and with some center p. Since C tiles $\mathbb{R}^n$ under the action of $\widetilde{\mathcal{W}}$, there exists a word $w \in \widetilde{\mathcal{W}}$ such that $w(C) \cap B_r(p)$ has positive measure. (Note here that $B_r(p)$ is the ball with the same center p but with smaller radius r.) Then $w(B_r(0)) \cap B_r(p) \neq \emptyset$. Since reflections (and hence words in $\widetilde{\mathcal{W}}$) preserve diameters of sets in $\mathbb{R}^n$, it follows that $w(B_r(0))$ is contained in $B_{3r}(p)$. Hence $w(E)$ is contained in $D^k(F)$, as required.

This establishes part (1) of Definition 2.8. Part (2) follows from the fact that $\theta = 0$ and D is multiplication by an expansive matrix in $M_n(\mathbb{R})$. $\qquad\square$

DEFINITION 7.3. Given an affine Weyl group $\widetilde{\mathcal{W}}$ acting on $\mathbb{R}^n$ with fundamental domain a foldable figure C, given a designated interior point θ of C, and given an expansive matrix A on $\mathbb{R}^n$, a dilation–reflection wavelet set for $(\widetilde{\mathcal{W}}, \theta, A)$ is a measurable subset E of $\mathbb{R}^n$ satisfying the properties:

    (1) E is congruent to C (in the sense of Definition 2.4) under the action of $\widetilde{\mathcal{W}}$, and

    (2) W generates a measurable partition of $\mathbb{R}^n$ under the action of the affine mapping $D(x) := A(x - \theta) + \theta$.

In the case where $\theta = 0$, we abbreviate $(\widetilde{\mathcal{W}}, \theta, A)$ to $(\widetilde{\mathcal{W}}, A)$.

THEOREM 7.4. *There exist $(\widetilde{\mathcal{W}}, \theta, A)$–wavelet sets for every choice of $\widetilde{W}$, θ, and A.*

PROOF. This is a direct application of Theorem 2.9. Let C be a fundamental domain for $\widetilde{W}$ which is a foldable figure, let θ be any interior point of C, and let A be any expansive matrix in $M_n(\mathbb{R})$. By Proposition 7.2, $(\mathcal{D}_\theta, \widetilde{\mathcal{W}})$ is an abstract dilation–translation pair with θ the dilation fixed point. Let C play the role of E in Theorem 2.9. As in the proof of Theorem 2.12, which is sketched above the statement of the theorem, let $F_A := A(B) \setminus B$, where B is the unit ball of $\mathbb{R}^n$, and let $F := F_A + \theta$. Then $\{D^k F \,|\, k \in \mathbb{Z}\}$ is a partition of $\mathbb{R}^n \setminus \theta$, where D is the

---

[2]Let $\{x_0, x_1, \ldots, x_n\}$ be a set of linearly independent points in $\mathbb{R}^n$. The set

$$\Sigma^n := \left\{ \sum_{i=0}^n \lambda_i \, x_i \,\middle|\, \sum_{i=0}^n \lambda_i = 1, \ \lambda_i \geq 0, \ i = 0, 1, \ldots n \right\}$$

is called a simplex.

affine map $D(x) := A(x_\theta) + \theta$. Since F has nonempty interior and is bounded away from θ, Theorem 2.9 applies yielding a measurable set W which is simultaneously congruent to C under the action of $\widetilde{W}$ and congruent ot F under the action of $\mathcal{D}_\theta$. Since F generates a measurable partition of $\mathbb{R}^n$ under $\mathcal{D}_\theta$, so must any set that is $\mathcal{D}_\theta$-congruent to F. Hence W satisfies (2) of Definition 7.3. Since it is also $\widetilde{W}-$congruent to C, this shows that it is a dilation–reflection wavelet set for $(\widetilde{\mathcal{W}}, \theta, A)$, as required. $\qquad\square$

REMARK 7.5. The role of C in the dilation–reflection wavelet theory is analogous to the role of the interval $[0, 2\pi)$ in the dyadic dilation–translation wavelet theory on the real line. For sake of exposition, let us recapture this role: The set of exponentials

$$(7.1) \qquad \left\{ \left. \frac{e^{i\ell s}}{\sqrt{2\pi}} \right|_{[0,2\pi)} \middle| \ell \in \mathbb{Z} \right\}$$

is an orthonormal basis for $L^2([0, 2\pi))$, hence if W is any set which is 2π–translation congruent to $[0, 2\pi)$, then

$$(7.2) \qquad \left\{ \left. \frac{e^{i\ell s}}{\sqrt{2\pi}} \right|_W \middle| \ell \in \mathbb{Z} \right\}$$

is an orthonormal basis for $L^2(W)$. A dyadic dilation–translation wavelet set on the line has this (spectral set) property, and also generates a measurable partition of $\mathbb{R}$ under dilation by 2, and consequently the union of the sets

$$(7.3) \qquad D^n \left\{ \left. \frac{e^{i\ell s}}{\sqrt{2\pi}} \right|_W \middle| \ell \in \mathbb{Z} \right\}$$

is an orthonormal basis for $L^2(\mathbb{R})$.

So, recapitulating, the role of $[0, 2\pi)$ is that it supports a "special" orthonormal basis for $L^2([0, 2\pi))$ induced by the translation group via the Fourier Transform, and thus W, being τ–congruent to $[0, 2\pi)$, also supports an orthonormal basis for $L^2(W)$ induced by the τ–congruence. The role of the fundamental domain C in the dilation-reflection theory is analogous to this.

For sake of exposition, it is natural to make the following somewhat abstract definition.

DEFINITION 7.6. An abstract wavelet set in $\mathbb{R}^n$ is a measurable set W that produces an orthonormal basis for $L^2(\mathbb{R}^n)$ under the action of two countable unitary systems acting consecutively, with the first system inducing an orthonormal basis for $L^2(W)$ by its action restricted to W, and the second is a system of dilations by a family of affine transformations on $\mathbb{R}^n$ whose action on W yield a measurable partition of $\mathbb{R}^n$ (and hence the dilation unitaries when applied to $L^2(W)$ yield a direct-sum orthogonal decomposition of $L^2(\mathbb{R}^n)$).

In the case of a dilation–translation wavelet set W, the two systems of unitaries are $\mathcal{D} := \{D_A^k \mid k \in \mathbb{Z}\}$, where $A \in M_n(\mathbb{R})$ is an expansive matrix, and $\mathcal{T} := \{T^\ell \mid \ell \in \mathbb{Z}^n\}$. An orthonormal wavelet basis of $L^2(\mathbb{R}^n)$ is then obtained by setting $\widehat{\psi}_W := (m(W))^{-1/2} \chi_W$ and taking

$$\left\{ \widehat{D}_A^k \widehat{T}^\ell \widehat{\psi}_W \mid k \in \mathbb{Z}, \ell \in \mathbb{Z}^n \right\}.$$

For the systems of unitaries $\mathcal{D} := \{D_A^k \mid k \in \mathbb{Z}\}$ and $\widetilde{\mathcal{W}}$, the affine Weyl group associated with a foldable figure C, one obtains as an orthonormal basis for $L^2(\mathbb{R}^n)$

$$\left\{ D_{\rtimes}^k \mathcal{B}_\phi \mid k \in \mathbb{Z} \right\},$$

where $\mathcal{B}_\phi = \left\{ \phi^a \circ r \mid a \in A,\, r \in \widetilde{\mathcal{W}} \right\}$ is a fractal surface basis as constructed in the previous section.

In [**DL**], two examples of wavelet sets are given in the plane for dilation–2 (i.e., the dilation matrix is $A := 2I$, where I is the 2×2 identity matrix) and 2π-translation (separately in each coordinate). Both examples are reproduced here and it is shown that the two dilation–translation wavelet sets are also dilation-reflection wavelet sets for $(\widetilde{W}, \theta, 2I)$, where $\widetilde{W}$ is the affine Weyl group generated by the reflections about the bounding hyperplanes of $C := [-\pi, \pi) \times [-\pi, \pi)$, and where $\theta = 0$. In other words, we take dilation to be exactly the same and replace the 2π–translation by the action of the Weyl group. We find it quite interesting that the same measurable set is a wavelet set in each of the two different theories because the Weyl group and the usual translation by 2π–group act completely differently as groups of transformations of $\mathbb{R}^2$. In some sense, this justifies our usage of the term "wavelet set" to denote our construction in the Coxeter/fractal surface theory. The other reason is the interpretation given in Remark 7.5.

EXAMPLE 7.7. Let $A := 2I$, where I denotes the identity matrix in $\mathbb{R}^2$. For $n \in \mathbb{N}$, define vectors $\vec{\alpha}, \vec{\beta} \in \mathbb{R}^2$ by

$$\vec{\alpha}_n := \frac{1}{2^{2n-2}} \begin{pmatrix} \frac{\pi}{2} \\ \frac{\pi}{2} \end{pmatrix};$$

$$\vec{\beta}_0 := 0, \quad \vec{\beta}_n := \sum_{k=1}^{n} \vec{\alpha}_k.$$

Define

$$G_0 := \left[0, \frac{\pi}{2} \right];$$

$$G_n := \frac{1}{2^{2n}} G_0 + \vec{\beta}_n;$$

$$E_1 := \bigcup_{k=1}^{\infty} G_k \subset 2G_0 \setminus G_0;$$

$$C_1 := G_0 \cup E_1 + \begin{pmatrix} 2\pi \\ 2\pi \end{pmatrix};$$

$$B_1 := 2G_0 \setminus (G_0 \cup E_1).$$

Finally, let

$$A_1 := B_1 \cup C_1;$$
$$A_2 := \{(-x, y) \mid (x, y) \in A_1\};$$
$$A_3 := \{(-x, -y) \mid (x, y) \in A_1\};$$
$$A_4 := \{(x, -y) \mid (x, y) \in A_1\};$$
$$W_1 := A_1 \cup A_2 \cup A_3 \cup A_4.$$

It is not hard to verify that W_1 is 2π-translation–congruent to $C = [-\pi, \pi) \times [-\pi, \pi)$ and a 2-dilation generator of a measurable partition for the two-dimensional plane $\mathbb{R}^2 \setminus \{0\}$. (Cf. [**DL**])

The set C is a fundamental domain of the (reducible) affine Weyl group $\widetilde{\mathcal{W}}$ that is generated by the (affine) reflections about the six lines, namely $L_\pm^x : x = \pm\pi$, $L_\pm^y : y = \pm\pi$, $L_0^x : x = 0$, and $L_0^y : y = 0$. (Cf. Left-hand side of Figure 6.) The roots for this Weyl group are given by $r_1 = -r_2 = (1, 0)^\top$ and $r_3 = -r_4 = (0, 1)^\top$. As a matter of fact, the *Coxeter group* associated with C is Klein's Four Group or the dihedral group D_4. Denote by B_j, C_j, and E_j, $j = 2, 3, 4$, the extension of B_1, C_1, and E_1, respectively, into the jth quadrant. Let ρ_-^x and ρ_-^y denote the affine reflection about the line L^x and L^y, respectively. Then it is easily verified that $\rho_-^y \rho_-^x (E_3) \cup B_1 = 2G_0$. Analogous arguments applied to E_1, E_2, and E_4 show that W_1 is $\widetilde{\mathcal{W}}$–congruent to C.

Since C is a foldable figure, there exists an orthonormal basis for $L^2(C)$ generated by fractal surfaces. Now, the sets $2^{-2n}G_0$ are copies of C under appropriate combinations of the maps u_i and we have, by Remark 5.2, for all $n \in \mathbb{N}$ an orthonormal basis. Hence, the sets G_n have such a basis for all $n \in \mathbb{N}$ and since they only intersect on a set of measure zero, so do the sets E_j and $G_0 \cup E_j$, $j = 1, \ldots, 4$. Since the sets C_j are obtained by applying elements of the affine Weyl groups, i.e., isometries, to $G_0 \cup E_j$, $j = 1, \ldots, 4$, one obtains an L^2-orthonormal basis $\mathcal{B}_\phi$ for W_1. Then, since $L^2(W_1)$ is wandering for D_2^k, $\{D_2^k \mathcal{B}_\phi : k \in \mathbb{Z}\}$ is an orthonormal basis for $L^2(\mathbb{R}^2)$.

The wavelet set W_1 is depicted in Figure 9 (See also [**DL**]).

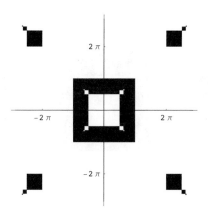

FIGURE 9. The wavelet set W_1 in $\mathbb{R}^2$.

EXAMPLE 7.8. Again, let $A := 2I$, where I denotes the identity matrix in $\mathbb{R}^2$. For $n \in \mathbb{N}$, define vectors $\vec{\alpha}, \vec{\beta} \in \mathbb{R}^2$ by

$$\vec{\alpha}_n := \frac{1}{2^{2n-2}} \begin{pmatrix} \frac{\pi}{2} \\ 0 \end{pmatrix};$$

$$\vec{\beta}_0 := 0, \quad \vec{\beta}_n := \sum_{k=1}^{n} \vec{\alpha}_k.$$

Define

$$G_0 := \left[0, \frac{\pi}{2}\right] \times \left[-\frac{\pi}{2}, \frac{\pi}{2}\right];$$

$$G_n := \frac{1}{2^{2n}} G_0 + \vec{\beta}_n;$$

$$E := \bigcup_{k=1}^{\infty} G_k \subset 2G_0 \setminus G_0;$$

$$D := G_0 \cup E + \begin{pmatrix} 2\pi \\ 0 \end{pmatrix};$$

$$B := 2G_0 \setminus (G_0 \cup E).$$

Define

$$A_1 := B \cup D;$$

$$A_2 := \{(-x, y) \,|\, (x, y) \in A_1\};$$

$$W_2 := A_1 \cup A_2.$$

That W_2 is a dilation–translation wavelet set was established in [**DL**]. To show that it is also a dilation–reflection wavelet set, note that C is the same foldable figure as in Example 7.7 above with the same (reducible) affine Weyl group $\widetilde{\mathcal{W}}$. Denote by $L_1 : x = -\pi$ and $L_2 : x = \pi$ the left and right bounding lines of C and by ρ_1, respectively, ρ_2 the corresponding reflections. Let $D^- := \{(-x, y) \,|\, (x, y) \in D\}$. Then it is easy to see that $\rho_2(D) \cup \rho_1(D^-) = C$, hence that W_2 is $\widetilde{\mathcal{W}}$–congruent to C. Arguments similar to those used in Example 7.7 show that there exists an orthonormal basis for $L^2(\mathbb{R}^2)$ consisting of fractal surfaces.

The wavelet set W_2 is shown in Figure 10.

FIGURE 10. The wavelet set W_2 in $\mathbb{R}^2$.

8. Generalitites and Open Problems

The two examples that were considered at the end of the previous section are representatives of what one might call "three-way tiling sets" of the Euclidean plane $\mathbb{R}^2$. To this end, observe that if that $\mathcal{G}$ is a group of transformation in $\mathbb{R}^n$ and some subset K tiles $\mathbb{R}^n$ under the action of $\mathcal{G}$, then any set L that is $\mathcal{G}$–congruent to K also tiles $\mathbb{R}^n$ under the action of $\mathcal{G}$. The sets W_i, $i = 1, 2$, are wavelet sets for both the affine reflection group $\widetilde{\mathcal{W}}$ and the translation group $\mathcal{T}$. Both clearly tile $\mathbb{R}^2$ under translation, but they also tile under reflections since both are $\widetilde{\mathcal{W}}$–congruent to a foldable figure, namely $C = [-\pi, \pi] \times [-\pi, \pi]$, which tiles the plane under reflections in its bounding hyperplanes.

The fact that W_1 and W_2 tile under both the dilation and translation group makes them dilation-translation wavelet sets, and the fact that they tile under both the dilation and reflection group makes them dilation-reflection wavelet sets. Indeed, more can be said about these two wavelets sets and the fundamental domain C.

The foldable figure C is in both situations a fundamental domain for the affine Weyl group $\widetilde{W}$ and for the standard translation-by-2π group $\mathcal{T}$. Recall that by Theorem 4.6, $\widetilde{W} = W \ltimes \Gamma$ where W is the stabilizer (isotropy group) of the origin and Γ the translation group generated by the coroots $r_j^\vee = 2\pi r_j$, $j = 1, \ldots, 4$ (cf. Example 7.7). (Note that the k in Equation 4.1 is πk in the two examples.) If we denote the intersection group $\widetilde{W} \cap \mathcal{T}$ by $\mathcal{J}$, then we have in this case $\mathcal{J} \subset \mathcal{T}$. Indeed, $\mathcal{J}$ is generated by translations of $(k \cdot 4\pi, \ell \cdot 4\pi)$, $k, \ell \in \mathbb{Z}$. To see this, we compute $\rho_{r,\pi k} \circ \rho_{s,\pi \ell}$, where r and s are any two of the roots r_j, $j = 1, \ldots, 4$, and find that for $s = \pm r$

$$\rho_{r,\pi k} \circ \rho_{\pm r,\pi \ell} = \mathrm{id}_{\mathbb{R}^2} \pm 2\pi(k - \ell)r, \qquad |k - \ell| \geq 2,$$

whereas for $s \perp r$,

$$\rho_{r,\pi k} \circ \rho_{s,\pi \ell} = -\mathrm{id}_{\mathbb{R}^2} + 2\pi(kr + \ell s).$$

Thus, every element of $\widetilde{W}$ is the product of a simple reflection and a translation by $(k \cdot 4\pi, \ell \cdot 4\pi)$, $k, \ell \in \mathbb{Z}$. Furthermore, $\mathcal{J}$ is clearly nontrivial but also big enough so that it together with the dilation-by-$2I$ group satisfies the axioms in the definition of "abstract dilation-translation pair" (Definition 2.8). In addition, both sets W_i, $i = 1, 2$, are actually congruent to C via the intersection group $\mathcal{J}$. In other words, for the translation-by-2π congruence only translations in $\mathcal{J}$ are used, and likewise for the $\widetilde{W}$–congruence.

These observations suggest now the validity of the following more general situation.

PROPOSITION 8.1. Suppose that C is any foldable figure which is a fundamental domain for both a translation group $\mathcal{T}$ and the affine Weyl group $\widetilde{W}$ for C, and which contains 0 in its interior. If the intersection group $\mathcal{J}$ of $\mathcal{T}$ and $\widetilde{W}$ is big enough, in the sense described above, so that for an expansive matrix $A \in M_n(\mathbb{R})$ the dilation group $\mathcal{D}$ for A and the intersection group $\mathcal{J}$ is an abstract dilation–translation pair, then there exist sets W which are simultaneously dilation-translation and dilation–reflection wavelet sets.

PROOF. Let B denote the unit ball of $\mathbb{R}^n$ and let W be the set whose existence is guaranteed by Theorem 2.9 (Theorem 1 of [**DLS1**]) and which is both dilation–congruent to $F := A(B) \setminus B$ and $\mathcal{J}$-congruent to C. (Here B is the unit ball of $\mathbb{R}^n$.) Then W is automatically congruent to C by the larger groups, namely the affine Weyl group $\widetilde{W}$ and the translation group $\mathcal{T}$. Now, dilation–congruence to F and translation-congruence to C makes W are dilation–translation wavelet set, whereas dilation–congruence to F and $\widetilde{W}$–congruence to C make it a dilation–reflection wavelet set.

The key is that W needs to tile $\mathbb{R}^n$ under the full translation group $\mathcal{T}$, and also the full affine Weyl group $\widetilde{W}$. Since this is true for C, the congruences of C to W under both groups guarantees that W tiles $\mathbb{R}^n$, too.   □

The two wavelets presented in Examples 7.7 and 7.8 are very special, for the reasons given above. In addition, the affine Weyl group associated with the foldable figure $C = [-\pi, \pi] \times [-\pi, \pi]$ is also rather specific in the sense that it is reducible. It consists of the bifold product of the Weyl group associated with the interval $[-\pi, \pi]$. Theorem 7.4 guarantees the existence of dilation-reflection wavelet sets in general, even when the Weyl group is irreducible. So it may be interesting to construct concrete examples of such sets. One example of a foldable figure whose affine Weyl group is irreducible is given in Figure 11. We therefore pose the following problem.

PROBLEM 1: Given the foldable figure C depicted in Figure 11, construct concrete examples of dilation–reflection wavelet sets W for C. In particular, are there any such examples which are bounded and bounded away from 0 (as are the sets in Examples 7.7 and 7.8). In principle, if one follows the proof of Theorem 2.6 in [DLS1] using that constructive proof as an algorithm for constructing a wavelet set, then dilation-reflection sets for the Weyl group associated with Figure 11 can easily be constructed. However, with that method, the sets constructed are not bounded subsets of the plane, and are also not bounded away from 0. In addition, they are difficult to work with.

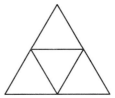

FIGURE 11. A triangular foldable figure.

Another interesting question connecting with the above arguments is the following.

PROBLEM 2: Let C be any foldable figure in $\mathbb{R}^n$ containing 0 in its interior and let $\widetilde{\mathcal{W}} = \mathcal{W} \ltimes \Gamma$ be the associated affine Weyl group. Suppose that A is any expansive matrix in $M_n(\mathbb{R})$, and $\mathcal{T}$ a translation group on $\mathbb{R}^n$ obtained by translating with respect to a basis $\mathcal{B} := \{b_1, b_2, ...b_n\}$ for $\mathbb{R}^n$, i.e., a vector x in $\mathbb{R}^n$ is mapped to $x - (k_1 b_1, k_2 b_2, ..., k_n b_n)$ for all n-tuples $(k_1, ..., k_n) \in \mathbb{Z}^n$. Give necessary and sufficient conditions for the existence a set W which is simultaneously

(1) $\widetilde{\mathcal{W}}$–congruent to C;

(2) D_A–congruent to the set $F := A(B) \setminus B$ (as in the proof of Theorem 7.4);

(3) $\mathcal{T}$–congruent to the set $[0, b_1) \times [0, b_2) \times \ldots \times [0, b_n)$. (Note that this last set is the simplest fundamental domain for $\mathcal{T}$).

Any W satisfying (1), (2), and (3) would be both a dilation–translation wavelet set for $(\mathcal{D}, \mathcal{T})$ and a dilation–reflection wavelet set for $(\widetilde{\mathcal{W}}, A)$. Conversely, any set which is both a dilation-translation wavelet set for $(\mathcal{D}, \mathcal{T})$ and a dilation–reflection wavelet set for $(\widetilde{\mathcal{W}}, A)$ must satisfy (1), (2), and (3). In particular, does there exist such a W for an irreducible Weyl group, such as the group corresponding to the foldable figure in Figure 11? We think that the answer is probably no. But in the topic of wavelet sets there are often surprises, so we would not be very surprised if the answer was yes.

Acknowledgments

The authors thank the anonymous referee for their detailed comments and recommendations improving the understanding of the paper and some interesting suggestions for further work. In particular, the referee observed that because of the semi-direct product structure of the affine Weyl group and the existence of three-way tiling sets, there may be connections between the dilation-reflection wavelet theory developed in this paper and the composite dilations developed by Krishtal, Robinson, Weiss and Wilson in [**KRWW**].

References

[BMM] L. Baggett, H. Medina, and K. Merrill, *Generalized multi-resolution analyses and a construction procedure for all wavelet sets in R^n*, J. Fourier Anal. Appl. **5** (1999).

[B] M. F. Barnsley, *Fractal functions and interpolation*, Constr. Approx. **2** (1986), 303 – 329.

[BD] M. F. Barnsely and S. Demko, *Iterated function systems and the global construction of fractals*, Proc. R. Soc. Lond. A **399** (1985), 243 – 275.

[BL] J. J. Benedetto and M. Leon, *The construction of single wavelets in D-dimensions*, J. Geom. Anal. **11** (2001), no. 1, 1 – 15.

[Bo] N. Bourbaki, *Lie Groups and Lie Algebras*, Chapters 4 - 6, Springer Verlag, Berlin 2002.

[C] H. S. M. Coxeter, *Regular Polytopes*, 3rd. ed., Dover, New York, 1973.

[G] L. C. Grove and C. T. Benson, *Finite Reflection Groups*, 2nd ed., Springer Verlag, New York, 1985.

[DL] X. Dai and D. Larson, *Wandering vectors for unitary systems and orthogonal wavelets*, Memoirs of the AMS, Vol. 134, No. 640, Providence, RI, 1998.

[DLS1] X. Dai, D. Larson and D. Speegle, *Wavelet Sets in $\mathbb{R}^n$*, J. Fourier Anal. and Appl. **3**(4) (1997), 451 – 456.

[DLS2] X. Dai, D. Larson and D. Speegle, *Wavelet sets in R^n - II*, Contemp. Math, **216** (1998), 15 – 40.

[DAU] I. Daubechies, *Ten Lectures on Wavelets*, SIAM, Philadelphia, PA, 1992.

[FW] X. Fang and X. Wang, Construction of minimally-supported frequency wavelets, J. Fourier Anal. Appl. **2** (1996), 315 – 327.

[GH] J. Geronimo and D. Hardin, *Fractal Interpolation Surfaces and a related 2-D Multiresolution Analysis*, J. Math. Anal. and Appl., **176**(2) (1993), 561 – 586.

[GHM1] J. Geronimo, D. Hardin and P. R. Massopust, *Fractal Surfaces, Multiresolution Analyses and Wavelet Transforms*, in Y. O, A. Toet, D. Foster, H. Heijmans, and P. Meer (eds.) *Shape in Picture*, NATO ASI Series, Vol. 126 (1994), 275 – 290.

[GHM2] J. Geronimo, D. Hardin and P. R. Massopust, *An Application of Coxeter Groups to the Construction of Wavelet Bases in $\mathbb{R}^n$*, in W. Bray, P.Milojević, and Č. Stanojević (eds.), *Fourier Analysis: Analytic and Geometric Aspects*, Lecture Notes in Pure and Applied Mathematics, Vol. 157, Marcel Dekker, New York 1994, 187 – 195.

[Gu] P. Gunnells, *Cells in Coxeter Groups*, in Notices of the AMS **53**(5) (2006), 528 – 535.

[HWW] E. Hernandez, X. Wang and G. Weiss, Smoothing minimally supported frequency (MSF) wavelets: Part I., J. Fourier Anal. Appl. 2 (1996), 329 – 340.

[H] J. Humphreys, *Reflection Groups and Coxeter Groups*, Cambridge University Press, Cambridge U.K. 1990.

[HM] D. Hardin and P. Massopust, *Fractal Interpolation Functions from $\mathbb{R}^n$ to $\mathbb{R}^m$ and their Projections*, Zeitschrift für Analysis u. i. Anw., **12**(1993), 535 – 548.

[ILP] E. Ionascu, D. Larson and C. Pearcy, *On the unitary systems affiliated with orthonormal wavelet theory in n-dimensions,* J. Funct. Anal. **157** (1998), no. 2, 413 – 431.

[Hu] J. Hutchinson, *Fractals and self-similarity*, Indiana Univ. J. Math., **30** (1981), 713 – 747.

[HW] M. Hoffman, W. D. Withers, *Generalized Chebyshev Polynomials Associated with Affine Weyl Groups*, Trans. Amer. Math. Soc., **308** (1) (1988), 91 – 104.

[KRWW] I. Krishtal, B. Robinson, G. Weiss, and E. Wilson, *Some simple Haartype wavelets in higher dimensions*, to appear in J. Geom. Anal. (2007).

[L96] D. R. Larson, *Von Neumann algebras and wavelets. Operator algebras and applications* (Samos, 1996), 267–312, NATO Adv. Sci. Inst. Ser. C Math. Phys. Sci., 495, Kluwer Acad. Publ., Dordrecht, 1997.

[L98] D. R. Larson, *Frames and wavelets from an operator-theoretic point of view,* Operator algebras and operator theory (Shanghai, 1997), 201 – 218, Contemp. Math., 228, AMS, Providence, RI, 1998.

[M1] P. Massopust, *Fractal Surfaces*, J. Math. Anal. and Appl., **151**(1) (1990), 275 – 290.

[M2] P. Massopust, *Fractal Functions, Fractal Surfaces, and Wavelets*, Academic Press, Orlando, 1995.

[M3] P. Massopust, *Fractal Functions and their Applications*, Chaos, Solitons & Fractals, Vol. 8, No. 2 (1997), 171 – 190.

[M4] P. Massopust, *Fractal Functions, Splines, and Besov and Triebel-Lizorkin Spaces*, in Fractals in Engineering: New trends and applications (J. Lévy-Véhel, E. Lutton, eds.), 21 – 32, Springer Verlag, London, 2005.

[S] D. Speegle, *The s-elementary wavelets are path-connected*, Proc. Amer. Math. Soc., **132** (2004), 2567 – 2575.

[SW] P. Soardi and D. Wieland, *Single wavelets in $\mathbb{R}^n$*, J. Fourier Anal. Appl. **4** (1998), 299 – 315.

[W] Wutam Consortium, *Basic properties of wavelets* J. Four. Anal. Appl., **4** (1998), 575 – 594.

DEPARTMENT OF MATHEMATICS, TEXAS A & M UNIVERSITY, COLLEGE STATION, TEXAS 77843
E-mail address: `larson@math.tamu.edu`

GSF - NATIONAL RESEARCH CENTER FOR ENVIRONMENT AND HEALTH, INSTITUTE OF BIOMATHEMATICS AND BIOMETRY, AND CENTRE OF MATHEMATICS M6, TECHNISCHE UNIVERSITÄT MÜNCHEN, GERMANY
E-mail address: `massopust@ma.tum.de`

Contemporary Mathematics
Volume **451**, 2008

CHARACTERIZATION OF MINIMIZERS OF CONVEX REGULARIZATION FUNCTIONALS

CHRISTIANE PÖSCHL AND OTMAR SCHERZER

ABSTRACT. We study variational methods of bounded variation type for the data analysis. Y. Meyer characterized minimizers of the Rudin-Osher-Fatemi functional in dependence of the G-norm of the data. These results and the follow up work on this topic are generalized to functionals defined on spaces of functions with derivatives of finite bounded variation. In order to derive a characterization of minimizers of convex regularization functionals we use the concept of generalized directional derivatives and duality. Finally we present some examples where the minimizers of convex regularization functionals are calculated analytically, repeating some recent results from the literature and adding some novel results with penalization of higher order derivatives of bounded variation.

1. INTRODUCTION

This paper is concerned with variational methods, consisting in minimization of the functional

$$\mathcal{T}_{\alpha,u^\delta}(u) := \mathcal{S}(u) + \alpha\mathcal{R}_l(u)\,, \quad \alpha > 0,\, l = 1, 2, \ldots,$$

for the analysis of data u^δ. Here

(1) $\mathcal{R}_l(u) := \left|D^l u\right|$ denotes the total variation of the $(l-1)$-th derivative of u and

(2) $\mathcal{S}(u)$ is a similarity measure. Typical examples are $\mathcal{S}_p(u) = \frac{1}{p}\int_\Omega \left|u - u^\delta\right|^p$.

Y. Meyer [Mey01] characterized minimizers of the ROF-functional (introduced in [RudOshFat92]), where $\mathcal{S}_2(u) = \frac{1}{2}\int_\Omega(u - u^\delta)^2$ and $l = 1$, in dependence of the G-norm of u^δ. This research has significant impact on the research in image analysis.

In this paper we use an alternative characterization based on *Fenchel's duality theorem* and *generalized directional derivatives* to generalize the results of Y. Meyer and the follow up work [OshSch04, SchYinOsh05]. Moreover, the results can also be applied to characterize minimizers of regularization functionals penalizing for derivatives with finite total variation. This generalizes the ideas in [ObeOshSch05]. Non-differentiable regularization functionals for higher order derivatives have attracted several research (see for instance [ChaLio95, Sch98, ChaMarMul00, SteDidNeu05, SteDidNeu, Ste06, HinSch06]).

Date: April 12, 2007.

1991 *Mathematics Subject Classification*. Primary 47A52; Secondary 65F22, 49M29.

Key words and phrases. Fenchel duality, TV, bounded Variation, bounded Hessian, G-norm.

CP was supported by DOC-FForte of the Austrian Academy of Sciences.

The work of OS is supported by the Austrian Science Foundation (FWF) Projects Y-123INF, FSP 9203-N12 and FSP 9207-N12.

The abstract results in this paper also allow to characterize minimizers of *metrical regularization* functionals, such as L^1-$BV(\Omega)$ regularization (see for instance Chan & Esedoglu [ChaEse05], Nikolova [Nik03a, Nik03b]) and showing a structure in regularization methods of this type.

Moreover, exploiting the Fenchel duality concept we exemplary derive explicit solutions for minimizers of the ROF-functional for denoising one-dimensional data (repeating the results of Strong & Chan [StrCha96] and Y. Meyer [Mey01]), the L^1-$BV(\Omega)$ regularization (repeating the results of Chan & Esedoglu [ChaEse05]), and also for novel metrical regularization techniques as well as regularization techniques with higher order penalization.

In Section 2 and the Appendix A we recall some basic facts on G-norms and bounded variation regularization. In Section 4 we recall the definition of the Fenchel dual of a functional, and establish a connection between the fenchel dual and the G-norm. In Section 3 we characterize minimizers of convex regularization functionals with differentiability concepts.

Finally in Section 5 we present some analytical examples of minimizers of regularization functionals.

Prerequisites for this paper:

- All along this paper we assume that Ω is a bounded, open, connected domain with Lipschitz boundary (bocL) or that $\Omega = \mathbb{R}^n$. $\vec{n}$ denotes the normal vector to the boundary of Ω.
- We denote by $|\cdot|$ the Euclidean norm.
- If $1 < p < \infty$, we denote by p_* the number $p/(p-1)$ such that $1/p + 1/p_* = 1$. For $p = 1$ we set $p_* = \infty$.
- For the sake of simplicity of presentation we identify in this paper functions in $L^{p_*}(\Omega)$ and functionals in $(L^p(\Omega))^*$. For $X = L^p$ we identify a functional $u_X^* \in (L^p(\Omega))^*$ with a function u^* in $L^{p_*}(\Omega)$ such that $\langle u_X^*, u \rangle_{p_*,p} := \langle u_X^*, u \rangle_{X^*,X} = \int_\Omega u^* u$. Moreover when we write $\|u^*\|_*$, $\langle u^*, u \rangle_{X^*,X}$, $\mathcal{S}^*(u^*)$, $\cdots$ we actually mean $\|u_X^*\|_*$, $\langle u_X^*, u \rangle_{X^*,X}$, $\mathcal{S}^*(u_X^*)$, $\cdots$ respectively, where $\|\cdot\|_*$ and $\mathcal{S}^*$ are defined later.
- The p-duality mapping $\mathcal{J}_p$ is defined by
$$\mathcal{J}_p : L^p(\Omega) \to L^{p_*}(\Omega),$$
$$u \to |u|^{p-2} u.$$
- If $U \subset X$ Banach spaces with norms $\|\cdot\|_U$ and $\|\cdot\|_X$, such that the inclusion $i : U \to X$ is continuous with respect to the norms $\|\cdot\|_U$ and $\|\cdot\|_X$. The adjoint $i^\# : X^* \to U^*$ of the inclusion mapping is defined implicitly by
$$\langle u^*, i(v) \rangle_{X^*,X} = \langle i^\#(u^*), v \rangle_{U^*,U}, \qquad u^* \in X^*, \ v \in U.$$
- For $l \in \mathbb{N}$ and $u \in L^1(\Omega)$ we define
$$\rho_u^l(x) := (-1)^l \int_{-1}^x \left(\ldots \int_{-1}^{t_2} u(t_1) dt_1 \ldots \right) dt_l, \quad x \in [-1,1].$$
- For $1 \le p < \infty$ and $l \in \{1, 2, \ldots\}$ $\mathcal{T}_{\alpha,u^\delta}^{l,p}$ is defined by
$$\mathcal{T}_{\alpha,u^\delta}^{l,p}(u) := \mathcal{S}_p(u) + \alpha \mathcal{R}_l(u) = \frac{1}{p} \int_\Omega |u - u^\delta|^p + \alpha |D^l u|.$$

Convention 1.1. *In the remainder of this paper we use the following convention. For $1 \leq p < \infty$ and $l \in \{1, 2, \ldots\}$ we assume that Ω is bocL or $\Omega = \mathbb{R}^n$.*

• *For Ω bocL we consider the Sobolev spaces*

$$\mathcal{W}^{l,p} := W^{l,p}_\diamond(\Omega) = \left\{ u \in W^{l,p}(\Omega) : \int_\Omega \partial^\gamma u = 0, \;\; \text{for all } |\gamma| \leq l - 1 \right\}$$

and the space of functions with derivatives of finite total variation

$$\mathcal{TV}^l := \mathrm{TV}^l_\diamond(\Omega) = \mathrm{TV}^l(\Omega) \cap W^{l-1,1}_\diamond(\Omega) .$$

• *For $\Omega = \mathbb{R}^n$ we consider the Sobolev space*

$$\mathcal{W}^{l,p} := \overline{\{u \in C_0^\infty(\Omega)\}} ,$$

the completion of $C_0^\infty(\Omega)$ with respect to $|\cdot|_{l,p}$. and the space of functions of finite total variation $\mathcal{TV}^l := \mathrm{TV}^l(\Omega)$.

Both Sobolev spaces are associated with the norm $\left\| \nabla^l u \right\|_p$. Both spaces of functions with derivatives of finite total variation are always associated with the norm $\left| D^l u \right|$.

2. G-Norm

Y. Meyer [Mey01] characterized minimizers of the ROF-functional

$$\mathcal{T}^{1,2}_{\alpha, u^\delta}(u) := \frac{1}{2} \int_{\mathbb{R}^2} (u - u^\delta)^2 + \alpha \left| Du \right| \quad (\alpha > 0)$$

using the dual norm of $W^{1,1}(\mathbb{R}^2)$, which he called the G-norm. Aubert & Aujol [AubAuj05] derived a characterization of minimizers of the ROF-functional defined on $\Omega \subseteq \mathbb{R}^2$ being bocL. Chan & Shen [ChaShe05] used a characterization of dual functions which applies both for bounded and unbounded domains. In [ObeOshSch05] we derived a characterization of minimizers of ROF-like functionals with penalization by the total variation of second order derivatives. In [OshSch04] we characterized minimizers of regularization functionals with anisotropic total variation regularization penalization term.

In [Mey01] Y. Meyer defined the G-space

$$G := \left\{ v : v = \nabla \cdot \vec{v}, \, \vec{v} \in L^\infty(\mathbb{R}^2; \mathbb{R}^2) \right\}$$

with the norm

$$\|v\|_G := \inf \left\{ \|\vec{v}\|_\infty : v = \nabla \cdot \vec{v} \right\} .$$

This definition was generalized by Aubert & Aujol [AubAuj05] to the case where the domain $\Omega \subset \mathbb{R}^2$ is bocL. The basic definition is the same, but boundary conditions have to be taken into account. Here,

(2.1) $$G := \left\{ v = (\nabla \cdot \vec{v}) \in L^2(\Omega) : \vec{v} \in L^\infty(\Omega), \, \vec{v} \cdot \vec{n} = 0 \text{ on } \partial\Omega \right\} ,$$

and again

$$\|v\|_G := \inf \left\{ \|\vec{v}\|_\infty : v = \nabla \cdot \vec{v} \right\} .$$

In both definitions the divergence $(\nabla \cdot \vec{v})$ has to be understood in a weak sense. Moreover, in (2.1) the *normal trace* is understood distributionally. More precisely, $v = \nabla \cdot \vec{v}$ with $\vec{v} \cdot \vec{n} = 0$ on $\partial\Omega$, if

$$\int_\Omega \vec{v} \cdot \nabla \phi = - \int_\Omega v \phi , \qquad \phi \in C_0^\infty(\mathbb{R}^n) .$$

Here we aim for a unified analysis. We rely on fundamental results in Adams [Ada75], which characterize the duals of $W^{l,p}(\Omega)$ for every $l \geq 1$ and Ω in any space dimension. Due to some structural properties of regularization functionals the results in Adams [Ada75] have to be slightly adapted.

Theorem 2.1. *Assume that Ω is bocL or $\Omega = \mathbb{R}^n$, $1 \leq p < \infty$, and $l \in \mathbb{N}_0$. Define $\mathcal{N}(l) := \#\{\vec{\gamma} \in \mathbb{N}_0^n : |\vec{\gamma}| = l\}$. For every $L \in (\mathcal{W}^{l,p})^*$ there exists $\vec{v} \in L^{p*}(\Omega; \mathbb{R}^{\mathcal{N}(l)})$ such that*

$$(2.2) \qquad \langle L, u \rangle = \langle \vec{v}, \nabla^l u \rangle_{p,p_*} = \sum_{|\gamma|=l} \int_\Omega \vec{v}_{\vec{\gamma}} \, \partial^{\vec{\gamma}} u, \qquad u \in \mathcal{W}^{l,p} \; .$$

Moreover,

$$(2.3) \qquad \|L\|_{(\mathcal{W}^{l,p})^*} = \min\{\|\vec{v}\|_{p_*} : \vec{v} \in L^{p*}(\Omega; \mathbb{R}^{\mathcal{N}(l)}) \text{ satisfies } (2.2)\} \; .$$

Proof. Let $L \in (\mathcal{W}^{l,p})^*$ and denote $X := L^p(\Omega; \mathbb{R}^{\mathcal{N}(l)})$. From the definition of $|\cdot|_{l,p}$ it follows that the operator $P := \nabla^l : \mathcal{W}^{l,p} \to X$ is a linear isometry from $\mathcal{W}^{l,p}$ into a linear subspace V of X. In particular, ∇^l is injective. Thus a linear functional $\tilde{L}$ on V is defined by

$$\tilde{L}(Pu) := L(u), \qquad u \in \mathcal{W}^{l,p} \; .$$

Since P is an isometry, it follows that

$$\sup\{\tilde{L}(Pu) : \|Pu\|_p = 1\} = \sup\{L(u) : |u|_{l,p} = 1\} = \|L\|_{(\mathcal{W}^{l,p})^*} \; .$$

In particular, $\tilde{L} : V \to \mathbb{R}$ is bounded. From the Hahn-Banach Theorem it follows that there exists an extension $\hat{L} : X \to \mathbb{R}$ such that $\left\|\hat{L}\right\|_{X^*} = \|L\|_{(\mathcal{W}^{l,p})^*}$. Since X^* can be identified with $L^{p_*}(\Omega; \mathbb{R}^{\mathcal{N}(l)})$, there exists $\vec{v} \in L^{p_*}(\Omega; \mathbb{R}^{\mathcal{N}(l)})$ such that $\|\vec{v}\|_{p_*} = \left\|\hat{L}\right\|_{X^*}$ and

$$\hat{L}(Pu) = \sum_{|\vec{\gamma}|=l} \int_\Omega \vec{v}_{\vec{\gamma}} \, \partial^{\vec{\gamma}} u \; .$$

This shows (2.2).

Now let $\vec{w} \in L^{p_*}(\Omega; \mathbb{R}^{\mathcal{N}(l)})$ be another function satisfying (2.2). Then

$$\|\vec{w}\|_{p_*} \geq \sup\left\{ \sum_{|\vec{\gamma}|=l} \int_\Omega \vec{w}_{\vec{\gamma}} \, \partial^{\vec{\gamma}} u : |u|_{l,p} \leq 1 \right\} = \|L\|_{(\mathcal{W}^{l,p})^*} \; .$$

Since on the other hand $\|\vec{v}\|_{p_*} = \|L\|_{(\mathcal{W}^{l,p})^*}$, equation (2.3) follows. $\qquad \square$

Definition 2.2. *Let $l \in \mathbb{N}$ and $1 \leq p < \infty$. The G-norm of $L \in (\mathcal{W}^{l,p})^*$ is defined as*

$$\|L\|_G := \min\left\{ \|\vec{v}\|_{p_*} : Lu = \int_\Omega \vec{v} \cdot \nabla^l u, \; u \in \mathcal{W}^{l,p} \right\} \; .$$

Remark 2.3. *From Theorem 2.1 it follows that $\|L\|_G$ is well-defined, and that it coincides with the dual norm on $(\mathcal{W}^{l,p})^*$. In particular,*

$$\|L\|_G = \sup\left\{ \langle L, u \rangle : u \in \mathcal{W}^{l,p}, \; \left\|\nabla^l u\right\|_p \leq 1 \right\} \; .$$

Remark 2.3 implies that the G-norm is useful for the analysis of regularization functionals with regularization term $\left\|\nabla^l u\right\|_p$. In the following we will generalize the concept to work with arbitrary norm-like regularization terms.

Definition of the $*$-Number. Let X be a linear space. Recall that a functional $\mathcal{R} : X \to \mathbb{R} \cup \{+\infty\}$ is *positively homogeneous*, if

$$\mathcal{R}(tu) = |t|\,\mathcal{R}(u)\,, \qquad u \in X,\ t \in \mathbb{R}\,.$$

Here, the product $0 \cdot (+\infty)$ is defined as 0.

Definition 2.4. *Let $(X, \mathcal{R})$ be a pair consisting of a locally convex space X and a positively homogeneous and convex functional $\mathcal{R} : X \to \mathbb{R} \cup \{+\infty\}$. We define the $*$-number of $u^* \in X^*$ with respect to $(X, \mathcal{R})$ by*

$$\boxed{\|u^*\|_* := \|u^*\|_{*,X,\mathcal{R}} := \sup\left\{\langle u^*, u\rangle_{X^*,X} : \mathcal{R}(u) \leq 1\right\}\,.}$$

Lemma 2.5. *Let $(X, \mathcal{R})$ be as in Definition 2.4, and let*

$$\mathcal{P} := \{p \in X : \mathcal{R}(p) = 0\}\,.$$

Since $\mathcal{R}$ is positively homogeneous, it follows that $\mathcal{P}$ is a linear subspace of X. Denote by

$$\mathcal{P}^\perp := \left\{u^* \in X^* : \langle u^*, p\rangle_{X^*,X} = 0 \text{ for all } p \in \mathcal{P}\right\}\,.$$

Then $\|u^*\|_* = +\infty$ for all $u^* \notin \mathcal{P}^\perp$.

Proof. Let $u^* \notin \mathcal{P}^\perp$. Then there exists $p \in \mathcal{P}$ such that $\langle u^*, p\rangle_{X^*,X} \neq 0$. Since $p \in \mathcal{P}$, it follows that $\mathcal{R}(p) = 0$. Consequently,

$$\|u^*\|_* = \sup\left\{\langle u^*, u\rangle_{X^*,X} : \mathcal{R}(u) \leq 1\right\} \leq \sup_{t\in\mathbb{R}}\langle u^*, tp\rangle_{X^*,X} = +\infty\,,$$

which proves the assertion. $\qquad\square$

The following results show that the $*$-number is a generalization of the G-norm.

Theorem 2.6. *Let U be a subspace of the normed linear space X. Assume that U is a Banach space with norm $\|\cdot\|_U$, and that the inclusion $i : U \to X$ is continuous with respect to the norms $\|\cdot\|_U$ and $\|\cdot\|_X$. Let*

$$\mathcal{R}(u) = \begin{cases} \|u\|_U & \text{if } u \in U\,, \\ +\infty & \text{if } u \in X \setminus U\,. \end{cases}$$

Then

$$\|u^*\|_* = \left\|i^\#(u^*)\right\|_{U^*}\,, \qquad u^* \in X^*\,.$$

Proof. Recall that the adjoint $i^\# : X^* \to U^*$ of the inclusion mapping is defined implicitly by

$$\langle u^*, u\rangle_{X^*,X} = \left\langle i^\#(u^*), u\right\rangle_{U^*,U}\,, \qquad u^* \in X^*,\ u \in U\,.$$

Thus, for $u^* \in X^*$

$$
\begin{aligned}
\|u^*\|_* &= \sup\left\{\langle u^*, v\rangle_{X^*,X} : v \in X,\ \mathcal{R}(v) \leq 1\right\} \\
&= \sup\left\{\langle u^*, u\rangle_{X^*,X} : u \in U,\ \|u\|_U \leq 1\right\} \\
&= \sup\left\{\left\langle i^\#(u^*), u\right\rangle_{U^*,U} : u \in U,\ \|u\|_U \leq 1\right\} \\
&= \left\|i^\#(u^*)\right\|_{U^*}\,.
\end{aligned}
$$

$\qquad\square$

Corollary 2.7. *Assume that either $n = 1$ and $1 \leq p < \infty$, or $n > 1$ and $1 \leq p \leq n/(n-1)$.*

Let $X = L^p(\Omega)$, and $\mathcal{R}(u) = \|\nabla u\|_1$ if $u \in \mathcal{W}^{1,1}$ and $\mathcal{R}(u) = +\infty$ else. Then

$$\|u^*\|_* = \left\|i^\#(u^*)\right\|_G = \left\|i^\#(u^*)\right\|_G$$

where $i^\# : X^* \to (\mathcal{W}^{1,1})^*$ denotes the adjoint of the inclusion $i : \mathcal{W}^{1,1} \to X$.

Proof. This is a consequence of Theorem 2.6 combined with the Sobolev Embedding Theorems (see [Ada75]).   □

Corollary 2.8. *Assume that either $n = 1$ and $1 \leq p < \infty$, or $n > 1$ and $1 \leq p \leq n/(n-1)$. Let $X = L^p(\Omega)$, and $\mathcal{R}(u) = |Du|$ if $u \in \mathcal{TV}^1$, and $\mathcal{R}(u) = +\infty$, else. Then*

$$\|u^*\|_* = \left\|i^\#(u^*)\right\|_G \ ,$$

where $i^\# : X^* \to (\mathcal{W}^{1,p})^*$ denotes the adjoint of the inclusion $i : \mathcal{W}^{1,p} \to X$.

Proof. We have to show that

$$\sup\left\{\int_\Omega u^*\, u : u \in X,\ R(u) \leq 1\right\} = \sup\left\{\int_\Omega u^*\, u : u \in \mathcal{W}^{1,1}, \|\nabla u\|_1 \leq 1\right\} .$$

This equality, however, is a direct consequence of the density result Theorem A.7.
                                                                                       □

3. Characterization of Minimizers of Convex Regularization Functionals with Differentiability Concepts

In the following we characterize the minimizers of the family of functionals

$$\boxed{\mathcal{T}_\alpha(u) := \mathcal{S}(u) + \alpha \mathcal{R}(u)\,, \quad \alpha > 0\,,}$$

where both $\mathcal{S}$ and $\mathcal{R}$ are proper and convex. For this purpose we use special differentiability concepts and make the following assumptions:

Definition 3.1. *Let $\mathcal{F} : X \to \mathbb{R} \cup \{+\infty\}$ be a functional defined on a normed space U. The directional derivative of $\mathcal{F}$ at $u \in \mathcal{D}(\mathcal{F})$ is defined by*

$$\mathcal{F}'(u; h) = \limsup_{t \to 0+} \frac{\mathcal{F}(u + th) - \mathcal{F}(u)}{t} \ .$$

Note that $\mathcal{F}'(u; h)$ can be $+\infty$.

Assumption 3.2. *Assume that X is a real Banach space and that $\mathcal{R}, \mathcal{S} : X \to \mathbb{R} \cup \{+\infty\}$ satisfy:*

(1) *$\mathcal{R}$ and $\mathcal{S}$ are convex, proper and uniformly bounded from below.*
(2) *Assume that $U := \mathcal{D}(\mathcal{S}) \cap \mathcal{D}(\mathcal{R}) \neq \emptyset$.*
(3) *For $u \in U$ and $h \in X$, $\mathcal{R}$ and $\mathcal{S}$ attain directional derivatives $\mathcal{R}'(u; h)$ and $\mathcal{S}'(u; h)$ at u in direction h. Note that the directional derivatives can be $+\infty$.*

Theorem 3.3. *Let $\mathcal{R}$ and $\mathcal{S}$ satisfy Assumption 3.2. Then $u = u_\alpha$ minimizes $\mathcal{T}_\alpha$ if and only if $u \in U$ satisfies*

(3.1) $$\boxed{-\mathcal{S}'(u; h) \leq \alpha \mathcal{R}'(u; h)\,, \quad h \in X\,.}$$

Proof. For $u \in X \backslash U$, by assumption either $\mathcal{R}(u) = +\infty$ or $\mathcal{S}(u) = +\infty$, showing that a minimizer u_α must be an element of U.

Moreover, from the minimality of u_α and the definition of the directional derivatives of $\mathcal{R}$ and $\mathcal{S}$ it follows that

$$0 \leq \liminf_{\varepsilon \to 0^+} \left(\frac{\mathcal{S}(u_\alpha + \varepsilon h) - \mathcal{S}(u_\alpha)}{\varepsilon} + \alpha \frac{\mathcal{R}(u_\alpha + \varepsilon h) - \mathcal{R}(u_\alpha)}{\varepsilon} \right)$$

$$\leq \limsup_{\varepsilon \to 0^+} \left(\frac{\mathcal{S}(u_\alpha + \varepsilon h) - \mathcal{S}(u_\alpha)}{\varepsilon} \right) + \alpha \limsup_{\varepsilon \to 0^+} \left(\frac{\mathcal{R}(u_\alpha + \varepsilon h) - \mathcal{R}(u_\alpha)}{\varepsilon} \right)$$

$$= \mathcal{S}'(u_\alpha; h) + \alpha \mathcal{R}'(u_\alpha; h), \quad h \in X,$$

showing (3.1).

To prove the converse direction we note that from the convexity of $\mathcal{S}$, $\mathcal{R}$ and (3.1) it follows that

$$\big(\mathcal{S}(u + h) - \mathcal{S}(u)\big) + \alpha\big(\mathcal{R}(u + h) - \mathcal{R}(u)\big) \geq \mathcal{S}'(u; h) + \alpha \mathcal{R}'(u; h) \geq 0, \quad h \in X.$$

Thus $u \in U$ satisfying (3.1) is a global minimizer. $\qquad\square$

The following result is an immediate consequence of Theorem 3.3.

Corollary 3.4. *Let Assumption 3.2 hold. Then*

$$-\mathcal{S}'(0; h) \leq \alpha \mathcal{R}'(0; h), \quad h \in X$$

if and only if $0 \in \arg\min \mathcal{T}_\alpha$.

Remark 3.5.   • *Let $\mathcal{R}$ be positively homogeneous. Then, the definition of u_α shows that*

$$\mathcal{S}(u_\alpha) + \alpha \mathcal{R}(u_\alpha) \leq \mathcal{S}(u_\alpha + \varepsilon(\pm u_\alpha)) + \alpha(1 \pm \varepsilon)\mathcal{R}(u_\alpha), \quad 0 < \varepsilon < 1,$$

and therefore

$$\mp \alpha \mathcal{R}(u_\alpha) \leq \liminf_{\varepsilon \to 0^+} \frac{1}{\varepsilon} \left(\mathcal{S}(u_\alpha + \varepsilon(\pm u_\alpha)) - \mathcal{S}(u_\alpha) \right).$$

The passage to the limit gives

$$-\mathcal{S}'(u_\alpha; u_\alpha) \leq \alpha \mathcal{R}(u_\alpha) \leq \mathcal{S}'(u_\alpha; -u_\alpha).$$

In particular, if $\mathcal{S}$ is Gâteaux-differentiable, then

(3.2) $-\mathcal{S}'(u_\alpha)u_\alpha = \alpha \mathcal{R}(u_\alpha).$

• *More general, if $\mathcal{R}$ satisfies*

$$\mathcal{R}((1 + \varepsilon)u_\alpha) \leq (1 + p\varepsilon)\mathcal{R}(u_\alpha) + o(\varepsilon),$$

then

$$-\mathcal{S}'(u_\alpha; u_\alpha) \leq \alpha p \mathcal{R}(u_\alpha) \leq \mathcal{S}'(u_\alpha; -u_\alpha).$$

In particular if $\mathcal{S}$ is Gâteaux-differentiable, then

$$-\mathcal{S}'(u_\alpha)u_\alpha = \alpha p \mathcal{R}(u_\alpha).$$

Remark 3.6. *Assume additionally that $\mathcal{S}'(u; \cdot)$ is positively homogeneous. Since $\mathcal{R}$ is convex, for all $u \in U$, $h \in X$ we have $\mathcal{R}'(u; h) \leq \mathcal{R}(u + h) - \mathcal{R}(u)$. Consequently it follows from (3.1) that*

(3.3) $\boxed{-\mathcal{S}'(u; h) \leq \alpha\big(\mathcal{R}(u + h) - \mathcal{R}(u)\big), \quad h \in X.}$

Replacing h by εh with $\varepsilon > 0$ it follows from (3.3) that

$$-\mathcal{S}'(u;h) \le \alpha \limsup_{\varepsilon \to 0^+} \frac{\mathcal{R}(u + \varepsilon h) - \mathcal{R}(u)}{\varepsilon} \le \alpha \mathcal{R}'(u;h), \quad h \in X.$$

Thus (3.1) and (3.3) are equivalent.

Analytical Examples. In the following we apply Theorem 3.3 to characterize minimizers of different regularization functionals. Before that we summarize derivatives of convex functionals used in this section:

- Assume that $p \ge 1$ and $X = L^p(\Omega)$. Let

$$\mathcal{S}_p : X \to \mathbb{R} \cup \{+\infty\}, \quad u \to \frac{1}{p}\int_\Omega |u - u^\delta|^p.$$

The directional derivative of $\mathcal{S}_p$ at $\tilde{u} \in X$ in direction $h \in X$ is given by

$$(3.4) \quad \begin{aligned} \mathcal{S}_p'(\tilde{u};h) &= \int_\Omega |\tilde{u} - u^\delta|^{p-1} \operatorname{sgn}(\tilde{u} - u^\delta)h \quad \text{if } p > 1, \\ \mathcal{S}_1'(\tilde{u};h) &= \int_{\{\tilde{u} \ne u^\delta\}} \operatorname{sgn}(\tilde{u} - u^\delta)h + \int_{\{\tilde{u} = u^\delta\}} |h|. \end{aligned}$$

- Let $l \in \mathbb{N}$, then the *directional derivative* of $\mathcal{R}_l$ at 0 satisfies

$$\mathcal{R}_l'(0;h) = \mathcal{R}_l(h), \qquad h \in X.$$

Example 3.7. *We consider the pair $(X = L^2(\Omega), \mathcal{R}_1)$. Let $u^\delta \in X$. Corollary 3.4 implies that $u_\alpha = 0 = \arg\min \mathcal{T}_{\alpha,u^\delta}^{2,1}$ if and only if*

$$\int_\Omega u^\delta h = -\mathcal{S}_2'(0;h) \le \alpha \mathcal{R}_1'(0;h) = \alpha |Dh|, \qquad h \in X,$$

which is equivalent to

$$(3.5) \quad \boxed{\left| \int_\Omega u^\delta h \right| \le \alpha |Dh|, \quad h \in X.}$$

Equation (3.5) is equivalent to $\|u^\delta\|_* \le \alpha$, where $\|\cdot\|_* = \|\cdot\|_{*,X,\mathcal{R}_1}$.
  From (3.2) it follows that

$$(3.6) \quad \boxed{\alpha |Du_\alpha| = -\int_\Omega (u_\alpha - u^\delta)u_\alpha.}$$

Taking into account inequality (3.3) and the triangle inequality it follows that

$$(3.7) \quad -\int_\Omega (u_\alpha - u^\delta)h \le \alpha(|D(u_\alpha + h)| - |D(u_\alpha)|) \le \alpha |Dh|, \quad h \in X.$$

Equation (3.7) implies that $\|u_\alpha - u^\delta\|_* \le \alpha$. Conversely, it follows from (3.6) that $\|u_\alpha - u^\delta\|_* \ge \alpha$. In [Mey01] it was shown that (3.7) and the condition $\|u_\alpha - u^\delta\|_G = \alpha$ uniquely characterize the minimizer u_α of the ROF-functional in the case $\|u^\delta\|_G > \alpha$.

The following example concerns L^1-TV minimization. In this case $\mathcal{T}_{\alpha,u^\delta}^{1,1}$ can have multiple minimizers since the functional is not strictly convex.

Example 3.8. *We consider the pair* $(X = L^1(\Omega), \mathcal{R}_1)$. *Let* $u^\delta \in X$. *From Theorem 3.3 and (3.4) it follows that* $u \in \arg\min \mathcal{T}_{\alpha,u^\delta}^{1,1}$ *if and only if*

$$(3.8) \qquad -\int_{\{u \neq u^\delta\}} \operatorname{sgn}(u - u^\delta)h - \int_{\{u = u^\delta\}} |h| \leq \alpha \mathcal{R}_1'(u; h), \quad h \in X.$$

In particular, $u_\alpha \equiv 0$ *if and only if*

$$\int_{\{0 \neq u^\delta\}} \operatorname{sgn}(u^\delta)h - \int_{\{0 = u^\delta\}} |h| \leq \alpha |Dh|, \quad h \in X.$$

Using this estimate both with h *and* $-h$ *it follows that*

$$\left(\left| \int_{\{0 \neq u^\delta\}} \operatorname{sgn}(u^\delta)h \right| - \int_{\{0 = u^\delta\}} |h| \right)^+ \leq \alpha |Dh|, \quad h \in X.$$

These results have been derived in [SchYinOsh05] *using a different mathematical methodology.*

In [ChaEse05] *minimizers of the functional* $\mathcal{T}_{\alpha,u^\delta}^{1,1}$ *with* $u^\delta = \chi_E$, $E \subset \Omega$ *have been calculated analytically. Some of the results can be reproduced from the considerations above. From Corollary 3.4 it follows that* $0 \in \arg\min \mathcal{T}_{\alpha,u^\delta}^{1,1}$ *if and only if*

$$(3.9) \qquad \left(\left| \int_E h \right| - \int_{\mathbb{R}^n \setminus E} |h| \right)^+ \leq \alpha |Dh|, \quad h \in X.$$

Taking $h = u^\delta$ *it follows from (3.9) that in this case* $\mathcal{L}^n(E)/\operatorname{Per}(E; \Omega) \leq \alpha$.

Using Remark 3.6 it follows from (3.8) that $u_\alpha = u^\delta = \chi_E$ *if*

$$-\int_E |h| \leq \alpha \left(|D(u^\delta + h)| - |Du^\delta| \right), \quad h \in X.$$

Taking $h = -\chi_E$ *shows that* $\alpha \leq \mathcal{L}^n(E)/\operatorname{Per}(E; \Omega)$.

The next example concerns an inverse problem.

Example 3.9. *Assume that* Ω *is bocL. We consider the pair* $(X = L^2(\Omega), \mathcal{R}_1)$ *and assume that* L *is a bounded linear operator on* X. *This implies that the functional* $\mathcal{S}(u) := \frac{1}{2} \|Lu - v^\delta\|_2^2$ *is convex. We consider minimization of*

$$\mathcal{T}_{\alpha,v^\delta}(u) := \mathcal{S}(u) + \alpha \mathcal{R}_1(u) = \frac{1}{2} \|Lu - v^\delta\|_2^2 + \alpha \mathcal{R}_1(u), \qquad u \in X.$$

From Theorem 3.3 it follows that $u_\alpha \equiv 0$ *if and only if* $\|L^*v^\delta\|_* \leq \alpha$, *where* L^* *is the adjoint of* L *on* X. *If* $v^\delta = Lu^\dagger$, *then this means that* $\|L^*Lu^\dagger\|_* \leq \alpha$.

Similarly as in the case of denoising zero is a minimizer if $\|L^*Lu^\dagger\|_* \leq \alpha$. *This property is not available for quadratic regularization. To see this, we consider the regularization functional*

$$\mathcal{T}_{\alpha,v^\delta}(u) := \|Lu - v^\delta\|_2^2 + \alpha \|\nabla u\|_2^2.$$

From Theorem 3.3 it follows that 0 *is a minimizing element of this* $\mathcal{T}_{\alpha,v^\delta}$ *if and only if* $L^*v^\delta = 0$. *In addition, if* $v^\delta = Lu^\dagger$, *then this means that* $u^\dagger$ *is an element of the null-space of* L. *Therefore, aside from trivial situations, it is* **not** *possible to completely remove data as it is the case for total variation regularization.*

The following example shows that variational regularization methods with a semi-norm penalization are capable of obtaining zero as a minimizer for non trivial data.

Example 3.10. *We consider the pair* $(X = L^2(\mathbb{R}^n), \mathcal{R})$ *with* $\mathcal{R}(u) = \|\nabla u\|_2$, *if* $u \in \mathcal{W}^{1,2}$, *and* $\mathcal{R}(u) = +\infty$ *else. Assume that* $L : \mathcal{W}^{1,2} \to X$ *is a bounded linear operator. For* $u \in \mathcal{W}^{1,2}$ *and* $h \in L^2(\mathbb{R}^n)$ *the directional derivative of* $\mathcal{R}$ *is given by*

$$\mathcal{R}'(u; h) = \begin{cases} \|\nabla h\|_2 & \text{if } u = 0, \\ \dfrac{1}{\|\nabla u\|_2} \int_\Omega \nabla u \, \nabla h & \text{if } u \neq 0. \end{cases}$$

From Theorem 3.3 it follows that zero is a minimizer of

$$\mathcal{T}_{\alpha, v^\delta}(u) := \frac{1}{2} \left\| Lu - v^\delta \right\|_2^2 + \alpha \mathcal{R}(u), \quad u \in X,$$

if and only if

$$\int_\Omega L^* v^\delta \, h \leq \alpha \left\| \nabla h \right\|_2, \quad h \in \mathcal{W}^{1,2}.$$

Here, L^* *denotes the adjoint of* L *on* X. *This is equivalent to stating that* $\left\| L^* v^\delta \right\|_* \leq \alpha$. *Applying Theorem 2.6 it follows that zero minimizes* $\mathcal{T}_{\alpha, v^\delta}$, *if and only if* $\left\| i^\#(L^*)(v^\delta) \right\|_{(\mathcal{W}^{1,2})^*} \leq \alpha$, *where* $i^\#$ *denotes the dual-adjoint of the inclusion* $i : \mathcal{W}^{1,2} \to L^2(\mathbb{R}^n)$.

4. FENCHEL DUALITY

In this Section we use Fenchel's duality theorem to characterize minimizers of convex regularization functionals. Below we review basic concepts from functional analysis (see for instance Ekeland & Temam [EkeTem76] and Aubin [Aub],[Aub91]).

Definition 4.1. *Assume that* X *is a locally convex space (for instance a Banach space). The* Fenchel transform *of a functional*

$$\mathcal{S} : X \to \mathbb{R} \cup \{+\infty\}, \quad u \mapsto \mathcal{S}(u)$$

is defined by

$$\mathcal{S}^* : X^* \to \mathbb{R} \cup \{+\infty\}, \quad u^* \mapsto \mathcal{S}^*(u^*) := \sup_{u \in X} \{\langle u^*, u \rangle - \mathcal{S}(u)\}$$

where $\langle \cdot, \cdot \rangle$ *denotes the bilinear pairing with respect to* X^* *and* X.

For a definition of the Fenchel transform in a finite dimensional space setting we refer to Rockafellar [Roc70] and for the infinite dimensional setting we refer to Ekeland & Temam [EkeTem76] and Aubin [Aub91].

Theorem 4.2. *Let* $\mathcal{S}, \mathcal{R}$ *be convex and lower semi continuous functionals from a locally convex space* X *into* $\mathbb{R} \cup \{+\infty\}$.
If $\tilde{u}$ *is a solution of*

(4.1)
$$\inf_{u \in X} \{\mathcal{S}(u) + \mathcal{R}(u)\},$$

$\tilde{u}^*$ *is a solution of*

(4.2)
$$\sup_{u^* \in X^*} \{-\mathcal{S}^*(u^*) - \mathcal{R}^*(-u^*)\}$$

and

$$(4.3) \qquad \inf_{u \in X} \{\mathcal{S}(u) + \mathcal{R}(u)\} = \sup_{u^* \in X^*} \{-\mathcal{S}^*(u^*) - \mathcal{R}^*(-u^*)\} < +\infty$$

then $\tilde{u} \in X$ and $\tilde{u}^* \in X^*$ satisfy the extremality relation

$$(4.4) \qquad \mathcal{S}(\tilde{u}) + \mathcal{R}(\tilde{u}) + \mathcal{S}^*(\tilde{u}^*) + \mathcal{R}^*(-\tilde{u}^*) = 0$$

which is equivalent to

$$\tilde{u}^* \in \partial \mathcal{S}(\tilde{u}) \ and \ -\tilde{u}^* \in \partial \mathcal{R}(\tilde{u})$$

or

$$\tilde{u} \in \partial \mathcal{S}^*(\tilde{u}^*) \ and \ -\tilde{u} \in \partial \mathcal{R}^*(\tilde{u}^*).$$

Conversely, if $u \in X$ and $u^* \in X^*$ satisfy (4.4), then u, u^* satisfy (4.1) and (4.2), respectively.

Proof. Follows from [EkeTem76, Proposition 2.4, Proposition 4.1, Remark 4.2 in Chapter 3] □

Theorem 4.3. *Assume that $\mathcal{S} + \mathcal{R}$ is convex, that $\inf(\mathcal{S} + \mathcal{R})$ is finite and that there exists $u_0 \in U$ such that $\inf(\mathcal{S}(u_0) + \mathcal{R}(u_0)) < \infty$, and the function $u^\delta \mapsto \mathcal{S}(u_0 - u^\delta) + \mathcal{R}(u_0)$ being continuous at u_0, then equation (4.3) holds.*

Proof. See [EkeTem76, Th. 4.1 in Chapter 3]. □

The next result summarizes some basic properties of the Fenchel transform:

Theorem 4.4. *Let $\mathcal{S} : U \to \mathbb{R} \cup \{+\infty\}$ be proper. Then:*

(1) *The functional $\mathcal{S}^*$ is weak* lower semi-continuous and convex.*
(2) *For every $\alpha > 0$*

$$(\alpha \mathcal{S})^*(u^*) = \alpha \mathcal{S}^*(u^*/\alpha) .$$

(3) *For every $t \in \mathbb{R}$*

$$(\mathcal{S} + t)^*(u^*) = \mathcal{S}^*(u^*) - t .$$

(4) *Let*

$$\mathcal{T}(u) = \mathcal{S}(u - u_0)$$

for some $u_0 \in U$. Then

$$\mathcal{T}^*(u^*) = \mathcal{S}^*(u^*) + \langle u^*, u_0 \rangle .$$

Proof. See [EkeTem76, Sec. 1.4]. □

Lemma 4.5. *Assume that U is a Banach space and that $\phi : \mathbb{R} \to \mathbb{R} \cup \{+\infty\}$ is a lower semicontinuous convex and proper function satisfying $\phi(-t) = \phi(t)$ for all $t \in \mathbb{R}$. Define*

$$\mathcal{S} : U \to \mathbb{R} \cup \{+\infty\}, \qquad \mathcal{S}(u) := \phi(\|u\|_U) .$$

The Fenchel transform of $\mathcal{S}$ is

$$\mathcal{S}^* : U^* \to \mathbb{R} \cup \{+\infty\}, \qquad \mathcal{S}^*(u^*) = \phi^*(\|u^*\|_{U^*}),$$

where $\phi^* : \mathbb{R} \to \mathbb{R} \cup \{+\infty\}$ is the Fenchel transform of ϕ.

Proof. See [EkeTem76, Ch. 1, Prop. 4.2]. □

We use Theorem 4.4 and Lemma 4.5 to compute the Fenchel transform in one simple but important case:

Example 4.6. *Let $1 \leq p < \infty$. We assume that Ω is bocL or $\Omega = \mathbb{R}^n$. We use Theorem 4.4 to calculate the Fenchel transform of*

$$\mathcal{S}_p : X = L^p(\Omega) \to \mathbb{R} \cup \{+\infty\}, u \to \frac{1}{p} \int_\Omega |u - u^\delta|^p =: \mathcal{T}_p(u - u^\delta), \quad p \geq 1.$$

For $u^* \in L^{p_*}(\Omega) \cong (L^p(\Omega))^*$

$$\mathcal{T}_p^*(u^*) := \sup \left\{ \int_\Omega uu^* - \frac{1}{p} \int_\Omega |u|^p : u \in L^p(\Omega) \right\}.$$

The supremum is attained for $u_\alpha \in L^p(\Omega)$ satisfying

$$u_\alpha^* = |u_\alpha|^{p-1} \, sgn(u_\alpha) \in L^{p_*}(\Omega).$$

Then for $p > 1$, the Fenchel transform of $\mathcal{T}_p$ is given by

$$\mathcal{T}_p^*(u^*) = \frac{1}{p_*} \int_\Omega |u^*|^{p_*}$$

and consequently

(4.5) $$\mathcal{S}_p^*(u^*) = \mathcal{T}_p^*(u^*) + \langle u^*, u^\delta \rangle_{L^{p_*}, L^p} = \frac{1}{p_*} \int_\Omega |u^*|^{p_*} + \int_\Omega u^* u^\delta.$$

For $p = 1$,

$$\mathcal{S}_1 : L^1(\Omega) \to \mathbb{R}, \quad u \to |u - u^\delta|_{L^1(\Omega)}$$

we have

$$\mathcal{S}_1^*(u^*) = \begin{cases} \int_\Omega u^\delta u^* & \text{if } \|u^*\|_{L^\infty} \leq 1, \\ +\infty & \text{else.} \end{cases}$$

$*$-Number and Fenchel Transform. Below we relate the $*$-number with the Fenchel transform of $\mathcal{R}_l$. We show that Fenchel duality is a more general concept than the $*$-number, and consequently by Corollary 2.7 is more general than the G-norm.

$$\boxed{\text{Fenchel-duality} \ \Rightarrow \ *\text{-Number} \Rightarrow G\text{-norm}.}$$

Theorem 4.7. *Assume that $l \in \mathbb{N}$, $1 \leq p < \infty$, and $\alpha > 0$. We consider the pair $(X = L^p(\Omega), \mathcal{R}_l)$. Then*

$$(\alpha \mathcal{R}_l)^*(u^*) = \begin{cases} 0 & \text{if } \|u^*\|_* \leq \alpha, \\ +\infty & \text{else.} \end{cases}$$

Moreover $\|u^*\|_* = +\infty$ if $\int_\Omega u^* p \neq 0$ for some polynomial $p \in \mathcal{P}_{l-1} \cap L^p(\Omega)$ of degree at most $l - 1$.

Proof. From the definition of the $*$-number it follows that $\|u^*\|_* \leq \alpha$ if and only if

$$\int_\Omega u^* u \leq \alpha, \qquad u \in X, \ \mathcal{R}_l(u) \leq 1.$$

This is equivalent to

$$\int_\Omega u^* u \leq \alpha \mathcal{R}_l(u), \qquad u \in X.$$

Now recall that

(4.6) $$(\alpha \mathcal{R}_l)^*(u^*) = \sup \left\{ \int_\Omega u^* u - \alpha \mathcal{R}_l(u) : u \in X \right\}.$$

Thus, if $\|u^*\|_* \leq \alpha$ it follows that $\int_\Omega u^* u - \alpha \mathcal{R}_l(u) \leq 0$ for all $u \in X$, and consequently $(\alpha \mathcal{R}_l)^*(u^*) \leq 0$. Choosing $u = 0$ in the right hand side of (4.6) shows that $(\alpha \mathcal{R}_l)^*(u^*) = 0$.

If on the other hand $\|u^*\|_* > \alpha$, then it follows that there exists $u_0 \in X$ with $\int_\Omega u^* u_0 - \alpha \mathcal{R}_l(u_0) > 0$. Consequently,

$$(\alpha \mathcal{R}_l)^*(u^*) \geq \sup\left\{ \int_\Omega u^* t u_0 - \alpha \mathcal{R}_l(t u_0) : t \in \mathbb{R} \right\} = +\infty \, .$$

The remaining part of the assertion follows from Lemma 2.5. □

Remark 4.8. *Assume that $1 \leq p < \infty$, $l \in \mathbb{N}$. Consider the pair $(X = L^p(\Omega), \mathcal{R}_l)$. Let u_α and u_α^* be extrema of $T_{\alpha,u^\delta}^{p,l} = S_p + \alpha \mathcal{R}_l$, $(T_{\alpha,u^\delta}^{p,l})^*$, respectively, where $(T_{\alpha,u^\delta}^{p,l})^*(u^*) := S_p^*(u^*) + \alpha \mathcal{R}_l^*(-u^*)$. Then from Theorem 4.3 it follows that*

$$\inf_{u \in X}\left\{ T_{\alpha,u^\delta}^{p,l} \right\} = \inf_{u^* \in X^*}\left\{ (T_{\alpha,u^\delta}^{p,l})^* \right\} \, .$$

Hence from Theorem 4.2, Theorem 4.7 and Example 4.6 it follows that for $p > 1$

$$(4.7) \quad S_p(u_\alpha) + (S_p)^*(u_\alpha^*) = \int_\Omega \left(\frac{1}{p}\left| u_\alpha - u^\delta \right|^p + \frac{1}{p_*}|u_\alpha^*|_*^p + u_\alpha^* u^\delta \right) = -\alpha \mathcal{R}_l(u_\alpha) \, ,$$

and for $p = 1$

$$(4.8) \qquad S_1(u_\alpha) + (S_1)^*(u_\alpha^*) = \int_\Omega \left(\left| u_\alpha - u^\delta \right| + u_\alpha^* u^\delta \right) = -\alpha \mathcal{R}_l(u_\alpha) \, .$$

Using that

$$(4.9) \qquad u_\alpha^* \in \partial S_p(u_\alpha) = \left| u_\alpha - u^\delta \right|^{p-1} \operatorname{sgn}(u_\alpha - u^\delta) = \mathcal{J}_p(u_\alpha - u^\delta)$$

we can simplify (4.7) and (4.8) and obtain

$$(4.10) \qquad -\alpha \left| D^l u_\alpha \right| = \int_\Omega u_\alpha^* u_\alpha \, .$$

This is a generalization of (3.6) for arbitrary $l \in \mathbb{N}$ and $p \geq 1$.

5. ANALYTICAL EXAMPLES

We apply duality arguments to analytically calculate minimizers of the functionals

$$T_{\alpha,u^\delta}^{l,p} : X = L^p(\Omega) \to \mathbb{R} \cup \{+\infty\} \, , \quad u \mapsto S_p(u) + \alpha \mathcal{R}_l(u)$$

where $\Omega = (-1, 1)$, $p = 1, 2$, and $l = 1, 2$ and set $(T_{\alpha,u^\delta}^{p,l})^*(u^*) := S_p^*(u^*) + \mathcal{R}_l(-u^*)$.

In the following we denote by u_α and u_α^* minimizers of $T_{\alpha,u^\delta}^{p,l}$ and $(T_{\alpha,u^\delta}^{p,l})^*$, respectively. For $u \in L^1(\Omega)$ we define

$$\rho_u^l(x) := (-1)^l \int_{-1}^x \left(\cdots \int_{-1}^{t_2} u(t_1) \, dt_1 \ldots \right) dt_l \, , \qquad x \in \Omega \, .$$

We show below that u_α is piecewise either a polynomial of order $l - 1$ or equals u^δ. Moreover, we prove that $\rho_{u_\alpha^*}^l$ shows structural behavior of u_α: if $\left| \rho_{u_\alpha^*}^l(x) \right| = \alpha$ and $\left| \rho_{u_\alpha^*}^l \right| < \alpha$ in a neighborhood of x (excluding the point x itself), then in the case $l = 1$, u_α is discontinuous and for $l = 2$, u_α bends (that is, the derivative is discontinuous) at x. Compare Figure 1.

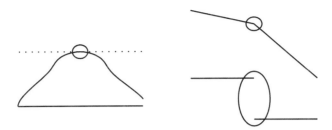

FIGURE 1.   *Left:* $\rho^l_{u^*_\alpha}$ touching the α-tube. *Right:* u_α is continuous (case $\mathcal{T}^{1,p}_{\alpha,u^\delta}$) or bends (case $\mathcal{T}^{2,p}_{\alpha,u^\delta}$) at that x-number where $\rho^l_{u^*_\alpha}$ touches the $\alpha-$tube.

Theorem 5.1. *Let* $1 \le p < \infty$. *Then* $(\alpha\mathcal{R}_l)^*(u^*) < +\infty$, *if and only if* $\rho^l_{u^*} \in W^{l,1}_0(\Omega)$, *and* $\left\|\rho^l_{u^*}\right\|_\infty \le \alpha$.

Proof. Note first that $\rho^l_{u^*} \in W^{l,1}_0(\Omega)$ if and only if $(\rho^l_{u^*})^{(i)}(-1) = (\rho^l_{u^*})^{(i)}(1) = 0$, $1 \le i \le l-1$. Moreover, from the definition of $\rho^l_{u^*}$ it directly follows that $(\rho^l_{u^*})^{(i)}(-1) = 0$ for all $1 \le i \le l-1$.

From Theorem 4.7 it follows that $(\alpha\mathcal{R}_l)^*(u^*) < +\infty$, if and only if $\|u^*\|_* = \|u^*\|_{*,X,\mathcal{R}_l} \le \alpha$. This is the case, if and only if

$$(5.1) \qquad \int_\Omega u^* u \le \alpha, \qquad u \in \mathrm{TV}^l(\Omega), \ \left|D^l u\right| \le 1 .$$

From Lemma A.9 it follows that for every $u \in \mathrm{TV}^l(\Omega)$ there exists a sequence $(u_k) \in C^\infty(\bar\Omega)$ with $\|u_k - u\|_1 \to 0$, and $\left|D^l u_k\right| \to \left|D^l u\right|$. Consequently, equation (5.1) is equivalent to

$$\int_\Omega u^* u \le \alpha, \qquad u \in C^\infty(\bar\Omega), \ \left|D^l u\right| \le 1 .$$

Inserting the definition of $\rho^l_{u^*}$ and integrating by parts shows that this in turn is equivalent to

(5.2)
$$+ \int_\Omega \rho^l_{u^*}\, u^{(l)} + \sum_{i=0}^{l-1}(-1)^{l+1}(\rho^l_{u^*})^{(i)}(1)\, u^{(l-1-i)}(1) \le \alpha, \qquad u \in C^\infty(\bar\Omega), \ \left\|u^{(l)}\right\|_1 \le 1 .$$

Now assume that $(\alpha\mathcal{R}_l)^*(u^*) < +\infty$. Then in particular inequality (5.2) holds for all polynomials $p \in \mathcal{P}_{l-1}$ of degree at most $l-1$. This implies that $(\rho^l_{u^*})^{(i)}(1) = 0$ for all $0 \le i \le l-1$, that is, $\rho^l_{u^*} \in W^{l,1}_0(\Omega)$. Consequently, inequality (5.2) reduces to

$$\int_\Omega \rho^l_{u^*}\, u^{(l)} \le \alpha, \qquad u \in C^\infty(\bar\Omega), \ \left\|u^{(l)}\right\|_1 \le 1 .$$

Since $\{u^{(l)} : u \in C^\infty(\bar\Omega)\}$ is dense in $L^1(\Omega)$, this shows that $\left\|\rho^l_{u^*}\right\|_{1_*} = \left\|\rho^l_{u^*}\right\|_\infty \le \alpha$.

The converse implication is a direct consequence of (5.2). $\square$

Theorem 5.2. *Let* $\Omega = (-1,1)$ *and* $p = 1,2$. *Assume that* u_α *and* u^*_α *are minimizers of* $\mathcal{T}^{l,p}_{\alpha,u^\delta}$, $(\mathcal{T}^{l,p}_{\alpha,u^\delta})^*$. *Then* u_α *and* $\rho^l_{u^*_\alpha}$ *satisfy the following relations:*

(a): If $(a, b) \subset \Omega$ such that $\left|\rho^l_{u^*_\alpha}(x)\right| < \alpha$ for all $x \in (a, b)$, then u_α is a polynomial of order $l - 1$ in (a, b).

(b): If $(a, b) \subset \Omega$ such that $\left|\rho^l_{u^*_\alpha}(x)\right| = \alpha$ for all $x \in (a, b)$, then $u_\alpha = u^\delta$ in (a, b).

(c): Let $-1 \leq x_1 < x_2 < x_3 < x_4 \leq 1$ be such that $\rho^l_{u^*_\alpha}(x) = \alpha$ for all $x \in [x_2, x_3]$ and $\left|\rho^l_{u^*_\alpha}(x)\right| < \alpha$ for $x \in (x_1, x_4) \backslash [x_2, x_3]$.

Then

$$(5.3) \qquad u_\alpha = \begin{cases} \displaystyle\sum_{i=0}^{l-1} c_i x^i \ \text{ in } (x_1, x_2), \\ \qquad u^\delta \ \text{ in } (x_2, x_3), \\ \displaystyle\sum_{i=0}^{l-1} d_i x^i \ \text{ in } (x_3, x_4). \end{cases}$$

Additionally we have

$$d_{l-1} \leq c_{l-1}.$$

Moreover ${u_\alpha}^{(l-1)} = u^{\delta\,(l-1)}$ is monoton decreasing in (x_2, x_3).

(c'): For $x_2 = x_3$ we have

$$u_\alpha(x) = \begin{cases} \sum_{i=0}^{l-1} c_i x^i & \text{for } x \in (x_1, x_2), \\ \sum_{i=0}^{l-1} d_i x^i & \text{for } x \in (x_2, x_4) \end{cases}$$

and

$$d_{l-1} \leq c_{l-1}.$$

Proof. (a) Since $\mathcal{C}_0(\Omega)$ is dense in $L^1(\Omega)$ we can approximate u^*_α by $\tilde{u}^*_\alpha \in \mathcal{C}_0(\Omega)$ with $\|u^*_\alpha - \tilde{u}^*_\alpha\|_{L^1} < \epsilon$. Define $\psi := \rho^l_{\tilde{u}^*_\alpha} + \omega$, such that $\|\psi\|_\infty \leq \alpha$ and $\omega \in C^l_0(\Omega)$ with $\omega(x) = 0$ for $x \in \Omega \setminus (a, b)$.

It follows from the definition of the total variation seminorm $\left|D^l u_\alpha\right|$ that

$$\int_\Omega \psi^{(l)} u_\alpha \leq \alpha \left|D u_\alpha\right| \ .$$

Using (4.10) it follows $\alpha \left|D^l u_\alpha\right| = -\int_\Omega u^*_\alpha u_\alpha = \int_\Omega {\rho^l_{u^*_\alpha}}' u_\alpha^{(l-1)}$ and consequently

$$-\int_\Omega (\psi' - {\rho^l_{u^*_\alpha}}') u_\alpha^{(l-1)} = -\left(\int_\Omega ({\rho^l_{\tilde{u}^*_\alpha}}' - {\rho^l_{u^*_\alpha}}') u_\alpha^{(l-1)} + \int_a^b \omega' u_\alpha^{(l-1)} \right) \leq 0 \ .$$

Since $u_\alpha^{(l-1)} \in TV(\Omega) \subset L^\infty(\Omega)$ we have:

$$(5.4) \qquad -\int_a^b \omega' u_\alpha^{(l-1)} \leq \left| \int_\Omega ({\rho^l_{\tilde{u}^*_\alpha}}' - {\rho^l_{u^*_\alpha}}') u_\alpha^{(l-1)} \right| \leq \left\| {\rho^l_{\tilde{u}^*_\alpha}}' - {\rho^l_{u^*_\alpha}}' \right\|_{L^1} \left\| u_\alpha^{(l-1)} \right\|_{L^\infty}$$

For $l = 1$, $\left\| \rho^l_{\tilde{u}^*_\alpha}{}' - \rho^l_{u^*_\alpha}{}' \right\|_{L^1} = \left\| \tilde{u}^*_\alpha - u^*_\alpha \right\|_{L^1} < \epsilon$, and for $l > 1$ we obtain

$$
\begin{aligned}
\left\| \rho^l_{\tilde{u}^*_\alpha}{}' - \rho^l_{u^*_\alpha}{}' \right\|_{L^\infty} &= \sup_{x \in (-1,1]} \left\{ \left| \int_{-1}^x \left(\cdots \int_{-1}^{t_2} (\tilde{u}^*_\alpha(t_1) - u^*_\alpha(t_1))\, dt_1 \cdots \right) dt_{l-1} \right| \right\} \\
&\leq \sup_{x \in (-1,1]} \left\{ \int_{-1}^x \left(\cdots \int_{-1}^{t_2} \left\| \tilde{u}^*_\alpha - u^*_\alpha \right\|_{L^1} dt_1 \cdots \right) dt_{l-1} \right\} \\
&\leq 2 \left\| \tilde{u}^*_\alpha - u^*_\alpha \right\|_{L^1} .
\end{aligned}
$$

Thus (5.4) becomes

$$
- \int_a^b \omega' u^{(l-1)}_\alpha \leq C\epsilon .
$$

for every $\omega \in C_0^l([a,b])$ and some constant C. Hence $\left| Du^{(l-1)}_\alpha \right|_{[a,b]} \leq C\epsilon$. Thus we conclude that $u^{(l-1)}_\alpha$ is constant.

(b) If $\left| \rho^l_{u^*_\alpha} \right| = \alpha$ in (a,b), then

$$
(-1)^l (\rho^l_{u^*_\alpha})^{(l)}(x) = u^*_\alpha(x) = 0 \quad \text{for } x \in (a,b) .
$$

The Kuhn-Tucker condition $u^*_\alpha \in \partial \mathcal{S}(u_\alpha)$ reads as follows

$$
\begin{aligned}
& u^*_\alpha(x) = -1 \Leftrightarrow u^\delta(x) \geq u_\alpha(x) , \\
(5.5) \quad & u^*_\alpha(x) = 1 \Leftrightarrow u^\delta(x) \leq u_\alpha(x) , \\
& u^*_\alpha(x) \in (-1,1) \Rightarrow u^\delta(x) = u_\alpha(x) \quad \wedge \quad u^\delta(x) = u_\alpha(x) \Rightarrow u^*_\alpha(x) \in [-1,1] ,
\end{aligned}
$$

for $p = 1$ and (4.5) for $p = 2$. Thus it follows in both cases that $u_\alpha = u^\delta$ in (a,b).

(c) From the Kuhn-Tucker condition $-u_\alpha \in \partial \mathcal{R}^* \left(\frac{u^*_\alpha}{\alpha} \right)$, it follows that for all $v^* \in L^{p_*}(\Omega)$

$$
(5.6) \qquad \mathcal{R}^* \left(\frac{v^*}{\alpha} \right) - \mathcal{R}^* \left(\frac{u^*_\alpha}{\alpha} \right) + \int_\Omega u_\alpha (v^* - u^*_\alpha) \geq 0 .
$$

Since by assumption $\left| \rho^l_{u^*_\alpha}(x) \right| < \alpha$ in $(x_1, x_4) \backslash [x_2, x_3]$ it follows from (a) and (b) that u_α is as in (5.3) and hence

$$
u^{(l-1)}_\alpha(x) = \begin{cases} c_{l-1} & \text{for } x \in (x_1, x_2) , \\ u^{\delta(l-1)}(x) & \text{for } x \in (x_2, x_3) , \\ d_{l-1} & \text{for } x \in (x_3, x_4) . \end{cases}
$$

For $x_2 < x_3$, $0 < \epsilon \leq |x_2 - x_1|$ define $\psi \in \mathcal{C}^{l-1}(\Omega)$ with

$$
\begin{aligned}
\psi(x) &= \rho^l_{u^*_\alpha}(x) & \text{for } x \notin (x_2 - \epsilon, x_3) , \\
\psi(x) &< \rho^l_{u^*_\alpha}(x) & \text{for } x \in (x_2 - \epsilon, x_3) , \\
\|\psi\|_{L^\infty} &\leq \alpha ,
\end{aligned}
$$

and define $\eta := \rho^l_{u^*_\alpha}(x_2) - \psi(x_2) > 0$. Then $w^* := (-1)^{(l)} \psi^{(l)} \in L^{p_*}(\Omega)$. With this choice of w^* and the fact that $\mathcal{R}^* \left(\frac{u^*_\alpha}{\alpha} \right) = \mathcal{R}^* \left(\frac{w^*}{\alpha} \right) = 0$ (see (4.7)), it follows from

(5.6) that

$$\int_\Omega u_\alpha(w^* - u_\alpha^*) = -\int_\Omega u_\alpha^{(l-1)}(\psi - \rho_{u_\alpha^*}^l)'$$

$$= -\left(\int_{x_2-\epsilon}^{x_2} u_\alpha^{(l-1)}(\psi - \rho_{u_\alpha^*}^l)' + \int_{x_2}^{x_3} u_\alpha^{(l-1)}(\psi - \rho_{u_\alpha^*}^l)'\right)$$

$$= c_{l-1}\eta - \int_{x_2}^{x_3}(u^\delta)^{(l-1)}(\psi - \rho_{u_\alpha^*}^l)' \geq 0 .$$

With the same argument we obtain

$$d_{l-1}\eta - \int_{x_2}^{x_3}(u^\delta)^{(l-1)}(\psi - \rho_{u_\alpha^*}^l)' \leq 0 ,$$

and thus

$$d_{l-1} \leq c_{l-1} .$$

For arbitrary $(a, b) \subset (x_2, x_3)$ we can find $\psi \in \mathcal{C}_0^{l-1}$ such that ψ' has only one zero x_0 in (a, b),

$$\rho_{u_\alpha^*}^l(x) = \psi(x) \qquad \text{for } x \notin (a, b) ,$$

$$\psi(x) < \rho(x) = \alpha \quad \text{for } x \in (a, b) .$$

Then $w^* := \psi^{(l)} \in L^{p_*}(\Omega)$. With this choice of w^* and the fact that $\mathcal{R}^*\left(\frac{u_\alpha^*}{\alpha}\right) = \mathcal{R}^*\left(\frac{w^*}{\alpha}\right) = 0$, it follows from (5.6) and the fact that $(\rho_{u_\alpha^*}^l)' = 0$ in (a, b) that

$$\int_\Omega u_\alpha(w^* - u_\alpha^*) = -\int_\Omega u_\alpha^{(l-1)}(\psi - \rho_{u_\alpha^*}^l)'$$

$$= -\left(\int_a^{x_0} u_\alpha^{(l-1)}\psi' + \int_{x_0}^b u_\alpha^{(l-1)}\psi'\right) \geq 0 .$$

Since

$$\int_a^{x_0}\psi' = \psi(x_0) - \alpha = -\int_{x_0}^b \psi' < 0$$

we have

$$\inf_{x\in(x_0,b)}(u^\delta)^{(l-1)} \leq \sup_{x\in(a,x_0)}(u^\delta)^{(l-1)}$$

for any $(a, b) \subset (x_2, x_3)$ and $x_0 \in (a, b)$. Thus $(u^\delta)^{(l-1)}$ is decreasing.

(c') In case $x_2 = x_3$ we define $\psi \in \mathcal{C}^{l-1}(\Omega)$ with

$$\psi(x) = \rho_{u_\alpha^*}^l(x) \qquad \text{for } x \notin (x_2 - \epsilon, x_2 + \epsilon) ,$$

$$\psi(x) < \rho_{u_\alpha^*}^l(x) \qquad \text{for } x \in (x_2 - \epsilon, x_2 + \epsilon) ,$$

$$\psi(x_2) = \rho_{u_\alpha^*}^l(x_2) - \eta ,$$

with $0 < \eta < \alpha$ and $\epsilon < \min\{|x_2 - x_1|, |x_4 - x_2|\}$. Then define $w^* := (-1)^l \psi^{(l)} \in L^{p_*}(\Omega)$. It follows from (5.6) that

(5.7) $$\int_\Omega u_\alpha(w^* - u_\alpha^*) = -\int_{x_2-\epsilon}^{x_2+\epsilon} u_\alpha^{(l-1)}(\psi' - \rho_{u_\alpha^*}^l{}') \geq 0 .$$

Since

$$-\int_{x_2-\epsilon}^{x_2}(\psi - \rho_{u_\alpha^*}^l)' = \eta = \int_{x_2}^{x_2+\epsilon}(\psi - \rho_{u_\alpha^*}^l)'$$

inequality (5.7) reduces to

$$(c_{l-1}\eta - d_{l+1}\eta) \geq 0.$$

□

Analytical Examples. In the following we present exact minimizers of $\mathcal{T}^{l,p}_{\alpha,u^\delta}$ for the cases $p = 1, 2$, $l = 1, 2$, $\Omega = (-1, 1)$, and $u^\delta = \chi_{[-\frac{1}{2}, \frac{1}{2}]} - \frac{1}{2}$. Then $\|\cdot\|_* = \|\cdot\|_{*,L^p,\mathcal{R}_l}$. Define

$$\Psi^{l,p}_\alpha := \{u^* \in L^{p*}(\Omega) : (\alpha\mathcal{R}_l)^*(u^*) = 0\}$$
$$= \{u^* \in L^{p*}(\Omega) : \|u^*\|_* \leq \alpha\}.$$

From Theorem 5.1 it follows that

$$\Psi^{l,p}_\alpha = \left\{u^* \in L^{p*}(\Omega) : \rho^l_{u^*} \in W^{l,1}_0(\Omega), \left\|\rho^l_{u^*}\right\|_\infty \leq \alpha\right\}.$$

Then

$$(\mathcal{T}^{p,l}_{\alpha,u^\delta})^*(u^*) = \begin{cases} (\mathcal{S}_p)^*(u^*) & \text{if } u^* \in \Psi^{l,p}_\alpha, \\ +\infty & \text{else}. \end{cases}$$

We construct the minimizing elements as follows: We start with $\alpha > \|u^\delta\|_*$, then $u_\alpha = 0$. Next we decrease the α-parameter until there exists an x such that $\left|\rho^l_{u^*_\alpha}(x)\right| = \alpha$. From Theorem 5.2(a) we know that u_α is a polynomial of order $(l-1)$ in the intervals where $\left|\rho^l_{u^*_\alpha}(x)\right| < \alpha$. Thus we set up according Ansatz functions u_C (piecewise polynomials of order $(l-1)$ in the connected intervals where $\left|\rho^l_{u^*_\alpha}\right| < \alpha$) and calculate the coefficients of u_C such that $u^*_C = \mathcal{J}_p(u_C - u^\delta)$ minimizes $(\mathcal{T}^{l,p}_{\alpha,u^\delta})^*$. We decrease the α-parameter again, until $\rho^l_{u^*}$ touches the α-tube again. Then the same procedure, but on more subintervals, is repeated.

Example 5.3 (L^1-TV minimization). *We derive the dual problem of of the minimization problem* $\arg\min(\mathcal{T}^{1,1}_{\alpha,u^\delta}) = \arg\min(\mathcal{S}_1 + \alpha\mathcal{R}_1)$. *From Example 4.6 we know that*

$$\mathcal{S}^*_1(u^*) = \begin{cases} \int u^\dagger u^* & \text{if } \|u^*\|_{L^\infty} \leq 1, \\ +\infty & \text{else}. \end{cases}$$

Thus the dual problem consists in minimizing $\int_\Omega u^\delta u^*$ *over the set* $\overline{\Psi}^{1,1}_\alpha := \Psi^{1,1}_\alpha \cap \{u^* \in L^\infty(\Omega) : \|u^*\|_\infty \leq 1\}$.

Using, that for $u^* \in \overline{\Psi}^{1,1}_\alpha$, $\rho^l_{u^*}(\pm 1) = 0$ *we obtain for* $u^* \in \overline{\Psi}^{1,1}_\alpha$ *following estimate*

(5.8)
$$(\mathcal{T}^{1,1}_{\alpha,u^\delta})^*(u^*) = \int_\Omega u^\delta u^* = \frac{1}{2}\left(-\int_{-1}^{-\frac{1}{2}} u^* + \int_{-\frac{1}{2}}^{\frac{1}{2}} u^* - \int_{\frac{1}{2}}^{1} u^*\right)$$
$$= \rho^l_{u^*}\left(-\frac{1}{2}\right) - \rho^l_{u^*}\left(\frac{1}{2}\right) \geq \min\{1, 2\alpha\},$$

and find that the minimizer of $(\mathcal{T}^{1,1}_{\alpha,u^\delta})^*$ *is*

$$u^*_\alpha := \begin{cases} -\min\{1, 2\alpha\} & \text{in } \left(-\frac{1}{2}, \frac{1}{2}\right) \\ \in [-1, 1] & \text{in } (-1, 1) \setminus \left(-\frac{1}{2}, \frac{1}{2}\right). \end{cases}$$

*Using the Kuhn-Tucker condition $u_\alpha^* \in \partial S_1(u_\alpha)$, we see that the minimizers u_α
and u_α^* of the functional and its dual are related by*

$$u_\alpha^* = \operatorname{sgn}(u_\alpha - u^\delta) = \mathcal{J}_1(u_\alpha - u^\delta)$$

(see also (5.5)). We distinguish between 3 cases:

- *$\alpha > \frac{1}{2}$: In this case $u_\alpha^* = -2u^\delta$ minimizes (5.8). Since $\rho_{u_\alpha^*}^l$ does not
  have contact with the α-tube, according to Theorem 5.2(a) u_α is constant
  in $(-1,1)$. From (5.5) it follows that*

$$u_\alpha(x) \geq u^\delta(x) = -\frac{1}{2} \quad for \ x \in \left(-1, -\frac{1}{2}\right) \cup \left(\frac{1}{2}, 1\right),$$

$$u_\alpha(x) \leq u^\delta(x) = \frac{1}{2} \quad for \ x \in \left(-\frac{1}{2}, \frac{1}{2}\right).$$

  Thus $u_\alpha = const$ with $const \in \left[-\frac{1}{2}, \frac{1}{2}\right]$ is a solution. Compare Figure 2.

- *$\alpha = \frac{1}{2}$: Here $u_\alpha^* = -2u^\delta \in \overline{\Psi}_\alpha^{1,1}$ and $\rho_{u_\alpha^*}^l(-\frac{1}{2}) = -\alpha = -\rho_{u_\alpha^*}^l(\frac{1}{2})$. From
  (5.5) it follows that*

$$-\frac{1}{2} = u^\delta \leq u_\alpha \quad in \ \left(-1, -\frac{1}{2}\right) \cup \left(\frac{1}{2}, 1\right),$$

$$\frac{1}{2} = u^\delta \geq u_\alpha \quad in \ \left(-\frac{1}{2}, \frac{1}{2}\right).$$

  *According to Theorem 5.2 (b) u_α is constant in the intervals $\left(-1, -\frac{1}{2}\right)$,
  $\left(-\frac{1}{2}, \frac{1}{2}\right)$, and $\left(\frac{1}{2}, 1\right)$. From Theorem 5.2 (c) we know that $a_1 = u_\alpha\left(-\frac{3}{4}\right) \leq
  a_2 = u(0)$ and $a_2 = u_\alpha(0) \geq a_3 = u\left(\frac{3}{4}\right)$. Thus the solutions of the L^1-TV
  minimization problem are*

$$u_\alpha = \begin{cases} a_1 & in \ \left(-1, -\frac{1}{2}\right), with \ -\frac{1}{2} \leq a_1, \\ a_2 & in \ \left(-\frac{1}{2}, \frac{1}{2}\right), with \ a_1 \leq a_2 \leq \frac{1}{2}, \\ a_3 & in \ \left(\frac{1}{2}, 1\right), with \ -\frac{1}{2} \leq a_3 \leq a_2. \end{cases}$$

  Compare Figure 3.

- *$\alpha < \frac{1}{2}$: The optimality condition in (5.8) fixes $\rho_{u_\alpha^*}^l$ for $x = \pm\frac{1}{2}$, with
  $\rho_{u_\alpha^*}^l(\frac{1}{2}) = \alpha = -\rho_{u_\alpha^*}^l(-\frac{1}{2})$. Exemplary we calculate u_α in $\left(-1, -\frac{1}{2}\right)$: Let*

$$A^+(u_\alpha^*) := \left\{ x \in \left(-1, -\frac{1}{2}\right) : u_\alpha^*(x) \geq 0 \right\},$$

$$A^-(u_\alpha^*) := \left\{ x \in \left(-1, -\frac{1}{2}\right) : u_\alpha^*(x) < 0 \right\}.$$

*Then from (5.5) we have $u^\delta \leq u_\alpha$ in $A^+(u_\alpha^*)$ and $u_\alpha \leq u_\alpha$ in $A^-(u_\alpha^*)$.
Moreover $\left(-1, -\frac{1}{2}\right) = A^+(u_\alpha^*) \cup A^-(u_\alpha^*)$. Define also the sets*

$$B^+(u_\alpha^*) := \left\{ x \in \left(-1, -\frac{1}{2}\right) : \rho_{u_\alpha^*}^l(x) = \alpha \right\} = \bigcup (b_{l_i}^+, b_{r_i}^+),$$

$$B^-(u_\alpha^*) := \left\{ x \in \left(-1, -\frac{1}{2}\right) : \rho_{u_\alpha^*}^l(x) = -\alpha \right\} = \bigcup (b_{l_j}^-, b_{r_j}^-),$$

$$B^0(u_\alpha^*) := \left\{ x \in \left(-1, -\frac{1}{2}\right) : \left|\rho_{u_\alpha^*}^l(x)\right| < \alpha \right\} = \bigcup (b_{l_k}^0, b_{r_k}^0).$$

For all $x \in B^0(u_\alpha^*)$ according to Theorem 5.2 u_α is constant. Then for every $(b_{l_i}^+, b_{r_i}^+)$ there exists an ϵ such that

$$(b_{l_i}^+ - \epsilon, b_{l_i}^+) \subset A^+ \cap B^0(u_\alpha^*) \quad and \quad (b_{r_i}^+, b_{r_i}^+ + \epsilon) \subset A^- \cap B^0(u_\alpha^*).$$

Then $u_\alpha = c_l^i \geq u^\delta = -\frac{1}{2}$ in $(b_{l_i}^+ - \epsilon, b_{l_i}^+)$ and $u_\alpha = c_r^i \leq u^\delta = -\frac{1}{2}$ in $(b_{r_i}^+, b_{r_i}^+ + \epsilon)$. Now according to Theorem 5.2 (c') $c_l^i \leq c_r^i$. Thus we have $c_l^i = -\frac{1}{2} = c_r^i$. For all $(b_{l_j}^-, b_{r_j}^-)$ we can argue analogously and obtain $u_\alpha = -\frac{1}{2}$ for all $x \in (-1, -\frac{1}{2})$. For $(-\frac{1}{2}, \frac{1}{2})$ and $(\frac{1}{2}, 1)$ we can argue analogously to show that $u_\alpha = u^\delta$. Compare Figure 4.

L^1-**TV Regularization**

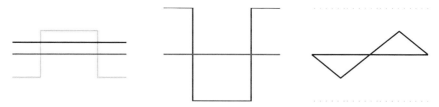

FIGURE 2. L^1-*TV minimization with $\alpha > 1/2$. Left: u_α, u^δ (gray). Middle: u_α^*. Right: $\rho_{u_\alpha^*}^1/\alpha$.*

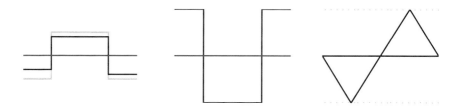

FIGURE 3. L^1-*TV minimization with $\alpha = 1/2$. Left: u_α, u^δ (gray). Middle: u_α^*. Right: $\rho_{u_\alpha^*}^1/\alpha$.*

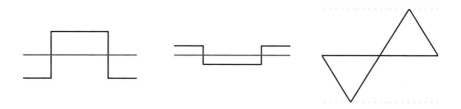

FIGURE 4. L^1-*TV minimization with $\alpha < 1/2$: Left: $u_\alpha = u^\delta$. Middle: u_α^*. Right: $\rho_{u_\alpha^*}^1/\alpha$. Note that $\rho_{u_\alpha^*}^1/\alpha$ is not unique.*

Example 5.4 (L^2-*TV minimization*). *We derive the dual problem of of the minimization problem* $\arg\min(\mathcal{T}_{\alpha, u^\delta}^{2,2}) = \arg\min(\mathcal{S}_2 + \alpha\mathcal{R}_2)$.

From Example 4.6 we know that

$$\mathcal{S}_2^*(u^*) = \int \frac{1}{2}(u^*)^2 + u^\delta u^* \ .$$

Thus the dual problem consists in minimizing $\mathcal{S}_2^*(u^*)$ over the set $\Psi_\alpha^{1,2}$.

Taking into account that the minimizer of $\int \frac{1}{2}(u^*)^2 + u^\delta u^*$ is the same as the minimizer of $\frac{1}{2}\int_\Omega (u^* + u^\delta)^2$ we see that u_α^* is the L^2-projection of $-u^\delta$ onto $\Psi_\alpha^{1,2}$.

Integrating u^δ once shows that the minimal α for which $u^\delta \in \Psi_\alpha^{1,2}$ is $\alpha = \left\|u^\delta\right\|_* = \frac{1}{4}$. Thus for $\alpha \geq \frac{1}{4}$ we have $u_\alpha^* = -u^\delta$. If $\alpha < \frac{1}{4}$ then $u_\alpha = -4\alpha\, u^\delta$. In summary

$$u_\alpha^* = -u^\delta \min\{1, 4\alpha\} \,,$$

hence, the Kuhn Tucker condition $u_\alpha \in \partial\mathcal{S}_2(u_\alpha^*)$ gives

$$u_\alpha = u^\delta + u_\alpha^* = \begin{cases} 0 & \text{for } \alpha \geq \frac{1}{4}, \\ (1 - 4\alpha)u^\delta & \text{for } 0 \leq \alpha \leq \frac{1}{4} \ . \end{cases}$$

See figures 5 and 6.

$L^2 - \text{TV}$ **Regularization**

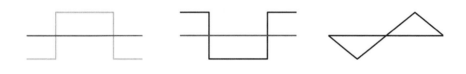

FIGURE 5. $\alpha \geq \frac{1}{4}$: *Left:* $u_\alpha = 0$, u^δ *(gray). Middle:* u_α^*. *Right:* $\rho_{u_\alpha^*}^1/\alpha$.

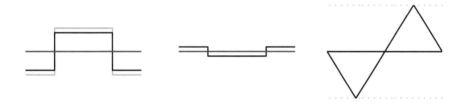

FIGURE 6. $\alpha < \frac{1}{4}$: *Left:* u_α, u^δ *(gray). Middle:* u_α^*. *Right:* $\rho_{u_\alpha^*}^1/\alpha$.

Example 5.5 (L^1-TV^2 *minimization*). *We derive the dual problem of of the minimization problem* $\arg\min(\mathcal{T}_{\alpha,u^\delta}^{1,2}) = \arg\min(\mathcal{S}_1 + \alpha\mathcal{R}_2)$. *From Example 4.6 and Theorem 4.7 we know that the dual problem consists in minimizing*

(5.9) $$\mathcal{S}_1^*(u^*) = \int u^\delta u^*$$

over the set $\overline{\Psi}_\alpha^{2,1} := \Psi_\alpha^{2,1} \cap \{u^* \in L^\infty : \|u^*\|_\infty \leq 1\}$.

We distinguish between the three cases $\alpha > \frac{1}{4} = \left\|u^\delta\right\|_* = \left\|u^\delta\right\|_{*,L^1,\mathcal{R}_2}$, $\alpha \in \left(\frac{3}{8} - \frac{\sqrt{2}}{4}, \frac{1}{4}\right]$ *and* $\alpha \leq \frac{3}{8} - \frac{\sqrt{2}}{4}$.

- Let $\alpha > \frac{1}{4}$, then $u_\alpha^* := -2u^\delta$ is an element of $\overline{\Psi}_\alpha^{2,1}$ which minimizes (5.9). Moreover $\left| \rho_{u_\alpha^*}^l(x) \right| < \alpha$ for all $x \in \Omega$. Thus according to Theorem 5.2 (a) u_α is a polynomial of order 1 on $(-1, 1)$. From (5.5) it follows that

$$
\begin{aligned}
0 \le u_\alpha \quad & in \ \left(-1, -\frac{1}{2} \right) \cup \left(\frac{1}{2}, 1 \right), \\
u_\alpha \le 1 \quad & in \ \in \left(-\frac{1}{2}, \frac{1}{2} \right).
\end{aligned}
$$

(5.10)

Thus the minimizers of $\mathcal{T}_{\alpha,u^\delta}^{2,1}$ are affine functions satisfying (5.10) (Figure 7).

- Let $\alpha \in (3/8 - \sqrt{2}/4, 1/4)$. If $\alpha \le 1/4$, $\rho_{u_\alpha^*}^l$ has at least one contact point $0 \le x_1 \le \frac{1}{2}$ with the α-tube and according to Theorem 5.2 u_α bends at $x = \pm x_1$. From Theorem 5.2 it follows that u_α is affine in $(-1, -x_1)$ and $(x_1, 1)$.

  Since u^δ is symmetric, there exists a symmetric minimizer of $\mathcal{T}_{\alpha,u^\delta}^{2,1}$ which satisfies $\left| \rho_{u_\alpha^*}^l(-x_1) \right| = \alpha$. In the following we concentrate on calculating symmetric minimizers.

  We calculate the minimizer u_{C_α} of $\mathcal{T}_{\alpha,u^\delta}^{2,1}$ in the class

$$
\mathcal{C} := \left\{ u_C : C = \{s, x_1, d\} \ with \ \frac{1}{2} \le -\frac{s}{d} \le 1 \right\},
$$

where

$$
u_C(x) = \begin{cases} \frac{1}{2} & x \in [-x_1, x_1], \\ s\,|x| + d & x \in (-1, -1) \backslash [-x_1, x_1], \end{cases}
$$

and prove afterwards that $u_\alpha = u_{C_\alpha}$.

  We set

$$
\mathcal{C}^* := \left\{ u_C^* \in \overline{\Psi}_\alpha^{2,1} : (u_C^*, u_C) \ satisfy \ (5.5) \right\}.
$$

One possibility of a function $u_C^* \in \mathcal{C}^*$ related to u_C by (5.5) is as follows:

$$
u_C^*(x) = \begin{cases} -1 & x \in (-1, s/d) \cup (-\frac{1}{2}, -x_1) \cup (x_1, \frac{1}{2}) \cup (-s/d, 1), \\ +1 & x \in (s/d, -\frac{1}{2}) \cup (\frac{1}{2}, -s/d), \\ 0 & x \in (-x_1, x_1). \end{cases}
$$

We determine $C_\alpha := \{s_\alpha, x_{1,\alpha}, d_\alpha\}$ as follows:

- u_α is continuous at $x = x_{1,\alpha}$. Thus

$$
u_{C_\alpha}(x_{1,\alpha}) = s_\alpha x_{1,\alpha} + d_\alpha = \frac{1}{2}.
$$

- Since u_α bends at $x_{1,\alpha}$, we aim for u_{C_α} which bends at $x = \pm x_{1,\alpha}$. Thus

$$
\rho_{u_{C_\alpha}^*}^2(x_{1,\alpha}) = \frac{4s_\alpha^2 - 8s_\alpha x_{1,\alpha} d_\alpha + 6x_{1,\alpha}^2 d_\alpha^2 - 8x_{1,\alpha} d_\alpha^2 - 3d_\alpha^2}{4d_\alpha^2} = \alpha.
$$

- $\rho_{u_{C_\alpha}^*}^2$ is maximal at $\pm x_{1,\alpha}$

$$
\left(\rho_{u_{C_\alpha}^*}^2 \right)'(x_{1,\alpha}) = \frac{x_{1,\alpha} d_\alpha - 2s_\alpha - 2d_\alpha}{d_\alpha} = 0.
$$

Solving the three equations above gives:

$$x_{1,\alpha} = \frac{1}{s_\alpha}\left(\frac{1}{2} - d_\alpha\right), \; d_\alpha = \frac{4 + \sqrt{1 + 12\alpha}}{3\sqrt{1 + 12\alpha}}, \; s_\alpha = \frac{-2}{\sqrt{1 + 12\alpha}} \, .$$

Since u_{C_α} and $u^*_{C_\alpha}$ satisfy (4.4) they are minimizers of $\mathcal{T}^{2,1}_{\alpha,u^\delta}, (\mathcal{T}^{2,1}_{\alpha,u^\delta})^*$ respectively.(Figure 8)

- *Let $\alpha \leq \frac{3}{8} - \frac{1}{4}\sqrt{2}$. We calculate the minimizers u_{C_α} of $\mathcal{T}^{2,1}_{\alpha,u^\delta}$ in*

$$\mathcal{C} := \{u_C : C := \{s_1, s_2, d_1, d_2, x_1, x_2\} \text{ with}$$

$$0 \leq x_1 \leq \frac{1}{2} \leq x_2 \leq 1, \, s_1 \leq 0 \text{ and } s_1 \leq s_2\}$$

with

$$u_C(x) = \begin{cases} s_2\,|x| + d_2 & x \in (-1,1)\backslash(-x_2, x_2) \\ s_1\,|x| + d_1 & x \in (-x_2, -x_1) \cup (x_1, x_2) \\ \frac{1}{2} & x \in (-x_1, x_1) \, . \end{cases}$$

Here the functions u_C can bend at least four times. Afterwards we verify that $u_\alpha = u_{C_\alpha}$.

Let

$$\mathcal{C}^* := \left\{u^*_C \in \Psi^{2,1}_\alpha : (u^*_C, u_C) \text{ satisfy (5.5)}\right\} \, .$$

One possible function $u^*_C \in \mathcal{C}^*$ related to u_C by (5.5) is

$$u^*_C(x) \mapsto \begin{cases} -c_1 & x \in (-1, -\frac{1}{2} - \frac{1}{2}x_2) \cup (\frac{1}{2} + \frac{1}{2}x_2, 1), \\ c_1 & x \in (-\frac{1}{2} - \frac{1}{2}x_2, -x_2) \cup (x_2, \frac{1}{2} + \frac{1}{2}x_2), \\ c_2 & x \in (-x_2, -\frac{1}{2}) \cup (\frac{1}{2}, x_2), \\ -c_2 & x \in (-\frac{1}{2}, -x_1) \cup (x_1, \frac{1}{2}), \\ 0 & x \in (-x_1, x_1) \end{cases}$$

with $0 \leq c_1, c_2 \leq 1$. Let $\rho^2_{u^*_C}$ be the second primitive of u^*_C.

Since u_α bends at $x = \pm x_1, \pm x_2$, we aim to find u_{C_α} that bends at x_1, x_2 as well and according to Theorem 5.2 enforcing the properties of $\rho^2_{u^*_\alpha}$ onto $\rho^2_{u^*_{C_\alpha}}$. We additionally require that $\rho^2_{u^*_{C_\alpha}}$ is extremal at $\pm x_1, \pm x_2$. From Theorem 5.2 (c) we know that $s_1 \leq 0$ and $s_1 \leq s_2$. Thus we solve following equations

$$\rho^2_{u^*_{C_\alpha}}(-x_{2,\alpha}) = \int_{-1}^{-x_{2,\alpha}}(\rho^2_{u^*_{C_\alpha}})' = -\alpha, \quad \rho^2_{u^*_{C_\alpha}}(-x_{1,\alpha}) = \int_{-1}^{-x_{1,\alpha}}(\rho^2_{u^*_{C_\alpha}})' = \alpha,$$

$$\rho^2_{u^*_{C_\alpha}}(x_{1,\alpha}) = \int_{-1}^{x_{1,\alpha}}(\rho^2_{u^*_{C_\alpha}})' = \alpha, \qquad \rho_C{}^*(x_{2,\alpha}) = \int_{-1}^{x_{2,\alpha}}(\rho^2_{u^*_{C_\alpha}})' = -\alpha \, .$$

Since $\rho^2_{u^*_{C_\alpha}}$ attains an extremum at $\pm x_{1,\alpha}, x_{2,\alpha}$, we have that

$$(\rho^2_{u^*_{C_\alpha}})'(\pm x_{2,\alpha}) = 0 \quad \text{and} \quad (\rho^2_{u^*_{C_\alpha}})'(\pm x_{1,\alpha}) = 0 \, .$$

Taking into account that $\rho^*_C \in \Psi^{2,2}_\alpha$ (thus is continuous and satisfies boundary conditions), we see that minimizing $\mathcal{T}^{2,2}_{\alpha,u^\delta}$ on $\mathcal{C}$ is equivalent to maximizing

$$(\rho^2_{u^*_C})'\left(-\frac{1}{2}\right) = \int_{-1}^{-\frac{1}{2}} u^*_C = \frac{c_2}{2}\left(x_2 - \frac{1}{2}\right)$$

Hence it follows that $c_{2,\alpha} = 1$. *Since* $\rho^2_{u^*_{C_\alpha}}(-x_{2,\alpha}) = -\alpha$ *and* $(\rho^2_{u^*_{C_\alpha}})'(-\frac{1}{2}) = 0$, $x_{2,\alpha}$ *has to satisfy*

$$\int_{-x_{2,\alpha}}^{-\frac{1}{2}} (\rho^2_{u^*_{C_\alpha}})' = \int_{-x_{2,\alpha}}^{-\frac{1}{2}} c_{2,\alpha}(x + x_{2,\alpha}) = \frac{1}{2}x^2_{2,\alpha} - \frac{1}{2}x_{2,\alpha} + \frac{1}{8} = \alpha .$$

Hence

$$x_{2,\alpha} = \frac{1}{2} + \sqrt{2\alpha} .$$

Analogous we find that

$$x_{1,\alpha} = \frac{1}{2} - \sqrt{2\alpha} .$$

Since $\rho^2_{u^*_{C_\alpha}}(-x_{2,\alpha}) = -\alpha$ *it follows that*

$$c_{1,\alpha} = \frac{16\alpha}{\left(2\sqrt{2\alpha} - 1\right)^2} \leq 1 .$$

Next we determine the coefficients such that u_{C_α} *and* $u^*_{C_\alpha}$ *are connected via* (5.5) *and obtain*

$$s_{2,\alpha} = 0, \qquad\qquad d_{2,\alpha} = 0,$$
$$k_{1,\alpha} = -\frac{1}{2\sqrt{2\alpha}}, \qquad\qquad d_{1,\alpha} = \frac{x_{2,\alpha}}{2\sqrt{2\alpha}}.$$

Then u_{C_α} *minimizes* $\mathcal{T}^{2,1}_{\alpha,u^\delta}$ *as can be shown by testing* (4.4). *See Figure 9.*

Example 5.6 (L^2-TV^2 *minimization*). *We derive the dual problem of the minimization problem* $\arg\min(\mathcal{T}^{2,2}_{\alpha,u^\delta}) = \arg\min(\mathcal{S}_2 + \alpha\mathcal{R}_2)$. *From Example 4.6 we know that* $\mathcal{S}^*_2(u^*) = \int u^\delta u^*$. *Thus the dual problem consists in minimizing* $\int_\Omega u^\delta u^*$ *over the set* $\Psi^{2,2}_\alpha$.

We investigate four different cases $\alpha > \frac{1}{8} = \|u^\delta\|_*$, $\alpha \in (\frac{1}{24}, \frac{1}{8})$, $\alpha \in (\alpha_m, \frac{1}{24})$, *and* $\alpha < \alpha_m$. *Here* α_m *denotes the largest* α-*value such that* $\rho^l_{u^*_\alpha}$ *takes the value* $-\alpha$ *for some* $x \in (-1, 1)$.

- *If* $\alpha > \|u^\delta\|_* = \frac{1}{8}$, *then* $-u^\delta \in \Psi^{2,2}_\alpha$. *Thus* $u^*_\alpha = -u^\delta$ *is the* L^2-*projection of* $-u^\delta$ *onto* $\Psi^{2,2}_\alpha$ *and* (4.9) *gives* $u_\alpha = u^*_\alpha + u^\delta = 0$. *Since* $(u_\alpha, u^*_\alpha = -u^\delta)$ *satisfy* (4.4) *they are minimizers of* $\mathcal{T}^{2,2}_{\alpha,u^\delta}, (\mathcal{T}^{2,2}_{\alpha,u^\delta})^*$ *respectively.*

- *For* $\alpha \in (\frac{1}{24}, \frac{1}{8}]$ *we calculate the minimizer* u_{C_α} *of* $\mathcal{T}^{2,2}_{\alpha,u^\delta}$ *in the class of functions*

$$\mathcal{C} := \{u_C : C := \{s, d\} \text{ and } u_C(x) = s|x| + d\} ,$$

and verify afterwards that $u_\alpha = u_{C_\alpha}$.

Let $u^*_C = u_C - u^\delta$ *and* $\rho^2_{u^*_C}$ *the second primitive of* u^*_C. *We determine* $C_\alpha := \{s_\alpha, d_\alpha\}$ *such that* $-u^*_{C_\alpha} \in \Psi_\alpha$, *that is*

$$\int_\Omega u^*_{C_\alpha} = s_\alpha + 2d_\alpha = 0 \qquad \Rightarrow d_\alpha = -\frac{1}{2}s_\alpha$$

$L^1 - \mathrm{TV}^2$ **Regularization**

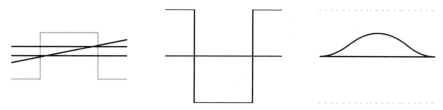

FIGURE 7. $\alpha > \frac{1}{4}$: Left: u_α (*bold*), u^δ (*gray*). Note that u_α is not unique. Middle: u_α^*. Right: $\rho_{u_\alpha^*}^2/\alpha$

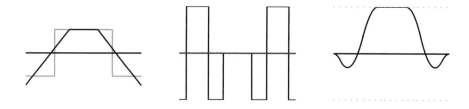

FIGURE 8. $\frac{3}{8} - \frac{1}{4}\sqrt{2} < \alpha < \frac{1}{4}$: Left: u_α bends at $x = \pm x_{1,\alpha}$, u^δ (*gray*). Middle: u_α^*. Right: $\rho_{u_\alpha^*}^2/\alpha$

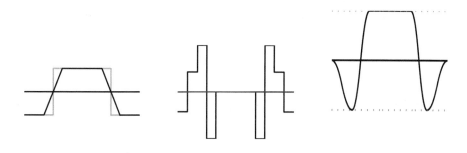

FIGURE 9. $\alpha < \frac{3}{8} - \frac{1}{4}\sqrt{2}$: Left: u_α bends at $x = \pm\left(1/2 \pm \sqrt{2\alpha}\right)$, u^δ (*gray*). Middle: $u^* = -(\rho_\alpha^*)''$. Right: Here ρ_α^* touches the α-tube at $x = \pm\left(1/2 \pm \sqrt{2\alpha}\right)$, where u_α bends

– u_{C_α} *bends at* $x = 0$, *which requires that*

$$\rho_{C_\alpha}^*(0) = \frac{1}{12}s_\alpha + \frac{1}{8} = \alpha, \Rightarrow s_\alpha = \frac{24\alpha - 3}{2}.$$

Then $(u_{C_\alpha}^* + u^\delta, u_{C_\alpha}^*)$ *satisfy* (4.4) *and thus are minimizers of* $\mathcal{T}_{\alpha,u^\delta}^{2,2}$, $(\mathcal{T}_{\alpha,u^\delta}^{2,2})^*$ *respectively.*

- *For* $\alpha \in (\alpha_m, \frac{1}{24}]$ *we calculate the minimizer* u_{C_α} *of* $\mathcal{T}_{\alpha,u^\delta}^{2,2}$ *in*

$$\mathcal{C} := \{u_C : C := \{s, d, x_1\}\}$$

with

$$u_C(x) = \begin{cases} s\,|x| + d & x \in (-1, -x_1) \cup (x_1, 1)\,, \\ u^\delta & \text{in } (-x_1, x_1)\,. \end{cases}$$

We verify afterwards that $u_\alpha = u_{C_\alpha}$. Let again $u_C^* = u_C - u^\delta$ and $\rho_{u_C^*}^2$ the second primitive of u_C^*. We determine $C_\alpha = \{s_\alpha, d_\alpha, x_{1,\alpha}\}$ such that
 – $u_{C_\alpha}^* \in \Psi_\alpha$, that is

$$\int_\Omega u_{C_\alpha}^* = s_\alpha + 2d_\alpha + s_\alpha\,x_{1,\alpha}^2 = 0\,.$$

This condition implies that

$$d_\alpha = -\frac{1}{2}\left(s_\alpha + s_\alpha\,x_{1,\alpha}^2\right).$$

 – $\rho_{u_{C_\alpha}^*}^2$ is maximal at $\pm x_1$

$$(\rho_{u_{C_\alpha}^*}^2)'(x_{1,\alpha}) = s_\alpha\left(x_{1,\alpha}^2 - \frac{1}{2}x_{1,\alpha} - \frac{1}{2}x_{1,\alpha}^3\right) - \frac{1}{2}x_{1,\alpha} = 0\,,$$

$$\rho_{u_{C_\alpha}^*}^2(x_{1,\alpha}) = -\frac{1}{6}x_{1,\alpha} + \frac{1}{24} = \alpha.$$

These two conditions are guaranteed if $s_\alpha = -\frac{1}{(1-x_{1,\alpha})^2}$ and $x_{1,\alpha} = 6\alpha - \frac{1}{4}$.
Since $u_{C_\alpha} = u_{C_\alpha}^* + u^\delta$ and $u_{C_\alpha}^*$ satisfy (4.4) they minimize $\mathcal{T}_{\alpha,u^\delta}^{2,2}, (\mathcal{T}_{\alpha,u^\delta}^{2,2})^*$ respectively. See Figure 11.
 • *For $\alpha \le \alpha_m$ we minimize $\mathcal{T}_{\alpha,u^\delta}^{2,2}$ on the set of piecewise affine functions*

$$u_C(x) = \begin{cases} s_2\,|x| + d_2 & x \in (-1, -x_2) \cup (x_2, 1)\,, \\ s_1\,|x| + d_1 & x \in (-x_2, -x_1) \cup (x_1, x_2)\,, \\ s_1 x_1 + d_1 & x \in (-x_1, x_1)\,, \end{cases}$$

with $C := \{x_2, x_1, k_2, k_1, d_1, d_2\}$ and proceed as above to determine C_α. See Figure 12.

Appendix A. Functions of Bounded Variation

In the following we highlight some properties of functions of bounded variation, which are collected from Evans & Gariepy [EvaGar92] and Ambrosio, Fusco & Pallara [AmbFusPal00].

Definition A.1. *For $1 \le p < n$ the Sobolev conjugate is*

$$p_n = \frac{np}{n - p}\,.$$

Definition A.2. *The space of functions of bounded variation (BV) consists of functions $u \in L^{p_n}(\Omega)$ satisfying*

$$|Du| := \left\{\int u\nabla \cdot \vec\phi : \vec\phi \in C_0^1(\mathbb{R}^n, \mathbb{R}^n), \left\|\vec\phi\right\|_{L^\infty(\Omega)} \le 1\right\} < \infty$$

The standard definition of BV requires $u \in L^1(\Omega)$ ([EvaGar92]). In this case $u \in L^{p_n}(\Omega)$ (which follows from the Gagliardo-Nirenberg-Sobolev inequality).

$L^2 - \mathrm{TV}^2$ **Regularization**

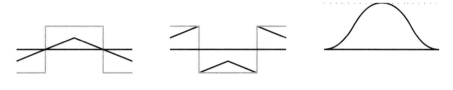

FIGURE 10. Left: u_α bends at $x = 0$, gray: u^δ. Middle: u_α^*. Right: $\rho_{u_\alpha^*}^2$.

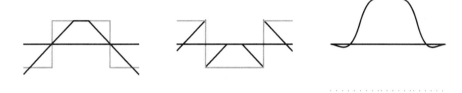

FIGURE 11. Left: u_α, gray: u^δ. Middle: u_α^*. Right: $\rho_{u_\alpha^*}^2$.

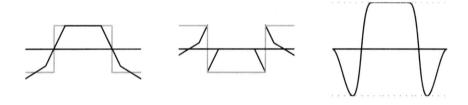

FIGURE 12. Left: u_α, gray: u^δ. Middle: u_α^*. Right: $\rho_{u_\alpha^*}^2$.

Definition A.3. *We define the set of functions with derivatives of bounded varia-tion (BV^l) as functions $u \in L^p(\Omega)$ (for some $p \geq 1$) satisfying*

$$\left| D^l u \right| := \sup \left\{ \int u \nabla^l \cdot \vec{\phi} : \vec{\phi} \in C_0^l(\mathbb{R}^n, \mathbb{R}^{\mathcal{N}(l)}), \left| \vec{\phi} \right| \leq 1 \right\} < \infty$$

where

$$\nabla^l \cdot \vec{\phi} = \sum_{\substack{i_l = 1, \ldots, n \\ l = 1, \ldots, l}} \frac{\partial^l \phi_{i_1, \ldots, i_l}}{\partial x_{i_1} \cdots \partial x_{i_l}}.$$

The gradient $Du = (D_1 u, \ldots, D_n u)$ of a function of bounded variation is repre-sentable by a finite Radon measure in Ω and satisfies

$$\int_\Omega u \frac{\partial \phi}{\partial x_i} \, dx = - \int_\Omega \phi \, dD_i u \quad \phi \in C_0^\infty(\Omega), i = 1, \ldots, N \, .$$

The following superspaces of $\mathrm{BV}^l(\Omega)$ are of importance for this work:

Definition A.4. *Functions with finite total variation. For $l \in \mathbb{N}$ let*

$$TV^l(\Omega) := \left\{ u \in C_0^l(\Omega)^* : \left| D^l u \right| < \infty \right\} .$$

In the definition above u has to be considered a distribution of order l, that is u is a linear operator on $C_0^l(\Omega; \mathbb{R}^{\mathcal{N}(l)})$ satisfying

$$u[\vec{\phi}] = (-1)^l \int_\Omega u \nabla^l \vec{\phi} \quad \text{for all } \vec{\phi} \in C_0^\infty(\Omega; \mathbb{R}^{\mathcal{N}(l)}) .$$

Theorem A.5. *Let $\Omega \subset \mathbb{R}^n$ be bocL. There exists a constant $C(\Omega, l, p)$ such that*

$$|u|_{l,p} \leq \|u\|_{l,p} \leq C(\Omega, l, p) |u|_{l,p} \qquad \text{for all } u \in W_\diamond^{l,p}(\Omega) .$$

In particular, $|\cdot|_{l,p}$ and $\|\cdot\|_{l,p}$ are equivalent norms on $W_\diamond^{l,p}(\Omega)$.

Proof. See [Zie89, Thm. 4.4.2]. $\qquad\square$

Theorem A.6. *For $l > 1$ and $u \in TV^l(\Omega)$, $\nabla^{(l-1)} u$ exists and*

$$\left| D^l u \right| = \left| D(\nabla^{(l-1)} u) \right| .$$

Proof. Since $TV^l(\Omega) \subset \left(C_0^l(\Omega) \right)^*$ and $C_0^l(\Omega) \subset C_0^{l-1}(\Omega)$ every $u \in \left(C_0^l(\Omega) \right)^*$ can be extended to a functional $\tilde{u} \in \left(C_0^{(l-1)}(\Omega) \right)^*$ such that

$$u[\phi] = \tilde{u}[\phi] \text{ for all } \phi \in C_0^l(\Omega).$$

Since $C_0^\infty(\Omega) \subset C_0^l(\Omega)$ it follows that

$$\int_\Omega u \nabla^l \cdot \psi = u[\phi] = \int_\Omega \nabla u \nabla^{l-1} \cdot \psi = \tilde{u}[\psi]$$

for all $\psi \in C_0^\infty(\Omega)$. Thus ∇u exists. According to Hahn-Banach, this extension is norm preserving, so that we have

$$\left| D^l u \right| = \left| D^{l-1}(\nabla u) \right| .$$

$\qquad\square$

Theorem A.7. *Let $\Omega \subset \mathbb{R}^n$ be open and $u \in \mathrm{BV}(\Omega)$. There exists a sequence $(u_k) \subset C^\infty(\Omega) \cap \mathrm{BV}(\Omega)$ strictly converging to u.*

Proof. See [AmbFusPal00, Thm. 3.9]. $\qquad\square$

Theorem A.8. *Let $u \in \mathrm{BV}(\Omega)$ with $\Omega = (a, b) \subset \mathbb{R}$. For every $x \in \Omega$ there exist $u^{(l)}(x)$ and $u^{(r)}(x)$, and they are equal outside of $\Sigma(u)$. In particular, u almost everywhere equals the function $\tilde{u}(x) := \left(u^{(l)}(x) + u^{(r)}(x) \right)/2$.*

Moreover, there exist $u^{(r)}(a)$ and $u^{(l)}(b)$. In particular, for every $\Omega \subset\subset \tilde{\Omega} \subset\subset \mathbb{R}$ the function

$$\tilde{u}(x) := \begin{cases} u^{(l)}(b), & x \geq b, \\ u(x), & x \in \Omega, \\ u^{(r)}(a), & x \leq a, \end{cases}$$

is in $\mathrm{BV}(\tilde{\Omega})$.

Proof. See [AmbFusPal00, Thm. 3.28]. $\qquad\square$

Lemma A.9. *Let $\Omega = (a, b) \subset \mathbb{R}$ be an open and bounded interval. For every $u \in \mathrm{TV}^l(\Omega)$ with $l \in \mathbb{N}$ there exists a sequence $(u_k) \subset C^\infty(\bar{\Omega})$ with $\|u - u_k\|_1 \to 0$ and $|D^l u_k| \to |Du|$.*

Proof. We prove the assertion by induction on l.

First let $l = 1$. From Theorem A.8 it follows that we can continue u by a function $\tilde{u} \in \mathrm{BV}(a-1, b+1)$ by setting $\tilde{u}(x) = u^{(r)}(a)$ for $x < a$, and $\tilde{u}(x) = u^{(l)}(b)$ for $x > b$. From Theorem A.7 it follows that there exists a sequence (u_k) in $\mathrm{BV}(a-1, b+1)$ such that $\|u_k - u\|_1 \to 0$ and $|Du_k| \to |D\tilde{u}| = |Du|$. Thus, the sequence $(u_k|_\Omega)$ has the desired properties.

Now assume that the assertion holds for $l - 1$. Then there exists a sequence $(\tilde{u}_k)$ in $\mathrm{TV}^{l-1}(\Omega)$ with $\|\tilde{u}_k - u'\|_1 \to 0$, and $|D\tilde{u}_k| \to |D^{l-1}u'| = |D^l u|$. Define

$$u_k(x) := (u')^{(r)}(a) + \int_a^x \tilde{u}_k \ .$$

Then $|D^l u_k| = |D^{l-1} u'_k| = |D^{l-1}\tilde{u}_k| \to |D^l u|$, and

$$\|u_k - u\|_1 = \int_a^b |u_k - u| \leq \int_a^b \int_a^x |\tilde{u}_k - u'| \leq (b-a) \|\tilde{u}_k - u'\|_1 \to 0 \ .$$

Thus the assertion follows. $\square$

ACKNOWLEDGMENTS

The authors would like to thank Markus Grasmaier for stimulating discusion, proof-reading and helpful remarks.

REFERENCES

[Ada75] R.A. Adams. *Sobolev Spaces.* Academic Press (New York), 1975.

[AmbFusPal00] L. Ambrosio, N. Fusco and D. Pallara. *Functions of bounded variation and free discontinuity problems.* The Clarendon Press Oxford University Press (New York), 2000.

[Aub91] J.-P. Aubin. *Optima and Equilibria. An Introduction to Nonlinear Analysis.* Springer-Verlag (Berlin, Heidelberg), Graduate Texts in Mathematics 140, editors: J.h: Ewing, F.W. Gehring, P.R.Halmos, 1991.

[Aub] J.-P. Aubin. *Explicit methods of optimization.* Dunod, 1985.

[AubAuj05] G. Aubert and J.-F. Aujol. *Modeling very oscillating signals. Application to image processing.* Appl. Math. Optim, vol. **51**, 163–182, 2005.

[ChaLio95] A. Chambolle and P.-L. Lions. *Image recovery via total variation minimization and related problems,* Numer. Math., vol. **76**, 167–188, 1997.

[ChaEse05] T.F. Chan, S. Esedoglu. *Aspects of total variation regularized L^1- function approximation.* SIAM J. Appl. Math. 65:5, pp. 1817–1837,2005.

[ChaMarMul00] T. Chan, A. Marquina and P. Mulet. *High-Order Total Variation-Based Image Restoration.* SIAM Journal on Scientific Computing, vol. **22** (2), 2000.

[ChaShe05] T. Chan and J. Shen. *Image Processing and Analysis - Variational, PDE, wavelet, and stochastic methods.* SIAM Publisher (Philadelphia), 2005.

[EvaGar92] L.C. Evans and R.F. Gariepy. *Measure Theory and Fine Properties of Functions.* CRC–Press (Boca Raton), 1992.

[EkeTem76] I. Ekeland and R. Temam. *Convex Analysis and Variational Problems.* North Holland (Amsterdam), 1976.

[HinSch06] W. Hinterberger and O. Scherzer. *Variational Methods on the Space of Functions of Bounded Hessian for Convexification and Denoising.* Computing 76, 109-133, 2006.

[Mey01] Y. Meyer. *Oscillating patterns in image processing and nonlinear evolution equations.* University Lecture Series, vol. **22**, Amer. Math. Soc., Providence, RI, 2001.

[Nik03a] M. Nikolova. *Minimization of Cost-Functions with Non-smooth Data-Fidelity Terms to Clean Impulsive Noise.* Int. workshop on Energy Minimization Methods in Computer Vision and Pattern Recognition, Lecture Notes in Computer Science, Springer-Verlag, pp. 391-406, 2003.

[Nik03b] M. Nikolova. *Efficient removing of impulsive noise based on an $L_1 - L_2$ cost-function.* IEEE Int. Conf. on Image Processing, 121–124, 2003.

[ObeOshSch05] A. Obereder, S. Osher and O. Scherzer. *On the use of dual norms in bounded variation type regularization.* Geometric Properties for Incomplete Data Series: Computational Imaging and Vision, vol. **31**, 2006.

[OshSch04] S. Osher and O. Scherzer. *G-norm properties of bounded variation regularization.* Comm. Math.Sci. vol. **2**, 2337–254, 2004.

[Roc70] R.T. Rockafellar. *Convex analysis.* Princeton Mathematical Series, No. **28**, Princeton University Press (Princeton, N.J.), 1970.

[RudOshFat92] L.I. Rudin, S. Osher and E. Fatemi. *Nonlinear total variation based noise removal algorithms.* Physica D, vol. **60**, 259–268, 1992.

[Sch98] O. Scherzer. *Denoising with higher order derivatives of bounded variation and an application to parameter estimation.* Computing 60, 1–27, 1998.

[SchYinOsh05] O. Scherzer, W. Yin and S. Osher. *Slope and G-set characterization of Set-valued Functions and Applications to Non-differentiable Optimization Problems.* Comm. Math. Sci., 3, 479–492, 2005.

[Ste06] *A note on the dual treatment of higher order regularization functionals* Computing 76, 135 – 148, 2006.

[SteDidNeu05] G. Steidl, S. Didas, and J. Neumann. *Relations Between Higher Order TV Regularization and Support Vector Regression.* Scale-Space 2005, LNCS 3459, Springer-Verlag (Berlin, Heidelberg), 515–527, 2005.

[SteDidNeu] G. Steidl, S. Didas and J. Neumann. *Splines in higher order TV regularization.* International Journal of Computer Vision, vol. **70**, 241–255, 2006.

[StrCha96] D. Strong and T. Chan. *Exact Solutions to Total Variation Regularization Problems* CAM Report, 41–96, 1996.

[Zie89] W.P. Ziemer. *Weakly differentiable functions. Sobolev spaces and functions of bounded variation..* Springer-Verlag (Berlin), 1989.

DEPARTMENT OF COMPUTER SCIENCE, LEOPOLD FRANZENS UNIVERSITY, INNSBRUCK, AUSTRIA
E-mail address: `Christiane.Poeschl@uibk.ac.at`

DEPARTMENT OF COMPUTER SCIENCE, LEOPOLD FRANZENS UNIVERSITY, INNSBRUCK, AUSTRIA
E-mail address: `Otmar.Scherzer@uibk.ac.at`

Contemporary Mathematics
Volume **451**, 2008

ERROR ESTIMATES FOR THE PSWF METHOD IN MRI

GILBERT G. WALTER AND TATIANA SOLESKI

ABSTRACT. The two-dimensional prolate spheroidal wave function (PSWF) method for MRI was proposed as an efficient method for determining the image intensity over a region of interest (ROI). Because of the optimization properties of the PSWFs it is expected that the intensity over the ROI can be approximated with a minimum of error when the data is taken from an appropriate subset of $k-$space. In this work we attempt to quantify this error and compare it to the error obtained with other methods. We first consider both the problem and its solutions in one dimension and then extend our results to two and higher dimensions.

1. INTRODUCTION

In a series of papers beginning with the work of Shepp and Zhang ([9],[18],[6]), a new method of estimating the amount of activity in a region of the brain from MRI data was proposed. This method was based on the use of multi-dimensional prolate spheroidal wave functions (PSWFs). It nicely avoids the need for FFT calculations since, for certain regions, the PSWFs are eigenfunctions of the Fourier transform.

These functions, although they have an ancient history as solutions to a Sturm-Liouville problem [13], were extensively studied at Bell Labs in the 1960's as solutions to an energy concentration problem ([10],[4],[5],[11]). More recently wavelet systems based on them were introduced [15] and shown to have some remarkable properties [14]. A simple method of calculating them based on a sampling theorem has been presented [16] as an alternative to the method based on the "lucky accident" of Slepian [10].

In the procedure of Shepp and Zhang based on these PSWFs, the intensity of the image obtained from an MRI scan over a region of interest (ROI) B must be approximated. That is, if $F(\mathbf{x})$ is the magnitude of the image in $\mathbb{R}^2$ the integral

$$\int_B F(\mathbf{x})d\mathbf{x}$$

must be estimated. The data consists of sampled values of the Fourier transform $f(\mathbf{k})$ of $F(\mathbf{x})$. Rather than use all possible values of $f(\mathbf{k})$, the above authors have suggested restricting the data used to a subset A of the Fourier transform $(\mathbf{k})$ space. Since only data on a small subset is used, the speed of computation can be increased.

Date: February 9, 2007.

2000 *Mathematics Subject Classification*. Primary 05C38, 15A15; Secondary 05A15, 15A18.

Key words and phrases. Prolate spheroidal wave functions, MRI, signal processing, imaging.

This paper is in final form and no version of it will be submitted for publication elsewhere.

1.1. **The maximization problem.** To attack this problem they proposed using two-dimensional PSWFs previously studied by Slepian [11]. These involve solving the maximization problem

$$(1) \qquad\qquad \max_{G \in \mathcal{B}_A} \frac{\int_B |G(\mathbf{x})|^2 d\mathbf{x}}{||G||^2}$$

where $\mathcal{B}_A$ is the Paley-Wiener space of all $A-$bandlimited functions, i.e., functions whose Fourier transform has support in A.

As observed by Slepian, this maximization problem is solved by means of the solution of an eigenvalue problem

$$(2) \qquad\qquad \int_B S_A(\mathbf{x} - \mathbf{t})\varphi(\mathbf{t})d\mathbf{t} = \lambda\varphi(\mathbf{x})$$

where $S_A(\mathbf{x})$ is the inverse Fourier transform of the indicator function of A (the function equal to one on A and zero elsewhere).

The eigenvalue problem has an infinite set of eigenpairs $\{\lambda_n, \varphi_n\}$ as solutions. If they are ordered in decreasing values of the eigenvalues, then we have

$$1 > \lambda_0 \geq \lambda_1 \geq \cdots > 0,$$

and the solution to the maximization problem (1) is given by φ_0. The value of the ratio in (1) is $\lambda_0 = \int_B |\varphi_0(\mathbf{x})|^2 d\mathbf{x}$ provided that the eigenfunctions are normalized ($||\varphi_n|| = 1$). These φ_n's constitute an orthonormal basis of the space $\mathcal{B}_A$. They also constitute an orthogonal basis of $L^2(B)$, but are not normalized in this space. To obtain an orthonormal basis we need merely divide by the eigenvalues, i.e., $\{\psi_n = \varphi_n/\sqrt{\lambda_n}\}$ is an orthonormal basis of $L^2(B)$.

1.2. **Estimating the intensity on B.** We now return to the problem of estimating the intensity on the set B given by $\int_B F(\mathbf{x})d\mathbf{x}$. The expansion of $F(\mathbf{x})$ in $L^2(B)$ in terms of the basis $\{\psi_n\}$ is used:

$$F(\mathbf{x}) = \sum_{n=0}^{\infty} \int_B F(\mathbf{t})\overline{\psi_n(\mathbf{t})}d\mathbf{t}\cdot\psi_n(\mathbf{x}), \quad \mathbf{x} \in B.$$

The desired intensity is therefore given exactly by

$$\int_B F(\mathbf{x})d\mathbf{x} = \sum_{n=0}^{\infty} \int_B F(\mathbf{t})\overline{\psi_n(\mathbf{t})}d\mathbf{t} \int_B \psi_n(\mathbf{x})d\mathbf{x}$$

$$= \int_B F(\mathbf{t})\overline{\psi_0(\mathbf{t})}d\mathbf{t} \int_B \psi_0(\mathbf{x})d\mathbf{x} + \sum_{n=1}^{\infty} \int_B F(\mathbf{t})\overline{\psi_n(\mathbf{t})}d\mathbf{t} \int_B \psi_n(\mathbf{x})d\mathbf{x}$$

$$(3) \qquad\qquad \approx \int_B F(\mathbf{t})\overline{\psi_0(\mathbf{t})}d\mathbf{t} \int_B \psi_0(\mathbf{x})d\mathbf{x}.$$

The last line in (3) is the first step in the approximation of the intensity on B. The first integral in this approximation is in turn approximated by the integral over all of the plane which is then expressed in terms of its Fourier transform,

$$
\int_B F(\mathbf{t})\overline{\psi_0(\mathbf{t})}d\mathbf{t} \approx \int_{\mathbb{R}^2} F(\mathbf{t})\overline{\psi_0(\mathbf{t})}d\mathbf{t}
$$
$$
= \frac{1}{(2\pi)^2}\int_{\mathbb{R}^2} f(\mathbf{k})\overline{\widehat{\psi}_0(\mathbf{k})}d\mathbf{k}
$$
(4)
$$
= \frac{1}{(2\pi)^2}\int_A f(\mathbf{k})\overline{\widehat{\psi}_0(\mathbf{k})}d\mathbf{k},
$$

where the last equality holds because the Fourier transform $\widehat{\psi}_0$ of ψ_0 has support on A. The other integral is approximated by

$$
(5) \qquad \int_B \psi_0(\mathbf{x})d\mathbf{x} \approx 1.
$$

These approximations are just the ones made in the works cited above [9]. They are not unreasonable since

(1) The solution φ_0 to the maximization problem has most of its energy ($= \lambda_0$) concentrated on B. This value can be made arbitrarily close to 1 by choosing A large enough.

(2) The energy of the other eigenfunctions $\varphi_1, \varphi_2, \cdots, \varphi_n, \cdots$ on B is given by

$$
\int_B |\varphi_n(\mathbf{x})|^2 d\mathbf{x} = \lambda_n, n = 1, 2, \ldots.
$$

By making A sufficiently small, these eigenvalues can be made close to zero. Thus the expansion coefficients of $F(\mathbf{x})$ (except the first) should be small.

(3) Since $\lambda_0 \approx 1$, the integrals in the last line of (3) may have ψ_0 replace by φ_0.

The second argument is not very convincing since the expansion coefficients are taken with respect to ψ_n and not φ_n. Even if $\int_B F(\mathbf{t})\overline{\varphi_n(\mathbf{t})}d\mathbf{t}$ is small $\int_B F(\mathbf{t})\overline{\psi_n(\mathbf{t})}d\mathbf{t}$ may not be. The authors in [9] argue instead that since the $\psi_n, n = 1, 2, \cdots$, have zeros in B while ψ_0 does not, the integral $\int_B F(\mathbf{t})\overline{\psi_n(\mathbf{t})}d\mathbf{t}$ should be small for non-negative $F(\mathbf{x})$.

A seeming contradiction occurs in these arguments: for λ_0 to be close to 1, A should be large, whereas for $\lambda_n, n > 0$, to be close to 0, A should be small. However, there is always a sharp drop in the eigenvalues for the PSWF from 1 to 0 as n increases, so it may be possible to satisfy both conditions simultaneously. In this work, we try to determine how good these approximations are and if they can be improved.

We have not mentioned a second part of the argument in these papers, viz., that among the sets A of measure a, the one which maximizes the energy on an ellipse B is a dual ellipse A. We defer an analysis of the error arising in this argument to a future work.

2. One-dimensional case

We first study the problem in one dimension to get a feel for the higher dimensional cases. This case is also quite well known and many of the calculations have already been made ([10],[4],[5]). It uses the fiirst prolate spheroidal wave function whose graph is shown in Figure 1.

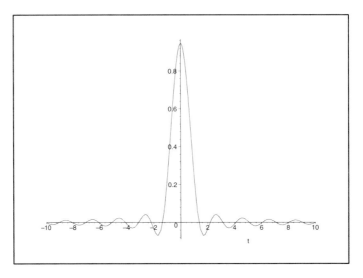

FIGURE 1. The first PSWF ψ_0 with parameter values $\sigma = \pi$ and $\tau = 1$.

In this case we set $B = [-\tau, \tau]$ and $A = [-\sigma, \sigma]$. We suppose again that $f(k)$ can be measured and try to estimate some of the errors introduced by the analogous approximations of the last section. We first look at the error arising from the approximations in (3), (4) and (5). We let ε_1 denote the error within the set B and ε_2 outside of the set B. Notice that ε_1 is the sum of the errors arising from the approximations in (3) and (5) and is expressed as

$$\varepsilon_1 = \int_B F(x)dx - \int_B F(t)\overline{\psi_0(t)}dt,$$

while ε_2 comes from the approximation in (4) and is

$$\varepsilon_2 = \int_B F(t)\overline{\psi_0(t)}dt - \int_{\mathbb{R}} F(t)\overline{\psi_0(t)}dt.$$

Then we have

$$\int_B F(x)dx = \varepsilon_1 + \int_B F(t)\overline{\psi_0(t)}dt$$
$$= \varepsilon_1 + \varepsilon_2 + \int_{\mathbb{R}} F(t)\overline{\psi_0(t)}dt$$
$$(6) \qquad\qquad = \varepsilon_1 + \varepsilon_2 + \frac{1}{2\pi}\int_A f(k)\widehat{\overline{\psi_0}}(k)dk$$

Hence all we need do is obtain small bounds on ε_1 and ε_2 to show that the method works in one dimension. Since the value of the last integral may be calculated directly from the data, the total error for this method is the sum of ε_1 and ε_2. But we would like to find a procedure independent of $F(x)$; to do so we use Schwarz's inequality:

$$\varepsilon_1^2 = |\int_B F(t)(1 - \overline{\psi_0(t)})dt|^2$$

$$\leq ||F||^2 \int_B |1 - \psi_0(x)|^2 dx$$

We denote by ε_3 the error of the factor independent of F,

(7) $$\varepsilon_3^2 = \int_B |1 - \psi_0(x)|^2 dx.$$

In order to get a lower bound on the estimate of ε_3, we use the fact that $|\int_{-\tau}^{\tau} \psi_0(x)dx|^2 \leq \int_{-\tau}^{\tau} 1 dx \int_{-\tau}^{\tau} |\psi_0(x)|^2 dx$ to get

$$\varepsilon_3^2 = \int_{-\tau}^{\tau} (1 - 2\psi_0(x) + \psi_0^2(x))dx$$

$$= 2\tau + 1 - 2\int_{-\tau}^{\tau} \psi_0(x)dx$$

$$\geq 2\tau + 1 - 2\sqrt{2\tau} = (\sqrt{2\tau} - 1)^2,$$

which is zero only at $\tau = 1/2$ and grows linearly in τ for large values of τ. The numerical estimates for the true values of ε_3 are given in Table 1.

In the same way a bound for the second error ε_2 may be found. In fact, we have

$$|\varepsilon_2| \leq \int_{\mathbb{R}-B} |F(t)\overline{\psi_0(t)}|dt$$

$$\leq \{\int_{\mathbb{R}-B} |F(t)|^2 dt \int_{\mathbb{R}-B} |\psi_0(t)|^2 dt\}^{1/2}$$

$$\leq ||F||\{\int_{\mathbb{R}} |\frac{\varphi_0(x)}{\sqrt{\lambda_0}}|^2 dx - \int_B |\psi_0(x)|^2 dx\}^{1/2}.$$

(8) $$= ||F|| \{1/\lambda_0 - 1\}^{1/2}$$

since φ_0 has been normalized on $\mathbb{R}$. We can denote by ε_4 the error of the factor independent of F given by

(9) $$\varepsilon_4^2 = \{1/\lambda_0 - 1\} = \int_{\mathbb{R}} |\psi_0(x)|^2 dx - 1.$$

If $\tau > 2$, for $\sigma = \pi$, then λ_0 is very close to 1 and this error is negligible. The values of λ_0 for $\sigma = \pi$ are shown in Table 1 for various values of τ. For example, for $\tau = 2$, this error will be less than 0.008. However for $\tau = 1$, the error will be about 0.14, hardly a negligible quantity.

Here we have used some of the well known properties of our PSWFs. We know [10] that ψ_0 has no zeros in the interval $[-\tau, \tau]$, is even and is monotone on $[0, \tau]$ with its max at 0. Its Fourier transform is

(10) $$\widehat{\psi_0}(k) = \sqrt{\frac{2\pi\tau}{\sigma\lambda_0}} \psi_0(\frac{\tau k}{\sigma})\chi_\sigma(k),$$

where χ_σ is the indicator function of the interval $[-\sigma, \sigma]$. This may then be used to calculate the final integral in (6).

τ	λ_0	ε_3	ε_4
0.2	0.38196	0.983	1.272
0.5	0.78314	0.176	0.526
1	0.98099	0.564	0.139
1.5	0.99889	0.975	3.363×10^{-2}
2	0.99994	1.301	7.746×10^{-3}
3	0.99999	1.771	3.162×10^{-3}
6	1.0000000000	2.907	0

Table 1. Bounds on the errors ε_3 and ε_4 for $\sigma = \pi$ and various values of τ.

In these calculations we have fixed σ at π while allowing τ to vary. We could just as easily have fixed τ and allowed σ to vary since the PSWFs depend only on the product of these two parameters. From the calculations we have made, it appears the best choice for this method for $\sigma = \pi$ is $\tau = 0.5$. But this will result in a total error of $\varepsilon_1 + \varepsilon_2 \leq 0.702\|F\|$. While this is an upper bound, it's the best we can do for all F, since Schwarz's inequality is exact for certain functions.

2.1. **An alternative.** We have seen that the method doesn't work well in the one dimensional case. The alternative to using ψ_0 in the approximations is to use the obvious choice, i.e., the indicator function χ_τ of the interval $[-\tau, \tau]$. Then both errors ε_1 and ε_2 are zero and the desired integral is given exactly by

$$(11) \qquad \int_B F(x)dx = \frac{\tau}{\pi} \int_{\mathbb{R}} f(k)s_\tau(k)dk,$$

where $s_\tau(k) = (\sin \tau k)/\tau k$. However another error is introduced if we are to approximate this last integral in (11) by the integral over a set A. It is

$$\varepsilon_5 = \frac{\tau}{\pi} \int_{\mathbb{R}-A} f(k)s_\tau(k)dk,$$

and is small when A is large, but converges to zero very slowly. Again an upper bound can easily be found by the same methods as in the last section. We have

$$\varepsilon_5^2 \leq \left(\frac{\tau}{\pi}\right)^2 \int_{\mathbb{R}-A} |f(k)|^2 dk \int_{\mathbb{R}-A} |s_\tau(k)|^2 dk$$

$$\leq \frac{4\tau^2}{\pi} \|F\|^2 \int_\sigma^\infty |\frac{\sin \tau k}{\tau k}|^2 dk$$

$$(12) \qquad = \frac{4\tau}{\pi} \|F\|^2 \int_{\tau\sigma}^\infty |\frac{\sin x}{x}|^2 dx = \varepsilon_6^2 \|F\|^2,$$

in which τ and σ are now decoupled. The last integral is bounded by 2π so that this error goes to zero as $\tau \to 0$. It also converges to 0 as $\sigma \to \infty$ independently of τ since we also have

$$(13) \qquad \varepsilon_6^2 = \frac{4}{\pi} \int_\sigma^\infty |\frac{\sin \tau k}{k}|^2 dk \leq \frac{4}{\pi\sigma}.$$

The convergence to 0 as $\tau \to \infty$ for fixed σ is more delicate but again follows from the above integrals. But we are interested in the case where B is small and A can be larger.

The question naturally arises as to whether the total error for this approach is smaller than that of the other. A plot of this error bound is shown in Figure 2 for $\sigma = \pi$ and 2π as a function of τ. Note that the error approaches zero as $\tau \to 0$, and hence the total error does as well since there is no other source of error.

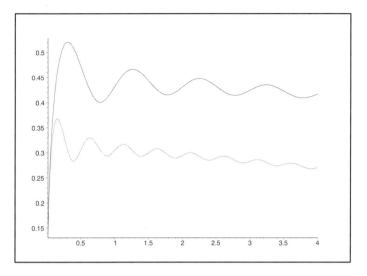

FIGURE 2. The error ε_6 as a function of τ for fixed $\sigma = \pi$ (top) and $\sigma = 2\pi$ (bottom).

If larger values of σ are chosen then the error will be even smaller. In any case, the error will always be smaller than the error in the best case in the last section which arose for $\sigma = \pi$, and $\tau = 1/2$. For these particular values we now have $\varepsilon_6 = 0.4720$. Hence this approach gives a slightly smaller error than the PSWF approach in this case, but does far better for other values of τ and σ as shown in Table 2.

τ	σ	ε_6
0.2	π	0.4954
0.2	2π	0.3580
0.2	4π	0.2005
0.5	π	0.4720
0.5	2π	0.3060
1	π	0.4328
1	2π	0.3051
2	4π	0.1911

Table 2. The error bounds associated with the alternate method for various values of τ and σ.

2.2. **A smoother alternative.** The apparent objection to this function is that the set B is known only approximately and hence the boundary should be a little fuzzy. This will not happen when the indicator function is used. But there appear to be many alternatives that can be used in this case, e.g., functions that are close to 1 in the interior of B but have smaller values close to the boundary. In fact there is an entire class of such functions: the Meyer wavelet scaling functions which can even be made infinitely differentiable [17].

One simple one, though not a C^∞ functon, is based on the raised cosine function [17, p. 51]

$$\theta(t) = (\frac{1}{2\varepsilon}(1 + \cos \pi \cdot /\varepsilon)\chi_\varepsilon * \chi_\tau)(t)$$

$$= \int_{t-\tau}^{t+\tau} \frac{1}{2\varepsilon}(1 + \cos \pi x/\varepsilon)\chi_\varepsilon(x)dx$$

$$= \begin{cases} 0, |t| > \tau + \varepsilon \\ 1, |t| < \tau - \varepsilon \\ \frac{1}{2} - \frac{(|t|-\tau)}{2\varepsilon} - \frac{1}{2\pi}\sin\frac{\pi(|t|-\tau)}{\varepsilon}, \tau - \varepsilon \le |t| \le \tau + \varepsilon \end{cases},$$

which is a good contiunous approximation to the indicator function of $[-\tau, \tau]$.

The use of this laterantive causes the errors ε_1 and ε_2 to be reintroduced, but reduces the error ε_5 since the Fourier transform of θ is

$$(14) \qquad \widehat{\theta}(k) = \frac{2\sin k\tau}{k}\{\frac{\sin k\varepsilon}{k\varepsilon} + \frac{1}{2\varepsilon}[\frac{\sin(k+\pi/\varepsilon)}{k+\pi/\varepsilon} + \frac{\sin(k-\pi/\varepsilon)}{k-\pi/\varepsilon}]\}$$

whose rate of decay is $O(k^{-2})$. This introduces another parameter ε as well which should be chosen to be small, to ensure that ε_3 and ε_4 are small. But ε will appear in the denominator of the expression for the error ε_5. The same inequality as in (13) shows that the error is dominated by a constant multiple of $1/\sigma^3\tau^3\varepsilon^2$. Thus by choosing $\sigma = 1/\varepsilon\tau$, say, the error ε_5 can also be made small, but requires that σ be large.

Other choices of θ are possible which give rates of decay $=O(k^{-p})$ for any p. They can be given by functions of the form

$$\theta(t) = \int_{t-\tau}^{t+\tau} h_\varepsilon(x)dx$$

where the positive function $h_\varepsilon \in C^\infty$, has compact support on $[-\varepsilon, \varepsilon]$, and has an integral over the real line equal to 1. This new θ has the same shape as the raised cosine but is infinitely differentiable, and thus its Fourier transform has more rapid decay. The two errors ε_1 and ε_2 are again small as in the raised cosine case, but the error ε_5 can now be made to converge to 0 more rapidly. We shall not explore these other alternatives in this work since the first alternative seems to have a smaller error than the PSWF approach, at least in one dimension.

3. Multidimensional alternatives

We can also adapt the one-dimensional alternatives to higher dimensions. The errors ε_1, ε_2, ε_3 and ε_4 are applicable to higher dimensions and the same short-comings hold as well: if B and A are chosen such that ε_1 is small then ε_2 will be large and vice-versa. We can find estimates in the same way, but only for certain regions. The formula (10) can be replaced by a similar formula if the regions satisfy certain symmetry conditions [11], but not otherwise.

Let A and B be scaled versions of each other and suppose that both are symmetric about the origin. We suppose that $A = cB$ and consider the eigenvalue problem

$$(15) \qquad \int_B e^{ic\mathbf{x}\cdot\mathbf{y}}\varphi(\mathbf{y})d\mathbf{y} = \alpha\varphi(\mathbf{x}).$$

This of course is not the same as (2), but is related to it as shown by Slepian [11]. In fact if ψ is a solution of (15) (normalized on B), it is also a solution to (2) and the eigenvalues are related by

$$(16) \qquad \lambda = (\frac{c}{2\pi})^d |\alpha|^2.$$

We again denote by ψ_0 and λ_0 the eigenpair corresponding to the largest eigenvalue. We see that (15) may then be expressed as

$$(2\pi)^{-d} \int_A e^{i\mathbf{x}\cdot\mathbf{t}} \varphi(\frac{\mathbf{t}}{c}) \frac{d\mathbf{t}}{c^d} = (2\pi)^{-d} \alpha \varphi(\mathbf{x}),$$

the inverse Fourier transform of $c^{-d}\varphi(\frac{\mathbf{t}}{c})\chi_A(\mathbf{t})$. Hence the Fourier transform of φ is

$$\widehat{\varphi}(\mathbf{k}) = \frac{(2\pi)^d}{\alpha c^d} \varphi(\frac{\mathbf{k}}{c}) \chi_A(\mathbf{k}),$$

and since the first eigenvalue of (2) is positive, we have

$$(17) \qquad \widehat{\psi_0}(\mathbf{k}) = \frac{(2\pi)^d}{\alpha_0 c^d} \psi_0(\frac{\mathbf{k}}{c}) \chi_A(\mathbf{k}) = \sqrt{\frac{(2\pi)^d}{\lambda_0 c^d}} \psi_0(\frac{\mathbf{k}}{c}) \chi_A(\mathbf{k})$$

which is the analog of (10).

Notice that the inequalities of the last section involving ε_2 still hold, i.e.

$$(18) \qquad |\varepsilon_2| \le ||F|| \{1/\lambda_0 - 1\}^{1/2}$$

whatever the dimension and whatever the two regions B and A. For ε_1, however, we have a slightly different notation. In terms of ε_3, it is

$$(19) \qquad \varepsilon_3^2 = \mu(B) + 1 - 2\int_B \psi_0(\mathbf{x})d\mathbf{x}$$
$$\ge \mu(B) + 1 - 2\sqrt{\mu(B)}.$$

The calculations of the eigenvalues and eigenfunctions cannot in general, be based on differential equations as in one dimension and usually must be based on the integral equation eigenvalue problem (2) or, in the case of symmetry, (15). The exceptions are the square and circular regions. In the former case we need merely take the tensor product of two one dimensional solutions; in the latter, Slepian [11] has introduced a procedure leading to a differential equation approach.

For certain regions we can find a formula for the kernel of the integral operator in closed form. This is clearly true for rectangles, but can also be found for disks and from it the formula for an ellipse. For the alternate method we need to find an expression for these kernels $S_B = \mathcal{F}(\chi_B)$ in $\mathbf{k}$-space. Hence our B sets should have one of these forms if we want a closed form expression.

Examples:
- The standard d-dimensional interval is $B = [-\tau, \tau]^d$; the kernel is $S_B(k_1, k_2, ..., k_d) = \frac{2\sin\tau k_1}{k_1} \cdot \frac{2\sin\tau k_2}{k_2} \cdot ... \cdot \frac{2\sin\tau k_d}{k_d}$.
- The ball in three dimensions $x_1^2 + x_2^2 + x_3^2 \le 1$ gives us $S_B(k_1, k_2, k_3) = \frac{4\pi}{|\mathbf{k}|^3}[\sin|\mathbf{k}| - |\mathbf{k}|\cos|\mathbf{k}|]$ where $|\mathbf{k}| = (k_1^2 + k_2^2 + k_3^2)^{1/2}$.
- The disk in two dimensions $x_1^2 + x_2^2 \le 1$ leads to the kernel $S_B(k_1, k_2) = 2\pi\int_0^1 J_0(|\mathbf{k}|r)rdr = 2\pi\frac{J_1(|\mathbf{k}|)}{|\mathbf{k}|}$ where $|\mathbf{k}| = (k_1^2 + k_2^2)^{1/2}$ and J_0, J_1 are the Bessel functions of order 0 and 1 respectively [8].

In these examples, the PSWFs turn out to be given by eigenfunctions of the integral operator with the same kernel (but with B replace by A) because of the symmetry and properties of the inverse Fourier transform [11] .

3.1. **The Square.** For both A and B given by squares (or, more generally, rectangles), no new calculations are needed since both the PSWFs and the kernel S_B are given by tensor products of the one dimensional functions. The errors arising when A is the square $[-\pi, \pi]^2$ and $B = [-\tau, \tau]^2$ and the estimations are based on the first PSWF, can be found by using the squares of the eigenvalues and the integrals for the one-dimensional case. For example, the error ε_3^2 in two dimensions is given by

$$(20) \quad (\varepsilon_3^2)^2 = (2\tau)^2 + 1 - 2(\int_{-\tau}^{\tau} \psi_0(x)dx)^2 = (2\tau)^2 + 1 - 2(\frac{2\tau + 1 - (\varepsilon_3^1)^2}{2})^2$$

where ψ_0 is the one-dimensional PSWF. For large values of τ, this error grows linearly in τ. Other values are given in Table 3.

τ	ε_3^2	ε_4^2
0.2	1.0643	2.4193
0.5	0.9597	0.7944
1	1.1848	0.1977

Table 3. A few values of the error bounds for the PSWF method in 2D for square regions.

The other error ε_4^2 can easily be calculated from the eigenvalues. It is given by

$$(21) \quad (\varepsilon_4^2)^2 = \{1/\lambda_0^2 - 1\}$$

and some values are also shown in Table 3. It's clear from Table 3, that the total error for this approach is quite large when the two regions B and A are both squares. If ε_3^2 is small then ε_4^2 is large and vise-versa just as in one dimension.

The alternate approach for these same squares again has no error in x-space but does have an error ε_6^2 in k-space as was the case in one dimension. Again we can express ε_6^2 in terms of similar error in one dimension by breaking up the region $R^2 - A$ into four infinite rectangles which are congruent to $[\sigma, \infty) \times (-\sigma, \infty)$. Hence from (13) we have

$$(22) \quad (\varepsilon_6^2)^2 = 4\{\frac{2}{\pi} \int_{\sigma}^{\infty} |\frac{\sin \tau k}{k}|^2 dk \cdot \frac{2}{\pi} \int_{-\sigma}^{\infty} |\frac{\sin \tau k}{k}|^2 dk\}$$
$$\leq 2(\varepsilon_6^1)^2(2\tau)$$

since by Parseval's equality $\frac{2}{\pi} \int_{-\infty}^{\infty} |\frac{\sin \tau k}{k}|^2 dk = \int_{-\tau}^{\tau} 1 dx = 2\tau$. Upper bounds are given in Table 4 for selected values of σ and τ.

τ	σ	ε_6^2
0.2	π	0.4431
0.2	2π	0.3202
0.2	4π	0.1793
0.5	2π	0.4327
1	2π	0.6102
2	4π	0.5405

Table 4. Bounds on the error for the alternate method in two dimensions with square regions.

Thus we see that for square regions the alternate method again has a smaller total error than the method based on the PSWF. The worst case in Table 4 is better than the best case in Table 3.

3.2. The Disk.

We now assume that both of the regions A and B are circular disks. We can no longer use the one dimensional results but must make new calculations. This requires that we use the function $S_B(k_1, k_2) = 2\pi \frac{J_1(|\mathbf{k}|)}{|\mathbf{k}|}$, $|\mathbf{k}| = \sqrt{k_1^2 + k_2^2}$ for B a unit disk, as our reproducing kernel in k−space and $S_A(x_1, x_2) = \frac{J_1(|\mathbf{x}|)}{2\pi|\mathbf{x}|}$ as the reproducing kernel in x-space when A is a unit disk. The latter is used to find the PSWFs and the former will be used in the alternate approach (We assume that the forward Fourier transform is from $x−$ space to $k−$space: $f(\mathbf{k}) = \int_{R^2} F(\mathbf{x}) e^{-i\mathbf{k}\cdot\mathbf{x}} d\mathbf{x}$ without the factor $(2\pi)^2$ in the denominator.)

In order to compare the results for the disk and the square we shall initially assume that the disk A has the same area as the square $[-\pi, \pi]^2$, i.e., has radius $r = 2\sqrt{\pi}$. Then the reproducing kernel becomes, by a change of scale, $S_A(x_1, x_2) = \frac{J_1(|2\sqrt{\pi}\mathbf{x}|)}{|\sqrt{\pi}\mathbf{x}|}$ and the problem of finding the first PSWF reduces to the eigenvalue problem

$$\lambda\varphi(\mathbf{x}) = \int_B \frac{J_1(|2\sqrt{\pi}(\mathbf{x} - \mathbf{y})|)}{\sqrt{\pi}|\mathbf{x} - \mathbf{y}|} \varphi(\mathbf{y}) d\mathbf{y}.$$

This may be put into polar form by taking $\mathbf{x} = r(\cos\theta, \sin\theta)$ and $\mathbf{y} = \rho(\cos\alpha, \sin\alpha)$ to get

$$\lambda\varphi(r\cos\theta, r\sin\theta)$$
$$= \int_0^b \int_0^{2\pi} \frac{J_1(|2\sqrt{\pi}((r\cos\theta, r\sin\theta) - (\rho\cos\alpha, \rho\sin\alpha)|)}{\sqrt{\pi}|(r\cos\theta, r\sin\theta) - (\rho\cos\alpha, \rho\sin\alpha)|} \varphi(\rho\cos\alpha, \rho\sin\alpha)\rho \, d\rho \, d\alpha$$
$$\text{(23)}$$
$$= \int_0^b \int_0^{2\pi} \frac{J_1(2\sqrt{\pi(r^2 + \rho^2 - 2r\rho\cos(\theta - \alpha))})}{\sqrt{\pi(r^2 + \rho^2 - 2r\rho\cos(\theta - \alpha))}} \varphi(\rho\cos\alpha, \rho\sin\alpha)\rho \, d\rho \, d\alpha$$

for a disk of radius b. The values of b that correspond to the squares given above, i.e., with the same area, are $\pi b^2 = 4\tau^2$ or $b = 2\tau/\sqrt{\pi}$. The error bounds are shown in Table 5 for the circular region; they may be compared to those for the square in Table 3 from which they do not differ markedly.

τ	λ_0	ε_3	ε_4
0.2	0.1482	0.9594	2.3976
0.5	0.6269	0.7017	0.7715
1	0.9744	1.1686	0.1621
1.5	0.9992	2.3135	0.0288
2	0.9999	3.3976	0.0069

Table 5. Error bounds for the PSWF method in 2D with circular regions.

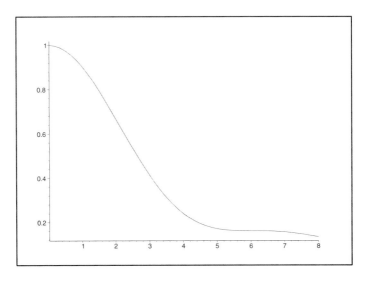

FIGURE 3. The error ε_6^2 as a function of σ with fixed $\tau = 0.5$.

The alternate approach again has no error in the $x-$space, but error is now introduced in $k-$space as before. The error ε_6 as in (22) is given by

$$
\begin{aligned}
\varepsilon_6^2 &= (2\pi)^{-2} \int_{R^2 - A} S_B^2(\mathbf{k}) d\mathbf{k} \\
&= \pi b^2 - (2\pi)^{-2} \int_0^a \int_0^{2\pi} \left(2\pi b \frac{J_1(b\rho)}{\rho} \right)^2 \rho \, d\rho \, d\theta \\
&= \pi b^2 - 2\pi b^2 \int_0^a \left(\frac{J_1(b\rho)}{\rho} \right)^2 \rho \, d\rho \\
&= \pi b^2 \left(1 - 2 \int_0^{ab} (J_1(r))^2 \frac{dr}{r} \right).
\end{aligned}
$$

Here a is the radius of the circle A and b is the radius of B. In order to compare them to the case of squares, we choose both circles to have the same area as the squares, i.e. $a = 2\sigma/\sqrt{\pi}$, $b = 2\tau/\sqrt{\pi}$. Then this error becomes

$$
\varepsilon_6^2 = 4\tau^2 \left(1 - 2 \int_0^{4\sigma\tau/\pi} (J_1(r))^2 \frac{dr}{r} \right).
$$

The plot of ε_6^2 as a function of σ with $\tau = 0.5$ is given in Figure 3. From this it seems apparent that for fixed τ, the error goes to 0 as $\sigma \to \infty$.

Similarly for fixed σ, the error goes to 0 as $\tau \to 0$ as in the case of the squares as can be seen in Figure 4 in which $\sigma = \pi$ and τ is allowed to vary.

The values of this error for various values of σ and τ are given in Table 6. These are again comparable to the error for square regions shown in Table 4.

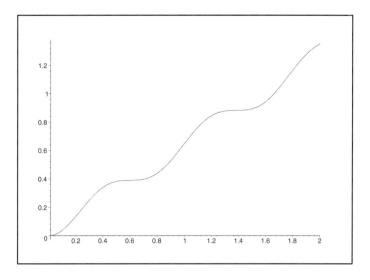

FIGURE 4. The error ε_6^2 as a function of τ with fixed $\sigma = \pi$.

τ	σ	ε_6
0.2	π	0.39968
0.2	2π	0.29180
0.2	4π	0.16532
0.5	2π	0.40260
1	2π	0.58144
2	4π	0.56246

Table 6. Error bounds alternate method on circular regions.

In both the case of the square and circular regions the error bounds have a similar pattern whether the PSWF method or the alternate method is used. The values for the disk in Table 6 and squares in Table 4 are within 10% of each other even though the estimate in Table 4 is rather crude. A comparison of Table 3 leads to a similar conclusion for both of the two errors found there for the square vs. the circular regions.

4. CONCLUSIONS

The object of these exercises has been to determine the best way of estimating the intensity of an image $\int_B F(\mathbf{x})d\mathbf{x}$ over ROI B when the size of B is small and the set A over which data is taken is also small. We have considered in this work only a method based on the PSFW and another based on the indicator function of the set B. In each of the three settings, the one-dimensional interval, two-dimensional square and two-dimensional disk, the latter method was found to have a smaller error. It is also simpler since it avoids the complex calculations associated with the PSWFs and uses functions which are given in closed form.

The reason the PSWF approach doesn't work very well is that if the size of B is chosen so small that the error within B is small, the eigenvalue will also be small and the error outside of B will be large. This can be compensated for by making A larger but because the error depends only on the product of the sizes of A and

B, this is equivalent to making B larger which results in error within B. If the error outside of B is to be small, the eigenvalue must be close to 1 and the error within B will be large since the other eigenvalues will be large as well. So we have a mathematical catch-22 for the PSWF method. This is not true for the alternate method since the two sizes of A and B are decoupled.

References

[1] Bouwkamp, C. J. (1947), On spheroidal wave functions of order zero, *Journal of Mathematics and Physics*, **26**, 79-92.

[2] Flammer, C. (1957), *Spheroidal Wave Functions*, Stanford University Press, Stanford, CA.

[3] Landau, H. J.and H. Widom (1980), The eigenvalue distribution of time and frequency limiting, *J. Math. Anal. Appl.* **77**, 469-481.

[4] Landau, H. J. and H.O. Pollak (1961), Prolate spheroidal wave functions, Fourier analysis and uncertainty, II, *Bell System Tech J.* **40**, 65-84.

[5] Landau, H. J. and H.O. Pollak (1962), Prolate spheroidal wave functions, Fourier analysis and uncertainty, III, *Bell System Tech J.* **41**, 1295-1336.

[6] Lindquist, M. (2003) Optimal data acquisition in fMRI using prolate spheroidal wave functions, *Int. J. Imaging Syst. Technol.,* **13,** 803-812.

[7] Meyer, Y. (1990), *Ondelettes et Op'erateurs* I. Herman, Paris.

[8] Papoulis, A. (1977), *Signal Analysis,* McGraw-Hill, New York.

[9] Shepp, L. and C.H. Zhang (2000), Fast functional magnetic resonance imaging via prolate wavelets, *Appl. Comp. Harmonic Anal.* **9,** 99-119.

[10] Slepian, D. and H.O. Pollak (1961), Prolate spheroidal wave functions, Fourier analysis and uncertainty, I, *Bell System Tech J.* **40,** 43-64.

[11] Slepian, D. (1964), Prolate spheroidal wave functions, Fourier analysis and uncertainty, IV, *Bell System Tech J.* **43**, 3009-3058.

[12] Slepian, D. (1983), Some comments on Fourier analysis, uncertainty,and modeling, *SIAM Review* **25**, 379-393.

[13] Volkmer, H. (2004), Spheroidal Wave Functions, in *Handbook of Mathematical Functions*, Nat. Bureau of Sts. Applied Math. Series.

[14] Walter, G. G. (2005), Prolate Spheroidal Wavelets: Differentiation, Translation, and Convolution Made Easy, *J. Fourier Analysis and Applications*, **11**, 73-84.

[15] Walter, G. G.and X. Shen (2004), Wavelets Based on Prolate Spheroidal Wave Functions, *J. Fourier Anal. and Appl.*, **10**, 1-26.

[16] Walter, G. G.and T. Soleski (2004), A friendly Method of Computing Prolate Spheroidal Wave Functions and Wavelets. *Preprint*

[17] Walter, G. G. and X. Shen (2001), *Wavelets and Other Orthogonal System, 2nd ed.,* CRC Press, Boca Raton, FL.

[18] Yang, Q.X., M. Lindquist, L. Shepp, C.H. Zhang, J. Wang, M.B. Smith (2002), Two dimensional prolate spheroidal wave functions for MRI, *J. Magn. Reson.* **158**, 43-51.

[19] Xiao, H, V. Rokhlin, N. Yarvin (2001), Prolate spheroidal wavefunctions, quadrature and interpolation, *Inverse Problems* **17**, 805–838.

[20] Zhang, S. and J.M. Jin (1996), *Computation of Special Functions*, Wiley, New York.

University of Wisconsin-Milwaukee
E-mail address: ggw@uwm.edu
URL: http://www.uwm.edu/~ggw

Current address: Marquette University
E-mail address: tigol@uwm.edu

Contemporary Mathematics
Volume **451**, 2008

STRUCTURE OF THE SET OF DYADIC PFW'S

H. ŠIKIĆ, D. SPEEGLE, AND G. WEISS

ABSTRACT. The purpose of this paper is to reveal the deep and rich structure of the set of Parseval frame wavelets. Two main directions are pursued. First, we study the reproducing properties of the translates of a Parseval frame wavelet within the closed linear span that the translates generate. In particular, we show that the translates need not have good reproducing properties, even though the translates and dilates form a Parseval frame. Second, we describe the effect of a semiorthogonalization procedure on the set of Parseval frame wavelets. Several examples illustrating the various possibilities are given.

1. INTRODUCTION

The study of various reproducing function systems has gained a lot of momentum in recent years. It is motivated on one side by the desire to understand and describe such systems; which is clearly related to some of the fundamental questions of mathematical analysis. On the other side, the tremendous success of orthonormal wavelets (which are particular examples of reproducing systems) in theory and practice has shown that such systems are more and more within our reach. At the same time the development of more general reproducing function systems has been well on its way.

The most basic case, which is the one studied in this paper, is the one of singly generated, one-dimensional systems where the generating groups are integer translations and dyadic dilations. More precisely, we shall analyze the systems of the form

$$\{\psi_{jk}(x)\} := \{2^{j/2}\psi(2^j x - k) : j, k \in \mathbb{Z}\}, \tag{1}$$

where $\psi \in L^2(\mathbb{R})$, and the system given in (1) satisfies the *reproducing property*, that is, for every $f \in L^2(\mathbb{R})$,

$$f = \sum_{j,k \in \mathbb{Z}} \langle f, \psi_{jk} \rangle \psi_{jk} \tag{2}$$

unconditionally in $L^2(\mathbb{R})$. This property is, of course, equivalent to the property that, for every $f \in L^2(\mathbb{R})$,

$$\|f\|_2^2 = \sum_{j,k \in \mathbb{Z}} |\langle f, \psi_{jk} \rangle|^2; \tag{3}$$

[0]The research of the first named author was supported by the MZOŠ grant of the Republic of Croatia and of the first and third named author by the US-Croatian grant NSF-INT-0245238. The second named author was supported by NSF DMS 0354957.

[0]*Keywords and Phrases.* Parseval frame wavelets, dimension function.

Date: December 28, 2007.

2000 *Mathematics Subject Classification.* 42C15.

which means that the system (1) is a normalized tight frame for $L^2(\mathbb{R})$ with upper and lower frame bounds 1. We denote by $\mathcal{P}$ the set of all functions $\psi \in L^2(\mathbb{R})$ such that ψ satisfies (2), and call an element of $\mathcal{P}$ a *Parseval frame wavelet (PFW)*. Many authors have studied various properties of PFW's, or of some subclasses of $\mathcal{P}$. Furthermore, studies of such systems have been generalized in many directions (multidimensional case, different generating groups, etc.). Despite this, some of the most fundamental questions about $\mathcal{P}$ are still unanswered. In particular, we do not know the full structure of the set $\mathcal{P}$, and we do not have the systematic classification of the various subclasses of $\mathcal{P}$. The purpose of this paper is to reveal the deep and rich structure of the set $\mathcal{P}$. In many ways, this is a continuation of [12, 13].

Already in [13], it was clear that the structure of $\mathcal{P}$ is potentially very rich, but we lacked some very basic answers to confirm our intuition. For example, it was not clear whether the various potential subclasses of $\mathcal{P}$ were empty. As it turned out, the structure is even more complex than we suspected, and in this paper we provide many examples (some of them highly non-trivial) to support this claim. In Section 2, we revisit (and somewhat reinterpret) the necessary basic theory, which we use in Section 3 to describe the rich structure of $\mathcal{P}$. In Section 4, we emphasize the MRA case even more. By formalizing the process of *semiorthogonalization*, we emphasize the limitations of the known theories and the importance of the filter based approach, which is grounded on ideas that go back to [18] and [11]. In particular, when we study the elements of $\mathcal{P}$ which are not semi-orthogonal, the shift-invariant spaces and MRA structure do not provide fine enough information, while the properties of the filters do.

We end this introduction with a figure that shows the structure of $\mathcal{P}$. The reader will not *a priori* understand all of the notation, but will need to go into sections 3 and 4. We do believe that the legend, though not being completely understandable to the reader without reading further, does give a description of what this work does. We will present not only results, but many examples. Some of these examples are technically complicated. Nevertheless, there are certain notions that we assume the reader is well acquainted with, such as MRA, non-MRA, semiorthogonality and Riesz bases. With this warning, we believe that the figure does indeed make the reader aware of the contents of the paper. Note that adjacency does not necessarily signify any relationship not already implied by inclusions, and that reflection about the middle line corresponds to toggling MRA and non-MRA.

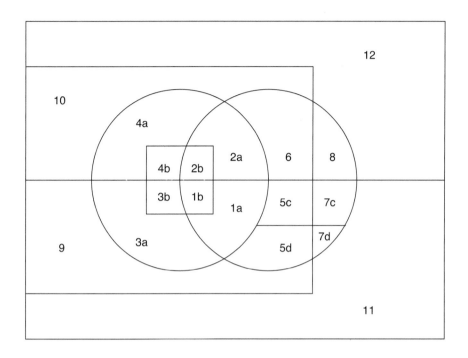

Legend

Odd numbers = MRA PFW's = $\mathcal{P}^{\mathrm{MRA}}$

Even numbers = $\mathcal{P}^{\mathrm{N}} = \mathcal{P} \setminus \mathcal{P}^{\mathrm{MRA}}$

1 = MRA orthonormal wavelets = $\mathcal{P}^{\mathrm{MRA}}_{tf,+}$

2 = non-MRA orthonormal wavelets = $\mathcal{P}^{\mathrm{N}}_{tf,+}$

3 = MRA semi-orthogonal, non-orthonormal PFW = $\mathcal{P}^{\mathrm{MRA}}_{tf,0}$

4 = semi-orthogonal non-MRA non-orthonormal PFW= $\mathcal{P}^{\mathrm{N}}_{tf,0}$

5 = MRA W_0 Riesz basis, non-orth. PFW = $\mathcal{P}^{\mathrm{MRA}}_{f,+}$

6 = non-MRA W_0 Riesz basis, non-orth. PFW = $\mathcal{P}^{\mathrm{N}}_{f,+}$

7 = MRA non W_0 frame, $\mathcal{I}$ 1-1 = $\mathcal{P}^{\mathrm{MRA}}_{0,+}$

8 = non-MRA non W_0 frame, $\mathcal{I}$ 1-1 = $\mathcal{P}^{\mathrm{N}}_{0,+}$

9 = MRA W_0 frame, non semi-orthogonal = $\mathcal{P}^{\mathrm{MRA}}_{f,0}$

10 = non-MRA W_0 frame, non semi-orthogonal = $\mathcal{P}^{\mathrm{N}}_{f,0}$

11 = MRA non W_0 frame, $\mathcal{I}$ not 1-1 = $\mathcal{P}^{\mathrm{MRA}}_{0,0}$

12 = non-MRA non W_0 frame, $\mathcal{I}$ not 1-1 = $\mathcal{P}^{\mathrm{N}}_{0,0}$

a $\cup$ b = semi-orthogonal PFW's

a = non-MSF semi-orthogonal PFW's

b = MSF PFS's

1 $\cup$ c $\cup$ d = MRA PFW's such that $\mathcal{I}$ is 1-1

1 $\cup$ c = $\mathcal{P}^{\mathrm{MRA}}_{-,1}$ = MRA PFW's whose semiorthogonalization yields o.n. wavelet

2. Preliminaries

In this section, we present the basic notions, known (and needed) facts, and some auxiliary results which are either of independent interest or are necessary for our main study. Our ambient space is $L^2(\mathbb{R})$, and for $\psi \in L^2(\mathbb{R})$, we shall often employ its Fourier transform $\hat{\psi}$, given by

$$\hat{\psi}(\xi) = \int_{\mathbb{R}} \psi(x) e^{-i\xi x}\, dx,$$

for $\psi \in L^1(\mathbb{R}) \cap L^2(\mathbb{R})$. Two basic operators are the translation operator T, defined by $T\psi(x) = \psi(x-1)$, and the dilation operator D, defined by $D\psi(x) = \sqrt{2}\psi(2x)$. They are both unitary operators on $L^2(\mathbb{R})$, and for $k \in \mathbb{Z}$, we denote their respective powers by T_k and D_k. It is often useful to have the translation projection τ on $\mathbb{R}$, defined by $\tau(\xi) = \eta$, where $\eta \in [-\pi, \pi)$ is such that $\xi - \eta = 2\pi k$ for some $k \in \mathbb{Z}$. The dilation projection d is defined on $\mathbb{R} \setminus \{0\}$ by $d(\xi) = \eta$, where η belongs to the so-called Shannon set, $S := [-2\pi, -\pi) \cup [\pi, 2\pi)$, and is such that $\eta/\xi = 2^k$ for some $k \in \mathbb{Z}$.

2.1. Shift-invariant spaces.
A closed subspace $V \subset L^2(\mathbb{R})$ is called *shift-invariant* if for every $f \in V$ and $k \in \mathbb{Z}$, we also have $T_k f \in V$. For a family $\mathcal{F} \subset L^2(\mathbb{R})$,

$$\langle \mathcal{F} \rangle := \overline{\text{span}\{T_k f : f \in \mathcal{F}, k \in \mathbb{Z}\}}$$

is the shift-invariant space generated by $\mathcal{F}$, i.e. the smallest shift-invariant space that contains $\mathcal{F}$. If $\mathcal{F} = \{f\}$ is a singleton, then $\langle f \rangle := \langle \{f\} \rangle$ is called a *principal shift-invariant space* (generated by f). The classical study of shift-invariant spaces in the context of separable Hilbert spaces goes back to H. Helson (see [9], for example). The techniques of shift-invariant spaces were used in the context of reproducing systems by C. de Boor, R. A. De Vore, and A. Ron (see, for example, [4]), and further developed in several papers by A. Ron and Z. Shen (for example, [14]). More recently, significant additions to the theory were obtained by M. Bownik [5], and by M. Bownik and Z. Rzeszotnik [6]. The ideas summarized by G. Weiss and E. Wilson [17] are closest to the spirit that we follow here. The results that we are going to present here (and use later in the article) are either in at least one of the articles mentioned here or are easily traceable to the results in these articles (and have become "folklore" by now).

Every shift-invariant space can be decomposed in terms of principal shift-invariant spaces. For a shift-invariant space V, there exists a countable family $\mathcal{F}$ (which is not unique) such that

$$V = \oplus_{f \in \mathcal{F}} \langle f \rangle \tag{4}$$

is the orthogonal sum of principal shift-invariant spaces [5][Theorem 3.3]. Hence, one would like to understand principal shift-invariant spaces. Since for $f \in L^2(\mathbb{R})$, $\langle f \rangle = \overline{\text{span}\{T_k f : k \in \mathbb{Z}\}}$, one obtains significant information on $\langle f \rangle$ by means of the periodization of $|\hat{f}|^2$

$$p_f(\xi) := \sum_{k \in \mathbb{Z}} |\hat{f}(\xi + 2k\pi)|^2, \qquad \xi \in \mathbb{R} \tag{5}$$

More precisely, let $L^2(\mathbb{T}, p_f)$ be the L^2 space on the torus $\mathbb{T} = \mathbb{R}/(2\pi\mathbb{Z})$ with the measure $p_f(\xi)d\xi$. For $f \in L^2(\mathbb{R})$, there is an isometric isomorphism

$$\mathcal{I} = \mathcal{I}_f : L^2(\mathbb{T}, p_f) \to \langle f \rangle, \tag{6}$$

given by $\mathcal{I}(t) := (t\hat{f})^{\vee}$, where we choose the inverse Fourier transform with the factor $\frac{1}{\sqrt{2\pi}}$, so that $(\hat{g})^{\vee} = g$ for every $g \in L^2(\mathbb{R})$. Consider also the set

$$U_f := \{\xi \in \mathbb{R} : p_f(\xi) > 0\}, \tag{7}$$

which is sometimes referred to as the spectrum of $\langle f \rangle$. Then, for $f, g \in L^2(\mathbb{R})$, we have that $\langle f \rangle = \langle g \rangle$ if and only if $g = \mathcal{I}_f(t)$ for some $t \in L^2(\mathbb{T}, p_f)$ such that $\text{supp}(t) = U_f$ a.e., where by supp we mean the ordinary set support of the function. In particular, $U_g = U_f$ a.e. Furthermore, for $f \in L^2(\mathbb{R})$, we define $h \in L^2(\mathbb{R})$ by

$$\hat{h}(\xi) := \frac{\hat{f}(\xi)}{\sqrt{p_f(\xi)}} \cdot \mathbf{1}_{U_f}(\xi), \, \xi \in \mathbb{R}, \tag{8}$$

and it follows that $\langle h \rangle = \langle f \rangle$ and that $\{T_k h : k \in \mathbb{Z}\}$ is a normalized (constant 1) tight frame for $\langle h \rangle$. In general, $\{T_k f : k \in \mathbb{Z}\}$ is a normalized (constant 1) tight frame for $\langle f \rangle$ if and only if

$$p_f = \mathbf{1}_{U_f} \text{ a.e.} \tag{9}$$

We emphasize and apply here yet another important function associated with shift-invariant spaces; the *dimension function*. This is a mapping $\dim_V : \mathbb{R} \to \{0\} \cup \mathbb{N} \cup \{\infty\}$, where V is a shift-invariant space, given by

$$\dim_V(\xi) := \text{dimension of } \overline{(\text{span}\{\hat{g}(\xi + 2k\pi) : k \in \mathbb{Z}) : g \in V\}}. \tag{10}$$

Observe that $(\hat{g}(\xi + 2k\pi) : k \in \mathbb{Z})$ is an element of $\ell^2(\mathbb{Z})$, so the definition above makes sense, and that actually we do not even have to put the span and the closure in (10), since it is well-known that $\{(\hat{g}(\xi + 2k\pi) : k \in \mathbb{Z}) : g \in V\}$ is a closed subspace of $\ell^2(\mathbb{Z})$. It is also known that $\dim_V$ is a measurable and 2π−periodic function. By (4) and the fact that

$$\dim_{\langle f \rangle} = \mathbf{1}_{U_f}, \tag{11}$$

it is very convenient (and possible) to choose the family $\mathcal{F}$ in (4) so that it forms a finite or infinite sequence $\{f_1, f_2, \ldots\}$ with the property that $U_{f_n} \supset U_{f_{n+1}}$ for all $n \in \mathbb{N}$. In this case, the dimension function of V has a particularly nice form. Observe that we also get that

$$Z_V := \{\xi : \dim_V(\xi) = 0\} = U_{f_1}^c, \tag{12}$$

and

$$I_V := \{\xi : \dim_V(\xi) = \infty\} = \cap_{n \in \mathbb{N}} U_{f_n}, \tag{13}$$

where in (13), the sets are decreasing.

The decomposition (4) has another interesting consequence. Since the dilation operator D is unitary, it follows that for every shift-invariant space V, the space DV is also shift-invariant, and that if V satisfies (4), then

$$DV = \oplus_{f \in \mathcal{F}} D\langle f \rangle. \tag{14}$$

Observe that, if we define $h(x) := \sqrt{2}f(2x - 1)$, then for every $k \in \mathbb{Z}$, $T_k Df = DT_{2k}f$, and $T_k h = DT_{2k+1}f$, i.e.,

$$D\langle f \rangle = \langle \{Df, h\} \rangle. \tag{15}$$

One then obtains a useful formula (for a much more general formula, see [6, Corollary 2.5])

$$\dim_{DV}(2\xi) = \dim_V(\xi) + \dim_V(\xi + \pi), \tag{16}$$

for a.e. $\xi \in \mathbb{R}$.

2.2. **Singly generated spaces.** Let $\psi \in L^2(\mathbb{R})$. It is useful to consider various "resolution levels" of ψ. Hence, for $j \in \mathbb{N} \cup \{0\}$, we define

$$W_j = W_j(\psi) := \overline{\text{span}\{\psi_{jk} : k \in \mathbb{Z}\}}, \tag{17}$$

and it is easy to see that all of them are shift-invariant space. Actually,

$$W_0 = \langle \psi \rangle, \qquad \text{and } D_j W_0 = W_j, \, j \geq 0. \tag{18}$$

Hence,

$$\dim_{W_0}(\xi) = \mathbf{1}_{U_\psi}(\xi) \, a.e.,$$

while (16) can be applied to find $\dim_{W_j}$. Observe that for $j < 0$, (17) does not define a shift-invariant spaces, so we consider

$$V_0 = V_0(\psi) := \langle \psi_{j,0} : j \in \mathbb{Z}, j < 0 \rangle, \tag{19}$$

which *is* a shift-invariant space. Related to it are $\ell^2(\mathbb{Z})$ vectors

$$\Psi_j(\xi) := \{\hat{\psi}(2^j(\xi + 2k\pi) : k \in \mathbb{Z}\}, \xi \in \mathbb{R}, \tag{20}$$

where $j \in \mathbb{Z}$ and $j \geq 0$; observe that the 2π-periodic, measurable function

$$\xi \to \|\Psi_j(\xi)\|_2^2 = \sum_{k \in \mathbb{Z}} |\hat{\psi}(2^j(\xi + 2k\pi))|^2$$

satisfies

$$\int_{-\pi}^{\pi} \|\Psi_j(\xi)\|_2^2 \, d\xi = \int_{\mathbb{R}} |\hat{\psi}(2^j u)|^2 \, du = \frac{1}{2^j} \|\hat{\psi}\|_2^2 < \infty. \tag{21}$$

The following useful formula is straightforward: for almost every $\xi \in \mathbb{R}$ and for every $j \in \mathbb{N} \cup \{0\}$,

$$\|\Psi_j(2\xi)\|_2^2 = \|\Psi_{j+1}(\xi)\|_2^2 + \|\Psi_{j+1}(\xi + \pi)\|_2^2. \tag{22}$$

By (10), it follows that

$$\dim_{V_0(\psi)}(\xi) = \dim \overline{\text{span}\{\Psi_j(\xi) : j \geq 1\}}, \tag{23}$$

where on the right hand side, we have the dimension function $\dim_\psi(\xi)$ used in [12, 13]. By $\dim_\psi$, we will mean $\dim_{V_0(\psi)}$, *not* the dimension function associated with W_0. Observe also that

$$p_\psi(\xi) = \|\Psi_0(\xi)\|_2^2, \xi \in \mathbb{R}, \tag{24}$$

and we will also use the function D_ψ, given by

$$D_\psi(\xi) := \sum_{j=1}^{\infty} \|\Psi_j(\xi)\|_2^2, \xi \in \mathbb{R}. \tag{25}$$

It is obvious that for a function $\psi \in L^2(\mathbb{R})$, the three functions $\dim_\psi$, D_ψ and p_ψ are measurable and 2π-periodic, and that they satisfy the following two properties:

$$D_\psi(2\xi) + p_\psi(2\xi) = D_\psi(\xi) + D_\psi(\xi + \pi), \, \xi \in \mathbb{R} \tag{26}$$

$$\int_{-\pi}^{\pi} D_\psi(\xi) \, d\xi = \int_{-\pi}^{\pi} p_\psi(\xi) \, d\xi = \|\hat{\psi}\|_2^2 < \infty. \tag{27}$$

In particular, D_ψ and p_ψ are finite a.e. The function $\dim_\psi$ can have infinite values, but in this case it can happen only with more restrictions than in the case of general shift-invariant spaces. Let us state and prove what we mean. By considering (22) coordinatewise, it is easy to get that for a.e. $\xi \in \mathbb{R}$,

$$\dim_\psi(2\xi) \geq \max(\dim_\psi(\xi) - 1, \dim_\psi(\xi + \pi) - 1). \tag{28}$$

Recall the following lemma (see [7], for example).

Lemma 2.1. *Let $E \subset \mathbb{R}$ be a measurable set such that $E + 2\pi = E$ and $2E \subset E$. Then, either $E = \mathbb{R}$ or $E = \emptyset$ a.e.*

Clearly, the set $I_{V_0(\psi)} = \{\xi : \dim_\psi(x) = \infty\}$ is 2π-periodic, while (28) implies that $2I_{V_0(\psi)} \subset I_{V_0(\psi)}$. By Lemma 2.1, we conclude:

Proposition 2.2. *If $\psi \in L^2(\mathbb{R})$, then either $I_{V_0(\psi)} = \mathbb{R}$ a.e. or $I_{V_0(\psi)} = \emptyset$ a.e.*

Let us also observe that

$$Z_{V_0(\psi)} = \{\xi : D_\psi(\xi) = 0\}. \tag{29}$$

Let us consider $DV_0(\psi)$; it is shift-invariant as we observed in (14). In order to describe it more precisely, we introduce the functions $h_j, j \in \mathbb{Z}, j < 0$ by

$$h_j(x) := \sqrt{2}\psi_{j,0}(2x - 1). \tag{30}$$

Since D is a unitary operator, we get $DV_0(\psi) = D(\overline{\text{span}\{T_k D_j \psi : j < 0, k \in \mathbb{Z}\}}) = \overline{\text{span}\{DT_k D_j \psi : j < 0, k \in \mathbb{Z}\}}$. Observe that for $k = 2\ell + 1$ odd, we get

$$DT_k D_j \psi = T_\ell h_j, \tag{31}$$

while for $k = 2\ell$ even, we get

$$DT_k D_j \psi = T_\ell D_{j+1} \psi. \tag{32}$$

Hence,

$$DV_0(\psi) = \langle\{\psi_{j,0} : j \leq 0\} \cup \{h_j : j < 0\}\rangle. \tag{33}$$

Since $\langle\psi_{j,0} : j \leq 0\rangle$ is equal to the sum of shift-invariant spaces $V_0(\psi) + W_0(\psi)$ (this sum is not necessarily orthogonal!), we get

$$V_0(\psi) \subset V_0(\psi) + W_0(\psi) \subset DV_0(\psi); \tag{34}$$

in particular, $V_0(\psi)$ is a *refinable* shift-invariant space (with respect to D). A consequence of (18) is then that

$$\overline{\text{span}\{\psi_{jk} : j, k \in \mathbb{Z}\}} \subset \overline{\cup_{j \in \mathbb{Z}} D_j V_0(\psi)}, \tag{35}$$

while (16) implies that

$$\dim_\psi(2\xi) \leq \dim_\psi(\xi) + \dim_\psi(\xi + \pi), \tag{36}$$

for a.e. $\xi \in \mathbb{R}$ (for a more general result, see [6, Theorem 3.2]).

2.3. **Parseval frame wavelets.** Suppose now that $\psi \in \mathcal{P}$, i.e. ψ is a PFW. Recall (see Chapter 7 in [10]) that for $\psi \in L^2(\mathbb{R})$, we have that $\psi \in \mathcal{P}$ if and only if

$$\sum_{j \in \mathbb{Z}} |\hat{\psi}(2^j \xi)|^2 = 1 \ a.e. \tag{37}$$

and

$$t_q(\xi) := \sum_{j \geq 0} \hat{\psi}(2^j \xi)\overline{\hat{\psi}(2^j(\xi + 2q\pi))} = 0 \ a.e. \tag{38}$$

whenever q is an odd integer. As a consequence, we have that $\|\psi\|_2 \leq 1$ and $|\hat{\psi}(\xi)| \leq 1$ a.e., for every $\psi \in \mathcal{P}$.

Furthermore, the reproducing property and (35) imply that for $\psi \in \mathcal{P}$, we have

$$\overline{\cup_{j \in \mathbb{Z}} D_j V_0(\psi)} = L^2(\mathbb{R}). \tag{39}$$

Remark 2.3. Since $V_0(\psi)$ is refinable, the question now becomes whether it is possible to have $\psi \in \mathcal{P}$ such that

$$V_0(\psi) = DV_0(\psi).$$

Observe that for such ψ, we have by (39) that $V_0(\psi) = L^2(\mathbb{R})$, and therefore, that $\dim_\psi = \infty$ a.e.

In the case that $\psi \in \mathcal{P}$, there are some additional useful properties. The results from the following proposition are essentially from [13, 15].

Proposition 2.4. *Suppose that $\psi \in \mathcal{P}$. Then*

(i) $p_\psi \leq 1$ *a.e.*
(ii) $D_\psi \leq \dim_\psi$ *a.e.*
(iii) $\liminf_{n \to \infty} D_\psi(2^{-n}\xi) \geq 1$ *for a.e. $\xi \in \mathbb{R}$.*

Part 1 of Proposition 2.4 enables us to consider another useful operator defined in the context of $\mathcal{I}_\psi$ and (6). Consider the operator $\tilde{\mathcal{I}} = \tilde{\mathcal{I}}_\psi : L^2(\mathbb{T}, dx) \to L^2(\mathbb{T}, p_\psi)$ given by $\tilde{\mathcal{I}}(f) = f$. Observe that by Proposition 2.4.1, we have that for $\psi \in \mathcal{P}$, the operator $\tilde{\mathcal{I}}_\psi$ is a bounded (norm less than 1) linear operator. The adjoint operator $\tilde{\mathcal{I}}^* : L^2(\mathbb{T}, p_\psi) \to L^2(\mathbb{T}, dx)$ is given by $\tilde{\mathcal{I}}^*(g) = p_\psi g$. Since both L^2 spaces involved contain bounded, measurable functions as a dense subset, it follows that the range of $\tilde{\mathcal{I}}$ is dense in $L^2(\mathbb{T}, p_\psi)$. Hence, the kernel of $\tilde{\mathcal{I}}^*$ is trivial; that is, $\tilde{\mathcal{I}}^*$ is injective. Consider the standard orthonormal basis $\{e_k(\xi) = \frac{e^{ik\xi}}{\sqrt{2\pi}} : k \in \mathbb{Z}\}$ of the space $L^2(\mathbb{T}, dx)$ and observe that, for every $k \in \mathbb{Z}$,

$$\mathcal{I}_\psi(\tilde{\mathcal{I}}_\psi(e_k)) = \psi_{0k} \in W_0, \tag{40}$$

and $\mathcal{I}_\psi \circ \tilde{\mathcal{I}}_\psi : L^2(\mathbb{T}, dx) \to W_0$. As we have already seen in [13], the properties of $\{\psi_{0k} : k \in \mathbb{Z}\}$ within W_0 play a major role in the analysis of $\mathcal{P}$. Recall that we say that $\psi \in \mathcal{P}$ is a W_0-*frame (W_0-Parseval frame, W_0-Riesz basis)* if $\{\psi_{0k} : k \in \mathbb{Z}\}$ is a frame (Parseval frame, Riesz basis; respectively) for W_0.

Let us emphasize yet another point; the basic idea most likely goes back to L. W. Baggett. For $\psi \in \mathcal{P}$, the space

$$\{f \in L^2(\mathbb{R}) : \|f\|_2^2 = \sum_{j \geq 0} \sum_{k \in \mathbb{Z}} |\langle f, \psi_{jk} \rangle|^2\} \tag{41}$$

is shift-invariant (since $j \geq 0$), and is the orthogonal complement of

$$\overline{\text{span}\{\psi_{jk} : k \in \mathbb{Z}, j < 0\}}.$$

It follows now easily, compare with (34), that for $\psi \in \mathcal{P}$,

$$V_0(\psi) = \overline{\text{span}\{\psi_{jk} : k \in \mathbb{Z}, j < 0\}} \tag{42}$$

and that

$$DV_0(\psi) = V_0(\psi) + W_0(\psi). \tag{43}$$

Remark 2.5. One should not conclude too much from this. Observe that the sum in (43) is not necessarily orthogonal and that we do not know immediately that $W_0(\psi)$ is not contained within $V_0(\psi)$. Hence, the question raised in Remark 2.3 remains; with some additional refinements. Namely, it follows that for $\psi \in \mathcal{P}$, the following statements are equivalent:

(i) $V_0(\psi) = L^2(\mathbb{R})$,
(ii) $V_0(\psi) = DV_0(\psi)$,

(iii) $\psi \in V_0(\psi)$,

(iv) the space in (41) is trivial

2.4. MRA Parseval frame wavelets.

Following [12], we add even more structure by assuming that $\mathcal{P}$ has a corresponding filter. A *generalized filter* is a measurable, 2π-periodic function $m : \mathbb{R} \to \mathbb{C}$ which satisfies

$$|m(\xi)|^2 + |m(\xi + \pi)|^2 = 1, \tag{44}$$

for a.e. $\xi \in \mathbb{R}$. A function $\varphi \in L^2(\mathbb{R})$ will be called a *pseudo-scaling function* if there exists a generalized filter m such that

$$\hat{\varphi}(2\xi) = m(\xi)\hat{\varphi}(\xi), \tag{45}$$

for a.e. $\xi \in \mathbb{R}$. A PFW ψ is an *MRA PFW* if there exist a pseudo-scaling function φ and an associated generalized filter m such that

$$\hat{\psi}(2\xi) = e^{i\xi}\overline{m(\xi + \pi)}\hat{\varphi}(\xi), \tag{46}$$

for a.e. $\xi \in \mathbb{R}$. We denote the set of all MRA PFW-s by $\mathcal{P}^{MRA}$. As it was proven in [12], if $\psi \in \mathcal{P}^{MRA}$ and m is its associated filter, then m has to be a *generalized low pass filter*, i.e. for a.e. $\xi \in \mathbb{R}$,

$$\lim_{n \to \infty} \prod_{j=1}^{\infty} |m(\frac{2^{-n}\xi}{2^j})| = 1. \tag{47}$$

It was proven in [11] that for a generalized filter, the limit in (47) always exists and is either 0 or 1. Furthermore, starting with any generalized low pass filter, we can build (using the multiplier techniques explained in [12]) an MRA PFW whose associated filter is the starting one.

One of the key results in [13] is the characterization of MRA PFW's; for $\psi \in \mathcal{P}$, we have

$$\psi \in \mathcal{P}^{MRA} \iff \dim_\psi \in \{0, 1\}. \tag{48}$$

Let us observe some consequences of this results. Obviously, for $\psi \in \mathcal{P}^{MRA}$, we have

$$0 \le D_\psi(\xi) \le \dim_\psi(\xi) = \mathbf{1}_{Z^c_{V_0(\psi)}}(\xi), \tag{49}$$

for a.e. $\xi \in \mathbb{R}$. It is also immediate that

$$I_{V_0(\psi)} = \emptyset \text{ a.e. and } \psi \notin V_0(\psi). \tag{50}$$

Furthermore, using the expression for the Fourier transform of ψ_{jk}, (46) and (6), it is clear that

$$V_0(\psi) \subset \langle \varphi \rangle,$$

while (48) shows that $V_0(\psi) = \langle f \rangle$, for some f. Observe that

$$|\hat{\varphi}(\xi)|^2 = \sum_{j=1}^{\infty} |\hat{\psi}(2^j\xi)|^2, \tag{51}$$

for a.e. $\xi \in \mathbb{R}$, which also shows that

$$D_\psi = p_\varphi. \tag{52}$$

It follows that

$$U^c_\varphi = Z_{V_0(\psi)}, \tag{53}$$

which implies the following result.

Proposition 2.6. *Suppose that $\psi \in \mathcal{P}$. Then $\psi \in \mathcal{P}^{MRA}$ if and only if $V_0(\psi)$ is a principal shift-invariant space. If $\psi \in \mathcal{P}^{MRA}$ and φ is the associated pseudo-scaling function, then*

$$V_0(\psi) = \langle \varphi \rangle.$$

Remark 2.7. One has to be somewhat careful in applying the above result. For example, we could have that $V_0(\psi) = \langle \varphi_1 \rangle$, but that does not necessarily mean that φ_1 is the pseudo-scaling function associated with ψ in the sense of (46). We will return to this issue in more detail in Section 4.

We complete this section with a few more simple and useful technical details. Observe that (49) and Proposition 2.4, part (3), imply that for every $\psi \in \mathcal{P}^{\mathrm{MRA}}$, we have

$$\lim_{n \to \infty} D_\psi(\frac{\xi}{2^n}) = \lim_{n \to \infty} \dim_\psi(\frac{\xi}{2^n}) = 1, \tag{54}$$

for a.e. $\xi \in \mathbb{R}$.

Lemma 2.8. *Suppose $\psi \in \mathcal{P}^{MRA}$ and φ and m are associated pseudo-scaling function and filter, respectively. Then, for a.e. $\xi \in \mathbb{R}$,*

    (i) $D_\psi(2\xi) = |m(\xi)|^2 D_\psi(\xi) + |m(\xi + \pi)|^2 D_\psi(\xi + \pi)$,
    (ii) $p_\psi(2\xi) = |m(\xi + \pi)|^2 D_\psi(\xi) + |m(\xi)|^2 D_\psi(\xi + \pi)$,
    (iii) $D_\psi(2\xi), p_\psi(2\xi) \in [\min\{D_\psi(\xi), D_\psi(\xi + \pi)\}, \max\{D_\psi(\xi), D_\psi(\xi + \pi)\}]$.

Proof The proof of (ii) goes along the same line as the proof of (i), while (iii) is an obvious consequence of (i) and (ii). Hence, we prove (i) using (45) and (46). Indeed,

$$\begin{aligned} D_\psi(2\xi) &= \sum_{k \text{ even}} |m(\xi + k\pi)|^2 |\hat{\varphi}(\xi + k\pi)|^2 + \sum_{k \text{ odd}} |m(\xi + k\pi)|^2 |\hat{\varphi}(\xi + k\pi)|^2 \\ &= |m(\xi)|^2 p_\varphi(\xi) + |m(\xi + \pi)|^2 p_\varphi(\xi + \pi). \end{aligned}$$

$\square$

3. STRUCTURE OF $\mathcal{P}$

We shall introduce some additional notation in order to be able to go through the various subclasses of $\mathcal{P}$ in a systematic way. We have already seen the first natural breaking point, a PFW ψ can be MRA, i.e., $\psi \in \mathcal{P}^{\mathrm{MRA}}$ or non-MRA. We denote the collection of non-MRA PFW's by $\mathcal{P}^{\mathrm{N}} := \mathcal{P} \setminus \mathcal{P}^{\mathrm{MRA}}$. We will also use two subscripts. The first, say x, will indicate which properties (of frames) the family $\{\psi_{0k} : k \in \mathbb{Z}\}$ has with respect to W_0, while the second, say y, will indicate which property $\tilde{\mathcal{I}}_\psi$ has, given in terms of p_ψ. Hence, we will always have

$$\mathcal{P}_{x,y} = \mathcal{P}_{x,y}^{\mathrm{N}} \cup \mathcal{P}_{x,y}^{\mathrm{MRA}}, \tag{55}$$

with the union being disjoint.

3.1. Non W_0-frames such that $\tilde{\mathcal{I}}_\psi$ is not injective. This is the class of PFW's with the least amount of structure, and its precise definition and our notation for it is as follows:

$$\mathcal{P}_{0,0} := \{\psi \in \mathcal{P} : \psi \text{ is not a } W_0\text{-frame and the kernel of } \tilde{\mathcal{I}}_\psi \text{ is not trivial}\}. \tag{56}$$

If we denote the Lebesgue measure of a measurable set $A \subset \mathbb{R}$ by $|A|$ and we use the known fact (see [13]) that for $\psi \in \mathcal{P}$, ψ is a W_0-frame if and only if

$$(\exists 0 < c \leq 1) \ (p_\psi \geq c\mathbf{1}_{U_\psi} \text{ a.e.}), \tag{57}$$

then we have the following characterization of the set $\mathcal{P}_{0,0}$ and its subsets $\mathcal{P}_{0,0}^N$ and $\mathcal{P}_{0,0}^{MRA}$.

Proposition 3.1. *Suppose $\psi \in \mathcal{P}$. Then, $\psi \in \mathcal{P}_{0,0}$ if and only if*

(i) $|U_\psi^c| > 0$, *and*

(ii) $|\{\xi : 0 < p_\psi(\xi) < \epsilon\}| > 0$, *for every $\epsilon > 0$.*

Furthermore, if $\psi \in \mathcal{P}_{0,0}$, then $\psi \in \mathcal{P}_{0,0}^{MRA}$ if and only if $\dim_\psi = \mathbf{1}_{Z_{V_0(\psi)}^c}$, which is equivalent to $\dim_\psi$ taking only the values of 0 and 1.

Observe that none of the conditions in Proposition 3.1 can be removed. However, it is not *a priori* clear that the classes $\mathcal{P}_{0,0}^N$ and $\mathcal{P}_{0,0}^{MRA}$ (and even $\mathcal{P}_{0,0}$ itself) are non-empty. The following examples show that they are.

Example 3.2. *We show $\mathcal{P}_{0,0}^N \neq \emptyset$.*

Consider the set $F := [\frac{3}{2}\pi, \frac{5}{3}\pi)$. It is easy to see that $d(F + 4\pi) = [\frac{11}{8}\pi, \frac{17}{12}\pi)$ and that

$$d|_{F \cup (F+4\pi)} \text{ is injective} \tag{58}$$

Observe also that $\tau(F) = [-\frac{1}{2}\pi, -\frac{1}{3}\pi)$, and $\tau(2F) = [-\pi, -\frac{2}{3}\pi)$. Hence,

$$0 \notin \overline{\tau(F \cup 2F)} \tag{59}$$

and

$$\tau|_{F \cup 2F} \text{ is injective.} \tag{60}$$

By (59), there exists $J \in \mathbb{N} \cup \{0\}$ such that for each $j \geq J$,

$$\left[-\frac{2\pi}{2^j}, \frac{2\pi}{2^j}\right] \cap \overline{\tau(F \cup 2F)} = \emptyset. \tag{61}$$

We define the set E by

$$E := \left(2^{-J} d(F \cup (F + 4\pi))\right)^c \cap \left(\left[-\frac{2\pi}{2^J}, -\frac{\pi}{2^J}\right) \cup \left[\frac{\pi}{2^J}, \frac{2\pi}{2^J}\right)\right).$$

Observe that E and F are two measurable sets of positive Lebesgue measure such that

$$d|_{E \cup F \cup (F+4\pi)} \text{ is a bijection,} \tag{62}$$

$$\tau|_{F \cup 2F \cup \tilde{E}} \text{ is injective, where } \tilde{E} := \cup_{j=0}^\infty 2^{-j}E, \text{ and} \tag{63}$$

$$2^{-j}E, j \geq 0; F; F + 4\pi; 2F; 2F + 8\pi \text{ are pairwise disjoint.} \tag{64}$$

It is easy to check (64) directly, (62) follows from (58) and the definition of E, while (63) follows from (60) and (61).

Consider now any $a \in (0, \frac{1}{2})$ and define $\psi \in L^2(\mathbb{R})$ such that

$$\hat{\psi}(\xi) = \begin{cases} -\sqrt{\frac{1}{2} + a}, & \xi \in 2F \\ \sqrt{\frac{1}{2} - a}, & \xi \in F \cup (2F + 8\pi) \\ \sqrt{\frac{1}{2} + a}, & \xi \in F + 4\pi \\ \sqrt{2^{-(j+1)}}, & \xi \in 2^{-j}E, j \geq 0 \\ 0, & \text{otherwise.} \end{cases} \tag{65}$$

We show that $\psi \in \mathcal{P}$. We need to show that $\hat{\psi}$ satisfies (37) and (38). Observe that (62) implies that it is enough to check (37) for $\xi \in E \cup F \cup (F + 4\pi)$. This, then, follows directly from (65).

In order to prove that (38) is satisfied, it is not difficult to see that the interesting cases are when $\xi \in \mathbb{R}$ and $q \in 2\mathbb{Z}+1$ such that there exists $j \geq 0$ with $2^j\xi \in \text{supp}(\hat{\psi})$ and $2^j(\xi + 2q\pi) \in \text{supp}(\hat{\psi})$. Without loss of generality, we consider the case $q > 0$. Since $\tau(2^j\xi) = \tau(2^j(\xi + 2q\pi))$, it follows by (63) that these conditions on j and q are satisfied only in the case that either $(2^j\xi \in F$ and $2^j(\xi + 2q\pi) \in F + 4\pi)$ or $(2^j\xi \in 2F$ and $2^j(\xi + 2q\pi) \in 2F + 8\pi)$. In the first case, we get $2^j q = 4$, which implies $q = 1$ and $j = 1$. In the second case, we get $2^j q = 8$, which implies that $q = 1$ and $j = 2$. Hence, in both cases, $\xi \in F/2$, so we get

$$t_q(\xi) = \sqrt{\frac{1}{2} - a}\sqrt{\frac{1}{2} + a} + \left(-\sqrt{\frac{1}{2} + a}\right)\sqrt{\frac{1}{2} - a} = 0.$$

It remains to prove that $\psi \in \mathcal{P}_{0,0}^N$; that is, to check that 1 and 2 from Proposition 3.1 hold, and that $\dim_\psi$ attains values bigger than 2 on a set of positive measure. Notice that

$$\tau(\text{supp}\hat{\psi}) = \tau(F \cup 2F \cup \tilde{E}) \subset [-\pi, -\frac{2}{3}\pi) \cup [-\frac{1}{2}\pi, -\frac{1}{3}\pi) \cup [-\frac{2\pi}{2^J}, \frac{2\pi}{2^J}).$$

Since $p_\psi = 0$ outside the set above (which does not include all of $[-\pi, \pi)$), condition i) is clearly fulfilled.

For $\xi \in 2^{-j}E$, we get $p_\psi(\xi) = |\hat{\psi}(\xi)|^2 = 2^{-(j+1)}$. Since $|2^{-j}E| > 0$, for every $j \geq 0$, and $\lim_{j \to \infty} 2^{-(j+1)} = 0$, we obtain ii).

For the last requirement, consider $\xi \in F/2$; recall that $|F/2| > 0$. Observe that among the vectors $\Psi_j(\xi)$ we have

$$\left(\ldots, 0, 0, \sqrt{\frac{1}{2} - a}, \sqrt{\frac{1}{2} + a}, 0, 0, \ldots\right)$$

and

$$\left(\ldots, 0, 0, -\sqrt{\frac{1}{2} + a}, \sqrt{\frac{1}{2} - a}, 0, 0, \ldots\right)$$

which are linearly independent. Hence, $\dim_\psi(\xi) \geq 2$ for $\xi \in F/2$. □

Example 3.3. *We show $\mathcal{P}_{0,0}^{MRA} \neq \emptyset$.*

Consider [13, Example 2.6] which give $\psi \in \mathcal{P}^{\text{MRA}}$ such that the associated scaling pseudo-scaling function φ is given by

$$\hat{\varphi}(\xi) = \begin{cases} 1, & |\xi| \leq \frac{\pi}{4} \\ -\frac{4}{\pi}|\xi| + 2, & \frac{\pi}{4} \leq |\xi| \leq \frac{\pi}{2} \\ 0 & \text{otherwise.} \end{cases}$$

It is not difficult to check then that the graphs of D_ψ and p_ψ are such that $\psi \in \mathcal{P}_{0,0}^{\text{MRA}}$.

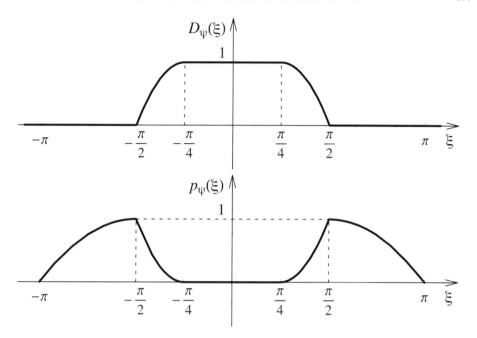

3.2. Non W_0-frames such that $\tilde{\mathcal{I}}_\psi$ is injective.

The notation for this collection of functions is

$$\mathcal{P}_{0,+} := \{\psi \in \mathcal{P} : \psi \text{ is not a } W_0\text{-frame and } \ker(\tilde{\mathcal{I}}_\psi) = \{0\}\} \qquad (66)$$

Observe that $\mathbf{1}_{U_\psi^c \cap [-\pi,\pi)}$ is equal to zero when considered in $L^2(\mathbb{T}, p_\psi)$. Then, it is straightforward to get

$$\ker(\tilde{\mathcal{I}}_\psi) = \{0\} \iff |U_\psi^c| = 0. \qquad (67)$$

Hence, we get the following characterization of $\mathcal{P}_{0,+}$ in which none of the conditions are redundant.

Proposition 3.4. *Suppose $\psi \in \mathcal{P}$. Then, $\psi \in \mathcal{P}_{0,+}$ if and only if*

  (i) $|U_\psi^c| = 0$ *and*
  (ii) $|\{\xi : 0 < p_\psi(\xi) < \epsilon\}| > 0$ *for every $\epsilon > 0$.*

The only construction of examples in $\mathcal{P}_{0,+}$ that we know of is somewhat delicate. We recall the following theorem, which is a special case of [8, Theorem 1] along the lines of [16, Theorem 1.1].

Theorem 3.5 (DLS)**.** *Let $E \subset [-\pi, \pi)$ and $F \subset [-2\pi, -\pi) \cup [\pi, 2\pi)$ be measurable sets such that $0 \in E^\circ$ and $E^\circ \neq 0$. Then, there exists measurable $G \subset \mathbb{R}$ such that $\tau|_G$ and $d|_G$ are injective functions with $\tau(G) = E$ and $d(G) = F$.*

Example 3.6. *We show $\mathcal{P}_{0,+} \neq \emptyset$.*

Consider sets $E := [-\frac{2\pi}{3}, \frac{\pi}{3})$ and $F := [-2\pi, -\pi)$. Apply Theorem 3.5 to obtain the set $G \subset \mathbb{R}$ such that $\tau : G \to E$ and $d : G \to F$ are measurable bijections. Observe that G is necessarily a subset of $(-\infty, 0)$. Consider $0 < \alpha < \frac{1}{2}$ and sets $H_1 := [\frac{\pi}{3}, \frac{2\pi}{3}), H_2 := 2H_1 = [\frac{2\pi}{3}, \frac{4\pi}{3})$. Observe that

$$\tau|_{H_1 \cup H_2 \cup G} : H_1 \cup H_2 \cup G \to [-\pi, \pi) \text{ is a bijection.} \qquad (68)$$

We define $\psi \in L^2(\mathbb{R})$ by

$$
\hat{\psi}(\xi) := \begin{cases}
\min(|\xi - \pi|^{\alpha}, \sqrt{\frac{1}{2}}) & \xi \in H_2 \\
\sqrt{1 - |\min(|2\xi - \pi|^{\alpha}, \sqrt{\frac{1}{2}})|^2} & \xi \in H_1 \\
1 & \xi \in G \\
0 & \text{otherwise.}
\end{cases} \tag{69}
$$

Note that on H_1, $\hat{\psi}(\xi) = \sqrt{1 - |\hat{\psi}(2\xi)|^2}$. See that $\hat{\psi}$ satisfies (37), since for $\xi > 0$ the dyadic orbit of ξ hits H_1 and H_2 exactly once, and for $\xi < 0$, it hits G only once. It is also easy to check (38), since $\hat{\psi}(\xi) \neq 0$ implies that $\hat{\psi}(\xi + 2k\pi) = 0$ for every $0 \neq k \in \mathbb{Z}$. Therefore, $\psi \in \mathcal{P}$.

Property (ii) from Proposition 3.4 follows from the definition of $\hat{\psi}$ on H_2, since $\hat{\psi}(\pi) = 0$ and $\hat{\psi}$ continuously approaches 0 around π (observe that $|\hat{\psi}(\xi)|^2$ equals $p_\psi(\xi)$ for $\xi \in H_1 \cup H_2 \cup G$).

We also have an even stronger condition than (i) from Proposition 3.4. It is easy to check that

$$
\frac{1}{p_\psi}\big|_{H_1 \cup H_2 \cup g} \text{ is in } L^1. \tag{70}
$$

Remark 3.7. An obvious question is if ψ from Example 3.6 belongs to $\mathcal{P}_{0,+}^{\mathrm{N}}$ or $\mathcal{P}_{0,+}^{\mathrm{MRA}}$. Although we have simple characterizations of these subclasses of $\mathcal{P}_{0,+}$, it is not easy to check the properties of $\dim_\psi$, since we do not have a detailed description of the set G.

Remark 3.8. By taking $\alpha \geq \frac{1}{2}$ in Example 3.6, we can adjust ψ so that $|U_\psi^c| = 0$, but $\frac{1}{p_\psi} \notin L^1$, which shows that the class $\mathcal{P}_{0,+}$ is even more interesting than one would expect. For example, when $\frac{1}{p_\psi} \in L^1$, as in our Example 3.6, then, for every h bounded in $L^2(\mathbb{T}, dx)$, we have that $\frac{h}{p_\psi} \in L^2(\mathbb{T}, p_\psi)$ and

$$
\tilde{\mathcal{I}}_\psi^*\left(\frac{h}{p_\psi}\right) = h. \tag{71}
$$

In particular, the sequence $\{y_k\}$, where $y_k := \frac{e_k}{p_\psi} \in L^2(\mathbb{T}, p_\psi)$, is biorthogonal to the sequence $\{x_k\}$, where $x_k := \tilde{\mathcal{I}}_\psi(e_k)$. Using $\mathcal{I}$, we get that $\{\psi_{0k}\}$ has a biorthogonal sequence in W_0. It is not difficult to prove that this fact is actually equivalent to $\frac{1}{p_\psi} \in L^1$. However, since such questions go beyond dimension one and the case of PFW's, we hope to address them in a separate article.

3.3. W_0-frames, but not more. As we have seen, when a PFW ψ has the property that the family $\{\psi(\cdot - k) : k \in \mathbb{Z}\}$ is a frame for W_0, we say that ψ is a W_0-frame. Recall that for a $\psi \in \mathcal{P}$, this is equivalent to the following property:

$$
(\exists 0 < c \leq 1) \text{ such that } p_\psi \geq c\mathbf{1}_{U_\psi} \ a.e. \tag{72}
$$

This property can be improved in at least two important ways; we could require that $c = 1$ and that $U_\psi = \mathbb{R}$ a.e. The first improvement would lead to semi-orthogonality and the second to W_0-Riesz bases. Here, we are interested in a subclass which does not have either of these improvements. Hence,

$$
\mathcal{P}_{f,0} := \{\psi \in \mathcal{P} : \psi \text{ is a } W_0\text{-frame}, \psi \text{ is not semi-orthogonal}, \tag{73}
$$
$$
\text{and } \psi \text{ is not a } W_0\text{-Riesz basis}\}.
$$

In the following characterization, none of the conditions are redundant.

Proposition 3.9. *Suppose that $\psi \in \mathcal{P}$ ($\psi \in \mathcal{P}^N, \psi \in \mathcal{P}^{MRA}$, respectively). Then, $\psi \in \mathcal{P}_{f,0}$ ($\psi \in \mathcal{P}_{f,0}^N$, $\psi \in \mathcal{P}_{f,0}^{MRA}$, respectively) if and only if (72) holds and*

> (i) $|U_\psi^c| > 0$,
> (ii) $|\{\xi : 0 < p_\psi(\xi) < 1\}| > 0$.

Furthermore, in the equivalence above, the condition (ii) can be replaced by either of the following conditions:

> (iii) p_ψ *is not integer valued,*
> (iv) D_ψ *is not integer valued.*

Proof: This is also more or less a straightforward application of the ideas from [12] and [13]. Being a W_0-frame is equivalent to (72). Adding $|U_\psi^c| = 0$ gives us a W_0-Riesz basis, while allowing c in (72) to be 1 would give us a semi-orthogonal PFW. Conditions (ii) and (iii) are clearly equivalent, while (iii) implies (iv) by (26). Finally, if (iv) holds, then D_ψ can not be equal to $\dim_\psi$; hence, ψ is not semi-orthogonal. Therefore, (ii) holds.

Observe that the fact that none of the conditions can be removed is actually proven by examples in this article.   □

Remark 3.10. As for the other classes, we know that for a $\psi \in \mathcal{P}_{f,0}$, we have that $\psi \in \mathcal{P}_{f,0}^{MRA}$ if and only if $\dim_\psi$ attains only the values 0 and 1. In particular, for $\psi \in \mathcal{P}_{f,0}^{MRA}$, we have $0 \leq D_\psi \leq 1$. It is natural to ask if within $\mathcal{P}_{f,0}$, the condition $0 \leq D_\psi \leq 1$ is also sufficient for ψ being in $\mathcal{P}_{f,0}^{MRA}$. The following example, which also shows that $\mathcal{P}_{f,0}^N \neq \emptyset$, shows that the answer is negative. Let us also mention that this example was announced in [13, Remark 3.7(b)].

Example 3.11. $\mathcal{P}_{f,0}^N \neq \emptyset$.

Proof: Let $\epsilon > 0$ be a small number to be specified later. Let $E := [\frac{\pi}{16} - \epsilon, \frac{\pi}{16} + \epsilon)$ so that $d(E) = [\pi, \pi + 16\epsilon) \cup [2\pi - 32\epsilon, 2\pi)$. For $\epsilon > 0$ small enough, we have that $d|_{E \cup (E+\pi)}$ is injective and

$$d(E \cup (E + \pi)) = [\pi, \pi + 16\epsilon) \cup [\frac{17}{16}\pi - \epsilon, \frac{17}{16}\pi + \epsilon) \cup [2\pi - 32\epsilon, 2\pi).$$

We define $\tilde{F} := S \setminus d(E \cup (E + \pi))$, where S is the Shannon set, and $F := 2^{-6}\tilde{F}$. Hence, $F \subset [-\frac{\pi}{32}, -\frac{\pi}{64}) \cup [\frac{\pi}{64}, \frac{\pi}{32})$.

We define $\psi \in L^2(\mathbb{R})$ by

$$\hat{\psi}(\xi) := \begin{cases} \frac{1}{\sqrt{2}} & \xi \in E \cup (E + \pi) \\ \frac{1}{\sqrt{6}} & \xi \in 2E \\ -\frac{1}{\sqrt{6}} & \xi \in 2E + 2\pi \\ \frac{1}{\sqrt{8}} - \frac{1}{\sqrt{24}} & \xi \in 4E \cup (8E + 8\pi) \\ \frac{1}{\sqrt{8}} + \frac{1}{\sqrt{24}} & \xi \in 8E \cup (4E + 4\pi) \\ 1 & \xi \in F \\ 0 & \text{otherwise.} \end{cases} \tag{74}$$

Let us show that ψ is a PFW. To check (37), observe that it is enough to do it for $\xi \in F \cup E \cup (E + \pi)$. For $\xi \in F$, the result is immediate, while for E and $E + \pi$, the calculations are essentially the same. For $\xi \in E$, the only elements in the dyadic orbit which are also in $\text{supp}(\hat{\psi})$ are ξ, 2ξ, 4ξ and 8ξ, and one checks directly that the corresponding sum is 1.

As usual, showing (38) is somewhat more delicate. Observe first that without loss of generality, we can assume that $q > 0$ in the q from t_q. Observe also that

$$\tau(F) \;=\; F \subset [-\frac{\pi}{32}, -\frac{\pi}{64}) \cup [\frac{\pi}{64}, \frac{\pi}{32}) \tag{75}$$

$$\tau(E) \;=\; E = [\frac{\pi}{16} - \epsilon, \frac{\pi}{16} + \epsilon) \tag{76}$$

$$\tau(E + \pi) \;=\; E - \pi = [\frac{15}{16}\pi - \epsilon, \frac{15}{16}\pi + \epsilon) \tag{77}$$

$$\tau(2E + 2\pi) \;=\; \tau(2E) = [\frac{\pi}{8} - 2\epsilon, \frac{\pi}{8} + 2\epsilon) \tag{78}$$

$$\tau(4E + 4\pi) \;=\; \tau(4E) = [\frac{\pi}{4} - 4\epsilon, \frac{\pi}{4} + 4\epsilon) \tag{79}$$

$$\tau(8E + 8\pi) \;=\; \tau(8E) = [\frac{\pi}{2} - 8\epsilon, \frac{\pi}{2} + 8\epsilon). \tag{80}$$

In particular, assuming $\epsilon > 0$ is small enough, we have that if for some $0 \neq k \in \mathbb{Z}$ that $\xi \in \text{supp}(\hat{\psi}) \cap (\text{supp}(\hat{\psi}) + 2k\pi)$, then $\xi \in 2E \cup (2E + 2\pi) \cup 4E \cup (4E + 4\pi) \cup 8E \cup (8E + 8\pi)$.

Take $\xi \in \mathbb{R}$. If for every $j \in \mathbb{N} \cup \{0\}$, we have

$$2^j \xi \notin \text{supp}(\hat{\psi}) \text{ or } 2^j(\xi + 2q\pi) \notin \text{supp}(\hat{\psi}),$$

then $t_q(\xi) = 0$. Otherwise, there exists $j \in \mathbb{N} \cup \{0\}$ such that $2^j \xi$ and $2^j(\xi + 2q\pi)$ are in $\text{supp}(\hat{\psi})$. It follows then by (75)-(80) and the assumption that $q > 0$ that we have three options:

$$\xi \in 2E \qquad j = 0, q = 1$$
$$\xi \in 2E \qquad j = 1, q = 1$$
$$\xi \in 2E \qquad j = 2, q = 1.$$

Hence, we get that

$$t_1(\xi) = \frac{1}{\sqrt{6}}(-\frac{1}{\sqrt{6}}) + 2(\frac{1}{\sqrt{8}} - \frac{1}{\sqrt{24}})(\frac{1}{\sqrt{8}} + \frac{1}{\sqrt{24}}) = 0.$$

We have proven that $\psi \in \mathcal{P}$. It is straightforward from (75)-(80) to conclude that for $\epsilon > 0$ small enough, we have (72) and (i) and (ii) from Proposition 3.9, i.e. $\psi \in \mathcal{P}_{f,0}$. It remains to show that ψ is not an MRA PFW and that $0 \leq D_\psi \leq 1$. The first claim follows from the fact that for $\xi \in 2E$ we will have among the vectors $\Psi_j(\xi), j \geq 0$, at least the vectors

$$(\ldots, 0, \frac{1}{\sqrt{8}} - \frac{1}{\sqrt{24}}, \frac{1}{\sqrt{8}} + \frac{1}{\sqrt{24}}, 0, \ldots)$$
$$(\ldots, 0, \frac{1}{\sqrt{8}} + \frac{1}{\sqrt{24}}, \frac{1}{\sqrt{8}} - \frac{1}{\sqrt{24}}, 0, \ldots),$$

which are linearly independent. Hence, $\dim_\psi(\xi) \geq 2$ and $\psi \in \mathcal{P}^N_{f,0}$.

In order to calculate D_ψ, we first consider the function

$$S_\psi(\xi) := \sum_{j=1}^{\infty} |\hat{\psi}(2^j \xi)|^2. \tag{81}$$

Observe that D_ψ is the periodization of S_ψ and that for every choice of j and k we have

$$2^j E \cap 2^k(E+\pi) = 2^j E \cap 2^k F = 2^j F \cap 2^k(E+\pi) = \emptyset.$$

Hence, we obtain

$$S_\psi(\xi) = \begin{cases} 1 & \xi \in \cup_{j\geq 1} 2^{-j}(E \cup (E+\pi) \cup F) \\ \frac{1}{2} & \xi \in E \cup (E+\pi) \\ \frac{1}{3} & \xi \in 2E \cup 2(E+\pi) \\ (\frac{1}{\sqrt{8}} + \frac{1}{\sqrt{24}})^2 & \xi \in 4E \\ (\frac{1}{\sqrt{8}} - \frac{1}{\sqrt{24}})^2 & \xi \in 4(E+\pi) \\ 0 & \text{otherwise.} \end{cases} \tag{82}$$

Recall that D_ψ is 2π-periodic, so it is enough to compute it on $[-\pi, \pi)$. It follows from (82) that for $\xi \in [-\pi, \pi)$, we obtain

$$D_\psi(\xi) = \begin{cases} 1 & \xi \in \cup_{j\geq 1} 2^{-j}(E \cup (E+\pi) \cup F) \\ \frac{1}{2} & \xi \in E \cup (E-\pi) \\ \frac{2}{3} & \xi \in 2E \\ \frac{1}{3} & \xi \in 4E \\ 0 & \text{otherwise,} \end{cases}$$

where we used the fact that $E - \pi$ is disjoint from all the other sets appearing above. Thus, in particular, $0 \leq D_\psi(\xi) \leq 1$ for every ξ, as desired. $\square$

Example 3.12. $\mathcal{P}_{f,0}^{MRA} \neq \emptyset$.

Proof: Consider $0 < a < \pi/2$. Since every generalized filter can be extended, through (44), from $[-\pi/2, \pi/2)$ to $[-\pi, \pi)$, it is obvious that there exists a generalized filter m such that the graph of the function $M(\xi) := |m(\xi)|^2$ on $[-a, a]$ is given by

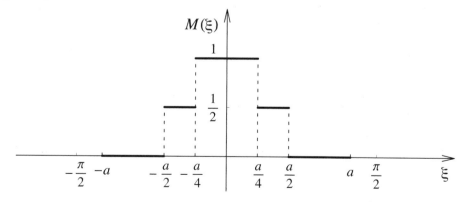

Since $m(\xi) = 0$ for $\xi \in [-a, -a/2] \cup [a/2, a)$, it follows by (45) that $\hat{\varphi}(\xi) = 0$ for $|\xi| > a$, where φ is the corresponding pseudo-scaling function. Furthermore, it follows that the graph of the function $\xi \to |\hat{\varphi}(\xi)|^2$ is as follows:

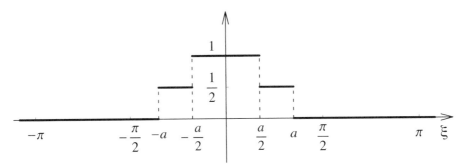

Observe that $M(\xi) = 1$ on $(-a/4, a/4)$, which ensures that (47) is valid, i.e., m is a generalized low pass filter, and by the well-known construction from [12], we know that it generates an MRA PFW ψ.

Since $M(\xi)$ and $|\hat{\varphi}(\xi)|^2$ attain only the values $0, 1/2$ and 1, by (46), the same is true of $|\hat{\psi}(\xi)|^2$. Hence, (72) is fulfilled. Again, by (46) and (3.3), we see that $\text{supp}(\hat{\psi}) \subset [-2a, 2a]$, which is a proper subset of $[-\pi, \pi)$. This shows (i) from Proposition 3.9. The second consequence is that, by (52), we get (iv) from Proposition 3.9. Hence, $\psi \in \mathcal{P}^{MRA}_{f,0}$.

3.4. W_0-Riesz bases, but not more. The next natural step is to consider W_0-frames for which $\tilde{\mathcal{I}}$ is injective, i.e.

$$\mathcal{P}_{f,+} := \{\psi \in \mathcal{P} : \psi \text{ is a } W_0\text{-frame}, \psi \text{ is not semi-orthogonal}, \tilde{\mathcal{I}}_\psi \text{ is injective}\}. \tag{83}$$

Recall that two of the conditions in (83), namely that ψ is a W_0-frame and $\tilde{\mathcal{I}}$ is injective, are actually equivalent to the single condition:

$$\left(\exists 0 < c \leq 1\right) \text{ such that } p_\psi \geq c, \text{ a.e.,} \tag{84}$$

which as we know from [13], is equivalent to ψ being a W_0-Riesz basis. Observe also that (84) is equivalent to

$$\frac{1}{p_\psi} \text{ is bounded,} \tag{85}$$

which implies that $\tilde{\mathcal{I}}_\psi$ is invertible and its inverse is a bounded operator. Hence, $\{\psi_{0k}\}$ is the image of the orthonormal basis via the regular operator. This is to be expected. Observe, however, that in this case both the orthonormal basis $\{e_k\}$ and the regular operator are explicitly given.

It is fairly straightforward to get the following characterization of $\mathcal{P}_{f,+}$ where none of the conditions are redundant.

Proposition 3.13. *Suppose that $\psi \in \mathcal{P}$ (resp. $\psi \in \mathcal{P}^N, \psi \in \mathcal{P}^{MRA}$). Then, $\psi \in \mathcal{P}_{f,+}$ (resp. $\psi \in \mathcal{P}^N_{f,+}, \psi \in \mathcal{P}^{MRA}_{f,+}$) if and only if (84) is valid and*

(i) $|\{\xi : p_\psi(\xi) \neq 1\}| > 0$.

Furthermore, in the equivalence above, the condition (i) can be replaced by either of the following conditions:

(ii) *p_ψ is not integer valued,*

(iii) *D_ψ is not integer valued.*

Remark 3.14. The elements of $\mathcal{P}_{f,+}$ are W_0-Riesz bases which are not semi-orthogonal. Example 2.5 in [13], which corresponds to the generalized low pass filter $m(\xi) = \frac{1}{2}(1 + e^{3i\xi})$, shows that $\mathcal{P}_{f,+}^{\mathrm{MRA}} \neq \emptyset$. In particular, $\mathcal{P}_{f,+} \neq \emptyset$. We will revisit the class $\mathcal{P}_{f,+}^{\mathrm{MRA}}$ in more detail in Section 4. The question whether $\mathcal{P}_{f,+}^{\mathrm{N}} = \emptyset$ remains open at this point.

3.5. Semiorthogonal PFW's, which are not orthonormal.

Let us consider all "resolution levels" $W_j(\psi)$, see (17). Recall that we say that a PFW ψ is *semi-orthogonal* if $W_j(\psi) \perp W_k(\psi)$ whenever $j \neq k$, $j, k \in \mathbb{Z}$. Observe that for such a ψ, we have $D_\psi = \dim_\psi$, so it is not possible that $\dim_\psi = \infty$. In particular, for a semi-orthogonal PFW ψ, we always have

$$\psi \notin V_0(\psi), \tag{86}$$

in particular, $V_0(\psi) \neq L^2(\mathbb{R})$.

It is useful to recall the various characterizations of semi-orthogonality within $\mathcal{P}$. Using (43) and [13, Theorem 2.7, Theorem 3.1, Corollary 3.2], we obtain directly the following theorem.

Theorem 3.15. *Suppose ψ is a PFW. The following are equivalent.*

   (i) *ψ is semi-orthogonal,*
  (ii) *ψ is a W_0-Parseval frame,*
 (iii) *p_ψ is integer valued,*
 (iv) *$p_\psi = \mathbf{1}_{U_\psi}$,*
  (v) *D_ψ is integer valued,*
 (vi) *$D_\psi = \dim_\psi$,*
(vii) *$\|\psi\|_2^2 = \sum_{k \in \mathbb{Z}} |\langle \psi, \psi_{0k} \rangle|^2$,*
(viii) *$DV_0(\psi) = V_0(\psi) \oplus W_0(\psi)$*

Remark 3.16. Condition (viii) in the previous theorem shows also that there is a close connection between the GMRA structure and semi-orthogonal PFW's. (See, for example, [1] for the definition and basics of GMRA's.) Indeed, if ψ is a semi-orthogonal PFW, then (viii) implies that

$$V_0(\psi) = \oplus_{j<0} W_j(\psi), \tag{87}$$

and $V_0(\psi)$ is the core space for a GMRA, assuming that ψ generated a GMRA to begin with. Observe also that in this case, the function S_ψ given in (81) is the spectral function of the shift-invariant space $V_0(\psi)$, see [6] for details.

Conversely, if $\{V_j\}$ is a GMRA and ψ is a Parseval frame for $V_1 \cap V_0^\perp$, then ψ is a semi-orthogonal PFW. Since these facts are part of folklore even in higher dimensions and for more general dilations, we do not say more here. Rather, we will emphasize that semi-orthogonal PFW's are most closely related to the theory of GMRA's, while when we go beyond semi-orthogonality, we obtain only very limited information through the GMRA approach. For example, if $\psi \in \mathcal{P}$ and $\psi \notin V_0(\psi)$, then we could consider

$$\psi_1 = \psi - \psi_0,$$

where ψ_0 is the orthogonal projection of ψ onto $V_0(\psi)$, and we can associate with ψ the corresponding semi-orthogonal PFW, which is going to have the same GMRA as ψ. However, since at this level of generality, we have a much more basic question given in Remark 2.5, we will address this issue in detail only in the MRA case in Section 4.

Let us also emphasize that a straightforward consequence of Theorem 3.15 is the following set of two analogous properties.

Corollary 3.17. *Suppose $\psi \in \mathcal{P}$ (resp. $\psi \in \mathcal{P}^{MRA}$). Then, ψ is a semi-orthogonal MRA PFW if and only if D_ψ (resp. p_ψ) has only values 0 and 1.*

Observe also that for a semi-orthogonal $\psi \in \mathcal{P}^{\mathrm{N}}$, we have $\dim_\psi = D_\psi$, and hence, $|\{\xi : D_\psi(\xi) \geq 2\}| > 0$. The integrability condition (27) then implies that the Lebesgue measure of the set of zeroes of D_ψ must be positive, too. Hence, we have proven the following proposition.

Proposition 3.18. *If $\psi \in \mathcal{P}^N$ is semi-orthogonal, then $|Z_{V_0(\psi)}| > 0$.*

It is now easy to characterize the class that interests us here, i.e.

$$\mathcal{P}_{tf,0} := \{\psi \in \mathcal{P} : \psi \text{ is a } W_0\text{-Parseval frame and } \tilde{\mathcal{I}}_\psi \text{ is not injective}\}. \tag{88}$$

Observe that for $\psi \in \mathcal{P}_{tf,0}$, we have that $p_\psi = \mathbf{1}_{U_\psi}$ and $|U_\psi^c| > 0$, so $L^2(\mathbb{T}, p_\psi)$ really becomes $L^2(\mathbb{T} \cap U_\psi, dx)$ and $\tilde{\mathcal{I}}_\psi$ is the orthogonal projection from $L^2(\mathbb{T}, dx)$ onto $L^2(\mathbb{T} \cap U_\psi, dx)$. Observe also that $\tilde{\mathcal{I}}_\psi^*$ then becomes the inclusion of $L^2(\mathbb{T} \cap U_\psi, dx)$ into $L^2(\mathbb{T}, dx)$.

Proposition 3.19.   (i) *Suppose $\psi \in \mathcal{P}$. Then, $\psi \in \mathcal{P}_{tf,0}$ if and only if ψ is semi-orthogonal and $|U_\psi^c| > 0$. Furthermore, the last condition can be replaced by $\|\psi\|_2 < 1$. The same is valid for the class $\mathcal{P}^N$.*

   (ii) *Suppose $\psi \in \mathcal{P}$. Then, $\psi \in \mathcal{P}_{tf,0}^{MRA}$ if and only if $D_\psi = \mathbf{1}_{Z_{V_0(\psi)}^c}$ and $|Z_{V_0(\psi)}| > 0$.*

   (iii) *Suppose $\psi \in \mathcal{P}^{MRA}$. Then, $\psi \in \mathcal{P}_{tf,0}^{MRA}$ if and only if $p_\psi = \mathbf{1}_{U_\psi}$ and $|U_\psi^c| > 0$.*

Observe that whenever $\psi \in \mathcal{P}_{tf,0}$, we always have $|Z_{V_0(\psi)}| > 0$. However, in Proposition 3.19 (i), the condition $|U_\psi^c| > 0$ can not be replaced by $|Z_{V_0(\psi)}| > 0$ (consider, for example, the Journé wavelet). In Proposition 3.19 (iii), it is possible to replace the condition $|U_\psi^c| > 0$ with $|Z_{V_0(\psi)}| > 0$. We omit the details.

In terms of the examples, we will revisit $\mathcal{P}_{tf,0}^{MRA}$ later, while at this point, let us consider $\mathcal{P}_{tf,0}^{\mathrm{N}}$.

Example 3.20. $\mathcal{P}_{tf,0}^{\mathrm{N}} \neq \emptyset$.

Proof: Consider the following graph of $\hat{\psi}$. Here, we have set a which will appear below equal to $\pi/3$. Any $0 < a \leq \pi/3$ will work, but for the remainder of this example we use a and $\frac{\pi}{3}$ interchangeably.

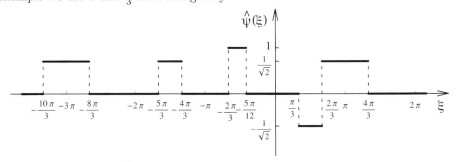

Observe that supp($\hat{\psi}$) consists of intervals of the form $[2a - 4\pi, 4a - 4\pi), [a - 2\pi, 2a - 2\pi), [a - \pi, a/4 - \pi/2), [a, 2a)$ and $[2a, 4a)$. It is easy to see that (37) is satisfied.

In order to check (38), observe that if $2^j\xi$ is in supp($\hat\psi$) for $j \geq 2$, then $\hat\psi(2^j(\xi + 2q\pi)) = 0$. Hence, if $\xi \in$ supp($\hat\psi$), then $[2a - 4\pi, 4a - 4\pi) \cup [2a, 4a)$ is resolved trivially, while the other possibilities are resolved in an analogous way. For example, $\xi \in [a - 2\pi, 2a - 2\pi)$ leads to

$$t_q(\xi) = \hat\psi(\xi)\hat\psi(\xi + 2\pi) + \hat\psi(2\xi)\hat\psi(2\xi + 4\pi) = \frac{1}{\sqrt{2}}\frac{-1}{\sqrt{2}} + \frac{1}{\sqrt{2}}\frac{1}{\sqrt{2}} = 0.$$

Hence, $\psi \in \mathcal{P}$.

It is easy to check directly that for $\xi \in [-\pi, \pi)$ we obtain

$$p_\psi(\xi) = \begin{cases} 1 & \xi \in [-\pi, -\frac{5\pi}{12}) \cup [\frac{\pi}{3}, \pi) \\ 0 & \xi \in [-\frac{5\pi}{12}, \frac{\pi}{3}). \end{cases}$$

Hence, ψ is semi-orthogonal and $|U_\psi^c| > 0$.

If $\xi \in [\frac{\pi}{6}, \frac{\pi}{3})$, then $2\xi \in [\frac{\pi}{3}, \frac{2\pi}{3})$ and $4\xi \in [\frac{2\pi}{3}, \frac{4\pi}{3})$. Therefore, we obtain

$$\Psi_1(\xi) = (\ldots, 0, \quad \tfrac{1}{\sqrt{2}}, -\tfrac{1}{\sqrt{2}}, \quad 0, \ldots)$$

$$\Psi_2(\xi) = (\ldots, 0, \quad 0, \tfrac{1}{\sqrt{2}}, \quad 0, \ldots).$$

This implies that $\dim_\psi(\xi) \geq 2$ for $\xi \in [\frac{\pi}{6}, \frac{\pi}{3})$. Therefore, $\psi \notin \mathcal{P}^{\mathrm{MRA}}$, and $\psi \in \mathcal{P}_{tf,0}^{\mathrm{N}}$. $\square$

3.6. Orthonormal wavelets.

The next natural class of interest is

$$\mathcal{P}_{tf,+} := \{\psi \in \mathcal{P} : \psi \text{ is a } W_0\text{-Parseval frame and } \tilde{\mathcal{I}}_\psi \text{ is injective}\}. \tag{89}$$

For this class, we have

$$p_\psi \equiv 1, \tag{90}$$

so $\tilde{\mathcal{I}}_\psi$ is the identity operator and ψ_{0k} is the orthonormal basis for W_0. Hence,

$$\psi \in \mathcal{P}_{tf,+} \iff \psi \text{ is an orthonormal wavelet.} \tag{91}$$

Observe also that $\psi \in \mathcal{P}_{tf,+}^{\mathrm{MRA}}$ if and only if ψ is an MRA orthonormal wavelet in the standard sense, as shown in, for example, [12]. Various examples of orthonormal wavelets are well known, and we know that both classes $\mathcal{P}_{tf,+}^{\mathrm{MRA}}$ and $\mathcal{P}_{tf,+}^{\mathrm{N}}$ are very rich.

For the sake of completeness, we recall various characterizations of orthonormal wavelets within $\mathcal{P}$, see [13], for example.

Proposition 3.21. *Suppose* $\psi \in \mathcal{P}$. *Then, the following are equivalent.*

(i) ψ *is an orthonormal wavelet,*

(ii) $\|\psi\|_2 = 1$,

(iii) $p_\psi \equiv 1$,

(iv) ψ *is a semi-orthogonal* W_0-*Riesz basis,*

(v) ψ *is a* W_0-*Parseval frame and* $\tilde{\mathcal{I}}_\psi$ *is injective.*

Furthermore, for $\psi \in \mathcal{P}$, *the following are equivalent:*

(i) ψ *is an MRA orthonormal wavelet,*

(ii) $D_\psi \equiv 1$.

Remark 3.22.   (i) Observe that, for $\psi \in \mathcal{P}$, $D_\psi \equiv 1$ already implies that ψ is an orthonormal wavelet, so the characterization of PFW's that are MRA orthonormal wavelets is the same as the characterization of orthonormal wavelets that come from an MRA.

(ii) It is not possible to add a third equivalence of $\dim_\psi \equiv 1$ to the characterization of MRA orthonormal wavelets given above. See Section 4 for details.

3.7. MSF Parseval frame wavelets.

An important class for the theory of PFW's is the one consisting of MSF PFW's. These are PFW's such that $|\hat\psi(\xi)| \in \{0,1\}$. By [12, Corollary 3.5], every MSF PFW is semi-orthogonal. If ψ is an MSF PFW and we denote by K the set where $|\hat\psi(\xi)| = 1$, then

$$p_\psi = \mathbf{1}_{\tau(K)}. \tag{92}$$

Obviously, MSF PFW's can be either within the class of orthonormal wavelets or outside of it. Clearly, an MSF PFW ψ is an orthonormal wavelet if and only if

$$\tau(K) = [-\pi, \pi). \tag{93}$$

The examples of various kinds of orthonormal wavelets which are or are not MSF are by this time well known. Hence, let us see what happens outside of orthonormal wavelets. If ψ is an MSF PFW and is not an orthonormal wavelet, then

$$\psi \in \mathcal{P}_{tf,0}^{N} \cup \mathcal{P}_{tf,0}^{\mathrm{MRA}} = \mathcal{P}_{tf,0}.$$

We shall see what happens within $\mathcal{P}_{tf,0}^{\mathrm{MRA}}$ in Section 4. Let us comment on $\mathcal{P}_{tf,0}^{N}$. Observe that the class of elements in $\mathcal{P}_{tf,0}^{N}$ which are not MSF is non-empty, since the PFW from Example 3.20 is not an MSF PFW.

Example 3.23. $MSF \cap \mathcal{P}_{tf,0}^{N} \neq \emptyset$.

Proof: Let $K = [-\frac{\pi}{2}, -\frac{\pi}{4}) \cup [\frac{\pi}{8}, \frac{7\pi}{32}) \cup [\frac{15\pi}{64}, \frac{\pi}{4}) \cup [7\pi, \frac{15\pi}{2})$. One checks that (37) holds, and that $\tau|_{\mathrm{supp}(\hat\psi)}$ is injective, but not onto. It follows that (38) holds, so ψ is an MSF PFW and ψ is not an orthonormal wavelet. It remains to show that ψ is not an MRA PFW. Since ψ is MSF, we have $D_\psi = \dim_\psi$. Consider $\xi \in [-\frac{\pi}{4}, -\frac{\pi}{8})$. In this case, $2\xi \in [-\frac{\pi}{2}, -\frac{\pi}{8})$ and $4(\xi + 2\pi) \in [7\pi, \frac{15\pi}{2})$, so $D_\psi(\xi) \geq 2$, so ψ is not MRA. $\qquad\square$

4. Semiorthogonalization

Results in this section are very much related to the ideas presented in Remark 3.16, but here we work on MRA PFW's. There are at least two strong facts that help us within $\mathcal{P}^{\mathrm{MRA}}$. The first is that for $\psi \in \mathcal{P}^{\mathrm{MRA}}$, we have $\dim_\psi \leq 1$, and therefore

$$\psi \notin V_0(\psi),$$

and $V_0(\psi)$ is the core space of a GMRA with respect to the dilation D. Second, within $\mathcal{P}^{\mathrm{MRA}}$ we can work on filters, which is very pleasing if we have in mind the construction from [12] that enables us to construct MRA PFW's from generalized low pass filters. Recall that this construction builds, from a given generalized low pass filter m, the associated pseudo-scaling function, which we denote φ_m, and the associated MRA PFW, which we denote ψ_m. Furthermore, every $\psi \in \mathcal{P}^{\mathrm{MRA}}$ can be constructed in this way. One needs to observe that, given m, its ψ_m is uniquely determined (as well as φ_m), but given ψ we can have several filters which are going to provide the same ψ.

The idea is now to modify the filter m in a minimal way, so as to obtain the new filter which corresponds to the semi-orthogonal MRA PFW. We define a map ζ, which we call the *semiorthogonalization map*, from the set of generalized low-pass filters into the set of generalized low-pass filters, by

$$\zeta(m)(\xi) := \begin{cases} \sqrt{\frac{D_{\psi_m}(\xi)}{D_{\psi_m}(2\xi)}}\, m(\xi) & D_{\psi_m}(2\xi) \neq 0 \\ m(\xi) & D_{\psi_m}(2\xi) = 0, \end{cases} \tag{94}$$

where m is the generalized low-pass filter. This semiorthogonalization procedure is similar in spirit to one outlined in the proof of Theorem 3.3 in [6], but here we get explicit formulas rather than expressing the procedure in terms of projections.

Theorem 4.1. *Suppose m is a generalized low-pass filter. Then $\zeta(m)$ is a generalized low-pass filter such that $\psi_{\zeta(m)}$ is a semi-orthogonal MRA PFW and*

$$V_0(\psi_{\zeta(m)}) = V_0(\psi_m); \tag{95}$$

in particular, $\dim_{\psi_{\zeta(m)}} = \dim_{\psi_m}$. Furthermore, there are direct formulas for $\varphi_{\zeta(m)}$ and $\psi_{\zeta(m)}$, i.e.,

$$\hat{\varphi}_{\zeta(m)} = \begin{cases} \frac{1}{\sqrt{D_{\psi_m}(\xi)}}\hat{\varphi}_m(\xi) & D_{\psi_m}(\xi) \neq 0 \\ 0 & D_{\psi_m}(\xi) = 0, \end{cases} \tag{96}$$

$$\hat{\psi}_{\zeta(m)}(\xi) = \begin{cases} \sqrt{\frac{D_{\psi_m}(\xi/2+\pi)}{D_{\psi_m}(\xi)D_{\psi_m}(\xi/2)}}\, \hat{\psi}_m(\xi) & D_{\psi_m}(\xi/2) \cdot D_{\psi_m}(\xi) \neq 0 \\ \frac{1}{\sqrt{D_{\psi_m}(\xi)}}\, \hat{\psi}_m(\xi) & D_{\psi_m}(\xi/2) \neq 0 \text{ and } D_{\psi_m}(\xi) = 0 \\ 0 & \textit{otherwise.} \end{cases} \tag{97}$$

Proof. Let us first show that m is a generalized filter. Observe that $D_{\psi_m}(2(\xi + \pi)) = D_{\psi_m}(2\xi)$. Hence, if $D_{\psi_m}(2\xi) \neq 0$, then

$$\begin{aligned} |\zeta(m)(\xi)|^2 + |\zeta(m)(\xi + \pi)|^2 &= \frac{D_{\psi_m}(\xi)}{D_{\psi_m}(2\xi)}|m(\xi)|^2 + \frac{D_{\psi_m}(\xi + \pi)}{D_{\psi_m}(2(\xi + \pi))}|m(\xi + \pi)|^2 \\ &= \frac{1}{D_{\psi_m}(2\xi)}\left(|m(\xi)|^2 D_{\psi_m}(\xi) + |m(\xi + \pi)|^2 D_{\psi_m}(\xi + \pi)\right) \\ &= \frac{D_{\psi_m}(2\xi)}{D_{\psi_m}(2\xi)} = 1, \end{aligned}$$

where in the second to last equality, we have used Lemma 2.8(i). If $D_{\psi_m}(2\xi) = D_{\psi_m}(2(\xi + \pi)) = 0$, then

$$|\zeta(m)(\xi)|^2 + |\zeta(m)(\xi + \pi)|^2 = |m(\xi)|^2 + |m(\xi + \pi)|^2 = 1.$$

The fact that $\zeta(m)$ is a generalized filter immediately shows that there is a corresponding pseudo-scaling function $\varphi_{\zeta(m)}$, as given by the multiplier construction from [12]. Hence, in order to prove that $\zeta(m)$ is a generalized low-pass filter, we need to prove that for a.e. $\xi \in \mathbb{R}$,

$$\lim_{n \to \infty} |\hat{\varphi}_{\zeta(m)}(2^{-n}\xi)|^2 = 1. \tag{98}$$

We also know from [11] that this limit is either 0 or 1, and it is 0 if and only if all the members of the sequence are 0. Observe that by (54) applied on D_{ψ_m}, we get that there exists $n_0 = n_0(\xi) \in \mathbb{N}$ such that for every $n \geq n_0$, $n \in \mathbb{N}$, we have

$$|\zeta(m)(2^{-n}\xi)|^2 = \frac{D_{\psi_m}(2^{-n}\xi)}{D_{\psi_m}(2^{-n+1}\xi)}\, |m(2^{-n}\xi)|^2.$$

It follows that, for every $n \geq n_0$, we have

$$
|\hat{\varphi}_{\zeta(m)}(2^{-n}\xi)|^2 = \frac{1}{D_{\psi_m}(2^{-n}\xi)} \cdot \lim_{k \to \infty} D_{\psi_m}(2^{-n-k}\xi) \cdot \Pi_{j=1}^k |m(2^{-n-j}\xi)|^2
$$

$$
= \frac{1}{D_{\psi_m}(2^{-n}\xi)} |\hat{\varphi}_m(2^{-n}\xi)|^2 \geq |\hat{\varphi}_m(2^{-n}\xi)|^2,
$$

where the second equality is from (54).

Since $\lim_{n \to \infty} |\hat{\varphi}_m(2^{-n}\xi)|^2 = 1$, by the assumption that m is a generalized low-pass filter, it follows that $\zeta(m)$ is a generalized low-pass filter, too. The consequence is that $\psi_{\zeta(m)} \in \mathcal{P}^{\mathrm{MRA}}$.

In order to check that $\varphi_{\zeta(m)}$ satisfies (96), it is enough to check that the function given by (96) is the pseudo-scaling function of $\zeta(m)$; recall the multiplier approach from [12] and the fact that $\varphi_{\zeta(m)}$ satisfies (98). Indeed, if $D_{\psi_m}(2\xi) \neq 0$, then by (94) and (96), we get

$$
\frac{1}{\sqrt{D_{\psi_m}(2\xi)}} \hat{\varphi}_m(2\xi) = \frac{1}{\sqrt{D_{\psi_m}(2\xi)}} m(\xi)\hat{\varphi}_m(\xi) = \frac{1}{\sqrt{D_{\psi_m}(2\xi)}} m(\xi)\sqrt{D_{\psi_m}(\xi)}\,h(\xi),
$$

where h is the function defined by (96) (observe that $\sqrt{D_{\psi_m}(\xi)}h(\xi) = \hat{\varphi}_m(\xi)$ irrespective of $D_{\psi_m}(\xi)$ being zero or not, by (52)). Hence, $h(2\xi) = \zeta(m)(\xi)h(\xi)$ if $D_{\psi_m}(2\xi) \neq 0$. If $D_{\psi_m}(2\xi) = 0$, then $h(2\xi) = 0$ and $\zeta(m)(\xi) = m(\xi)$. Furthermore, by (52), we also have $\hat{\varphi}_m(2\xi) = 0$, which implies that either $m(\xi) = 0$ or $\hat{\varphi}_m(\xi) = 0$. In any case, we get $\zeta(m)(\xi)h(\xi) = 0$. This proves (96), which by Proposition 2.6 and (8), shows that

$$
V_0(\psi_{\zeta(m)}) = \langle \varphi_{\zeta(m)} \rangle = \langle \varphi_m \rangle = V_0(\psi_m).
$$

Obviously, then $\dim_{\psi_{\zeta(m)}} = \dim_{\psi_m}$. Moreover, (8) and (9) applied on (96) show that

$$
p_{\varphi_{\zeta(m)}} = \mathbf{1}_{Z^c_{V_0(\psi_m)}} = \mathbf{1}_{Z^c_{V_0(\psi_{\zeta(m)})}},
$$

which by (52) and Corollary 3.17 proves that $\psi_{\zeta(m)}$ is semi-orthogonal. Finally, we leave it to the reader to check that (97) follows from (46), (94) and (96).   □

Let us clarify some aspects of the semiorthogonalization.

Remark 4.2.   (i) It is easy to see that if m is a generalized low-pass filter such that ψ_m is semi-orthogonal, then $\zeta(m) = m$, $\varphi_{\zeta(m)} = \varphi_m$ and $\psi_{\zeta(m)} = \psi_m$.

(ii) It is possible to consider (97) directly as the definition of the semiorthogonalization given on MRA PFW's, and we obtain the same results. The reader can check the details. Whichever way one considers it, the semiorthogonalization always maps MRA PFW's into semi-orthogonal MRA PFW's.

(iii) One way to be tempted by (ii) is to extend (97) to arbitrary PFW's. However, this does not work, as one can check on various examples. Hence, to extend semiorthogonalization, one needs to use ideas from Remark 3.16. The problem is that they work only if $\psi \notin V_0(\psi)$, and we do not know when this is the case.

(iv) One consequence of $\dim_{\psi_{\zeta(m)}} = \dim_{\psi_m}$ is that the sets of zeroes of $D_{\psi_{\zeta(m)}}$ and D_{ψ_m} are equal. Observe that (96) gives us even more, i.e.,

$$
\mathrm{supp}(\hat{\varphi}_{\zeta(m)}) = \mathrm{supp}(\hat{\varphi}_m). \tag{99}
$$

One also easily obtains from (97) that

$$
\mathrm{supp}(p_{\psi_{\zeta(m)}}) = \mathrm{supp}(p_{\psi_m}). \tag{100}
$$

Since $\dim_{\psi_{\zeta(m)}} = D_{\psi_{\zeta(m)}}$, and (26) is valid, we obtain that for every $\psi \in \mathcal{P}^{\mathrm{MRA}}$,

$$\dim_{\psi}(\xi) + \dim_{\psi}(\xi + \pi) = \dim_{\psi}(2\xi) + \mathbf{1}_{U_{\psi}}(2\xi), \qquad (101)$$

for a.e. $\xi \in \mathbb{R}$.

(v) Sometimes it may not be trivial to see what we get by the semiorthogonalization procedure, despite having explicit formulas (94), (96) and (97). For example, apply them to the filter $m(\xi) = \frac{1}{2}(1 + e^{3i\xi})$, and you may not immediately see the result. E. Hernandez and F. Soria proved in unpublished notes that for every nonnegative integer n, the semiorthogonalization procedure applied to the filter $m_n(\xi) := \frac{1}{2}(1 + e^{(2n+1)i\xi})$ gives that $\psi_{\zeta(m_n)}$ is the Haar wavelet. Recall that ψ_{m_1} is an example of a W_0-Riesz basis MRA PFW which is not semi-orthogonal.

Remark 4.3. We think that the semiorthogonalization procedure, even considered only on $\mathcal{P}^{\mathrm{MRA}}$, raises an important issue with respect to the GMRA approach to the analysis of $\mathcal{P}$. More precisely, one can (as several authors do) say that a $\psi \in \mathcal{P}$ is associated with a GMRA if $V_0(\psi)$ is the core space for the GMRA. This notion can be somewhat misleading. Namely, if a semi-orthogonal PFW is associated with its GMRA in this way, everything is fine in the sense that all the crucial information about the PFW is given in the GMRA. However, things are different when we go outside of semi-orthogonal PFW's. If $\psi \in \mathcal{P}^{\mathrm{MRA}}$, then it is always associated with a GMRA (in the above sense). What are all the possible GMRA's that we can get from $\mathcal{P}^{\mathrm{MRA}}$? By Theorem 4.1, these are exactly those that we can get from semi-orthogonal MRA PFW's, i.e., the ones studied by J. J. Benedetto and S. Li [2] and by J. J. Benedetto and O. M. Treiber [3]. But, the associated GMRA's are not going to tell us anything about some important features of PFW's, as soon as it is not semi-orthogonal. Take the filters $m_0(\xi) = \frac{1}{2}(1 + e^{i\xi})$ and $m_1(\xi) = \frac{1}{2}(1 + e^{3i\xi})$ from Remark 4.2 (v). They will both generate exactly the same GMRA, but one of them, ψ_{m_0} is the orthonormal Haar wavelet, while the other, ψ_{m_1}, is a PFW which is a W_0-Riesz basis but is not even semi-orthogonal.

The semiorthogonalization procedure points to yet another interesting class of MRA PFW's. Through the semiorthogonalization procedure, an MRA PFW will end either in $\mathcal{P}_{tf,0}^{\mathrm{MRA}}$ or in $\mathcal{P}_{tf,+}^{\mathrm{MRA}}$ (the orthonormal MRA wavelets). Clearly, from Theorem 4.1, MRA PFW's which are going to end in $\mathcal{P}_{tf,+}^{\mathrm{MRA}}$ are precisely the elements of the class

$$\mathcal{P}_{-,1}^{\mathrm{MRA}} := \{\psi \in \mathcal{P} : \dim_{\psi} = 1\}. \qquad (102)$$

Remark 4.4. Other authors have encountered this class in related situations. In particular, in [6], the authors consider Parseval frame wavelets (in higher dimensions with an expansive dilation) for which the space $V_0(\psi)$ is the core space of a GMRA. Then, they ask which ones come from a classical MRA, and they prove that these are precisely the ones for which $D_{\psi} > 0$ a.e. Observe that in the case of dyadic, one-dimensional wavelets, this is exactly our class $\mathcal{P}_{-,1}^{\mathrm{MRA}}$. Indeed, for $\psi \in \mathcal{P}^{\mathrm{MRA}}$, this is obvious, and for $\psi \in \mathcal{P}^{\mathrm{N}}$ and such that it is associated with a GMRA, we will have $\psi \notin V_0(\psi)$, so we can apply Remark 3.16, and then using the fact that for semi-orthogonal elements in $\mathcal{P}^{\mathrm{N}}$, $|Z_{V_0(\psi)}| > 0$, conclude that $|\{\xi : D_{\psi}(\xi) = 0\}| > 0$. Hence, the class $\mathcal{P}_{-,1}^{\mathrm{MRA}}$ is clearly of importance, and we shall analyze it further.

The following result is both useful and of independent interest.

Lemma 4.5. *Suppose $\psi \in \mathcal{P}^{MRA}$. If $|U_\psi^c| > 0$, then $|Z_{V_0(\psi)}| > 0$.*

Proof. By Lemma 2.8 (iii), whenever $p_\psi(2\xi) = 0$, then either $D_\psi(\xi) = 0$ or $D_\psi(\xi + \pi) = 0$. The result now follows easily. $\square$

Proposition 4.6. *The following holds:*

$$\mathcal{P}_{tf,+}^{MRA} \subset \mathcal{P}_{-,1}^{MRA} \subset \mathcal{P}_{tf,+}^{MRA} \cup \mathcal{P}_{f,+}^{MRA} \cup \mathcal{P}_{0,+}^{MRA}.$$

Proof. The first inclusion is obvious, while for the second inclusion, we apply Lemma 4.5 together with the obvious fact that for $\psi \in \mathcal{P}_{-,1}^{\text{MRA}}$, we must have $|Z_{V_0(\psi)}| = 0$. Indeed, if $\psi \in \mathcal{P}_{tf,0}^{\text{MRA}} \cup \mathcal{P}_{f,0}^{\text{MRA}} \cup \mathcal{P}_{0,0}^{\text{MRA}}$, then $|U_\psi^c| > 0$, that is $\psi \notin \mathcal{P}_{-,1}^{\text{MRA}}$. $\square$

We have

$$\mathcal{P}_{-,1}^{\text{MRA}} \cap \mathcal{P}_{f,+}^{\text{MRA}} \neq \emptyset, \tag{103}$$

since the MRA PFW generated by the filter $m(\xi) = \frac{1}{2}(1 + e^{3i\xi})$ is in both classes. One would expect that the semiorthogonalization of a W_0-Riesz basis is always going to produce an orthonormal wavelet, i.e. that $\mathcal{P}_{f,+}^{\text{MRA}} \subset \mathcal{P}_{-,1}^{\text{MRA}}$. However, this is not true, as the following example shows. Observe that it also shows that the converse of Lemma 4.5 is not true.

Example 4.7. $\mathcal{P}_{f,+}^{MRA} \not\subset \mathcal{P}_{-,1}^{MRA}$.

Proof. Our ψ is an MRA PFW which is generated by the filter m, whose function $M(\xi) = |m(\xi)|^2$ has the following graph (clearly, such an m exists and it is a generalized low-pass filter).

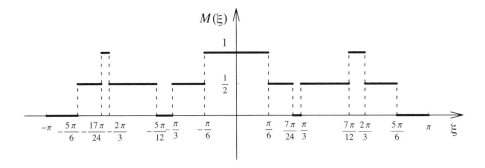

In order to analyze the key properties of this ψ, it is enough to look into the function $F(\xi) = |\hat{\varphi}_{|m|}(\xi)|^2 = \Pi_{j=1}^\infty M(2^{-j}\xi)$. We claim that the graph of F is as follows.

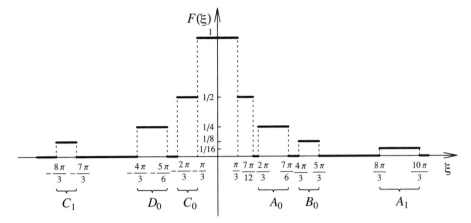

We divide $\mathrm{supp}(F) \setminus [-\frac{\pi}{3}, \frac{7\pi}{12})$ into four classes of intervals, two on a positive side and two on the negative side. Consider the positive side first. We define, for $j \geq 0$,

$$A_j \quad := \quad \mathrm{supp}(F) \cap 4^j [\frac{2\pi}{3}, \frac{4\pi}{3})$$

$$B_j \quad := \quad \mathrm{supp}(F) \cap 4^j [\frac{4\pi}{3}, \frac{8\pi}{3}).$$

We can prove by induction that

$$A_j \quad = \quad [\frac{4^j 2\pi}{3}, \frac{(4^{j+1}+3)\pi}{6}), \ \tau(A_j) = \tau(A_0)$$

$$\tau(A_0) \quad = \quad [\frac{2\pi}{3}, \pi) \cup [-\pi, -\frac{5\pi}{6}), \ F|_{A_j} = (\frac{1}{4})^{j+1}$$

$$B_j \quad = \quad [\frac{4^j 4\pi}{3}, \frac{(4^{j+1}+1)\pi}{3}), \ \tau(B_j) = \tau(B_0)$$

$$\tau(B_0) \quad = \quad [-\frac{2\pi}{3}, -\frac{\pi}{3}), \ F|_{B_j} = \frac{1}{8}(\frac{1}{4})^j.$$

Similarly, we get two classes on the negative side, i.e. for $j \geq 0$,

$$C_j \quad = \quad [-\frac{4^j 2\pi}{3}, -\frac{(4^j \cdot 2 - 1)\pi}{3}), \ \tau(C_j) = [-\frac{2\pi}{3}, -\frac{\pi}{3})$$

$$D_j \quad = \quad [-\frac{4^j 4\pi}{3}, -\frac{(4^{j+1} \cdot 2 - 3)\pi}{6}), \ \tau(D_j) = [\frac{2\pi}{3}, \pi) \cup [-\pi, -\frac{5\pi}{6}).$$

Observe that $D_\psi(\xi) = \sum_{k \in \mathbb{Z}} F(\xi + 2k\pi)$, i.e. we get the graph of D_ψ:

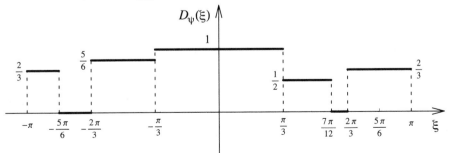

Since $p_\psi(\xi) = D_\psi(\xi/2) + D_\psi(\xi/2 + \pi) - D_\psi(\xi)$, it is easy to check that $p_\psi(\xi) \geq 1/2$ a.e. Hence, ψ is an MRA PFW, it is a W_0-Riesz basis, but $|Z_{V_0(\psi)}| = |\{\xi : D_\psi(\xi) = 0\}| > 0$. In other words, $\psi \in \mathcal{P}_{f,+}^{\mathrm{MRA}} \setminus \mathcal{P}_{-,1}^{\mathrm{MRA}}$.

Observe also that the semiorthogonalization of ψ provides us with an example in $\mathcal{P}_{tf,0}^{\mathrm{MRA}}$ which is not MSF. $\qquad\square$

In order to complete the picture about MSF PFW's, let us mention that it is indeed not difficult to construct MRA PFW's which are MSF, but are not orthonormal. Let us also mention that if the generalized pow-pass filter m has the property that $|m(\xi)|$ attains only values 0 and 1, then the associated ψ must be an MSF.

Acknowledgment: The authors wish to thank Marcin Bownik for pointing out a mistake in an earlier version of this paper.

REFERENCES

[1] L. Baggett and K. Merrill, *Abstract harmonic analysis and wavelets in* $\mathbf{R}^n$, The functional and harmonic analysis of wavelets and frames (San Antonio, TX, 1999), 17–27, Contemp. Math., **247, Amer. Math. Soc., Providence, RI, 1999.**

[2] J. Benedetto and S. Li, *The theory of multiresolution analysis frames and applications to filter banks,* Appl. Comput. Harmon. Anal. **5** (1998), no. 4, 389–427.

[3] J. Benedetto and O. Treiber, *Wavelet frames: multiresolution analysis and extension principles,* Wavelet transforms and time-frequency signal analysis 3–36, Appl. Numer. Harmon. Anal., Birkhäuser Boston, Boston, MA, 2001.

[4] C. de Boor, R. DeVore, and A. Ron, *The structure of finitely generated shift-invariant spaces in* $L_2(\mathbb{R}^d)$, J. Funct. Anal. **119** (1994), no. 1, 37–78.

[5] M. Bownik, *The structure of shift-invariant subspaces of* $L^2(\mathbb{R}^n)$, J. Funct. Anal. **177** (2000), no. 2, 282–309.

[6] M. Bownik and Z. Rzeszotnik, *On the existence of multiresolution analysis for framelets,* Math. Ann. **332** (2005), 705-720.

[7] M. Bownik, Z. Rzeszotnik, and D. Speegle, *A characterization of dimension functions of wavelets,* Appl. Comput. Harmon. Anal. **10** (2001), no. 1, 71–92.

[8] X. Dai, D. R. Larson, and D. Speegle, *Wavelet sets in* $\mathbb{R}$, J. Fourier Anal. Appl. **3** (1997), no. 4, 451–456.

[9] H. Helson, *Lectures on invariant subspaces,* Academic Press, New York-London 1964 xi+130 pp.

[10] E. Hernández and G. Weiss, *A first course on wavelets,* With a foreword by Yves Meyer. Studies in Advanced Mathematics. CRC Press, Boca Raton, FL, 1996. xx+489 pp.

[11] M. Papadakis, H. Šikić, and G. Weiss, *The characterization of low pass filters and some basic properties of wavelets, scaling functions and related concepts,* J. Fourier Anal. Appl. **5** (1999), no. 5, 495–521.

[12] M. Paluszyński, H. Šikić, G. Weiss, and S. Xiao, *Generalized low pass filters and MRA frame wavelets,* J. Geom. Anal. **11** (2001), no. 2, 311–342.

[13] M. Paluszyński, H. Šikić, G. Weiss, and S. Xiao, *Tight frame wavelets, their dimension functions, MRA tight frame wavelets and connectivity properties,* Adv. Comput. Math. **18** (2003), no. 2-4, 297–327.

[14] A. Ron and Z. Shen, *Frames and stable bases for shift-invariant subspaces of* $L_2(\mathbb{R}^d)$, Canad. J. Math. **47** (1995), 1051–1094.

[15] A. Ron and Z. Shen, *The wavelet dimension function is the trace function of a shift-invariant system,* Proc. Amer. Math. Soc. **131** (2003), no. 5, 1385–1398.

[16] D. Speegle *The s-elementary wavelets are path connected,* Proc. Amer. Math. Soc. **227** (1999), no. 1, 223–233.

[17] G. Weiss and E. Wilson, *The mathematical theory of wavelets,* Twentieth century harmonic analysis—a celebration (Il Ciocco, 2000), 329–366, NATO Sci. Ser. II Math. Phys. Chem., **33**, Kluwer Acad. Publ., Dordrecht, 2001.

[18] The Wutam Consortium, *Basic properties of wavelets,* J. Fourier Anal. Appl. **4** (1998), no. 4-5, 575–594.

DEPARTMENT OF MATHEMATICS, UNIVERSITY OF ZAGREB, BIJENICKA 30, HR-10 000 ZAGREB, CROATIA

E-mail address: hsikic@cromath.math.hr

DEPARTMENT OF MATHEMATICS AND COMPUTER SCIENCE, SAINT LOUIS UNIVERSITY, 221 NORTH GRAND BLVD., ST. LOUIS, MO 63103, USA

E-mail address: speegled@slu.edu

DEPARTMENT OF MATHEMATICS, WASHINGTON UNIVERSITY , CAMPUS BOX 1146, ONE BROOKINGS DRIVE, ST. LOUIS, MO 63130-4899

E-mail address: guido@math.wustl.edu

Titles in This Series

TITLES IN THIS SERIES

For a complete list of titles in this series, visit the
AMS Bookstore at **www.ams.org/bookstore/**.